Coordination Chemistry of Aluminum

Coordination Chemistry of Aluminum

EDITOR

Gregory H. Robinson

Howard L. Hunter Chemistry Laboratory
Department of Chemistry
Clemson University
Clemson, South Carolina

VCH

Gregory H. Robinson
Howard L. Hunter Chemistry Laboratory
Department of Chemistry
Clemson University
Clemson, South Carolina 29634-1905

Library of Congress Cataloging-in-Publication Data
CIP pending.

Printed in the United States of America
ISBN 1-56081-059-9 VCH Publishers, Inc.
ISBN 1-56081-656-2 VCH Publishers, Inc. (paper cover)
ISBN 3-527-28149-5 Verlagsgesellschaft
ISBN 3-527-89656-2 Verlagsgesellschaft (paper cover)

Printing History:
10 9 8 7 6 5 4 3 2 1

Published jointly by

VCH Publishers, Inc.
220 East 23rd Street
New York, New York 10010

VCH Verlagsgesellschaft mbH
P.O. Box 10 11 61
D-6490 Weinheim
Germany

VCH Publishers (UK) Ltd.
8 Wellington Court
Cambridge CB1 1HZ
United Kingdom

To Cynthia and Corbin

Preface

The chemistry of aluminum has been a fertile area of research for decades. The past decade has, arguably, been the most interesting and productive as a number of seminal discoveries in this large and varied areea have been made. This point in particular has served as an impetus for this volume. *Coordination Chemistry of Aluminum* consists of six chapters, authored by some of the recognised experts in the field. Each chapter concerns a timely and relevant topic. Indeed, these diverse topics underscore the breath of the chemistry of aluminum. It is hoped that this volume may be utilized by both organic and inorganic chemists alike as a valuable resource.

Certainly it is true that the author (or contributing editor) of any book is indebted to many others. I owe a great deal of gratitude to the contributors without whom this project would not have been possible. A special acknowledgement is extended to my research group (Jaye A. Burns, Jameica J. Byers-Hill, Mary Dillingham, Sheila Eng, Kevin Gripper, Lagette Kennedy, Baosheng Lee, Kris Scaboo, Steven J. Schauer, and Terry Talley) which has been forced to "make do" a bit more than they should have during the duration of this project.

Gregory H. Robinson
Clemson, South Carolina
November 1992

Contents

Contributors

JERRY L. ATWOOD. Department of Chemistry, The University of Alabama, Tuscaloosa, Alabama 35487

ANDREW R. BARRON. Department of Chemistry, Harvard University, Cambridge, Massachusetts 02138

ARNE HAALAND. Department of Chemistry, University of Oslo, Blindern 0315 Oslo (Norway)

RAJESH KUMAR. Department of Chemistry, Wayne State University, Detroit, Michigan 48202

JOHN P. OLIVER. Department of Chemistry, Wayne State University, Detroit, Michigan 48202

CHRIS ORVIG. Department of Chemistry, University of British Columbia, 2036 Main Mall, Vancouver, British Columbia V6T 1Z1

GREGORY H. ROBINSON. Howard L. Hunter Chemistry Laboratory, Department of Chemistry, Clemson University, Clemson, South Carolina 29634-1905

MAJID TAGHIOF. Department of Chemistry, Wayne State University, Detroit, Michigan 48202

CHAPTER

1

Normal and Dative Bonding in Neutral Aluminum Compounds

Arne Haaland

Department of Chemistry, University of Oslo, Blindern 0315 Oslo (Norway)

1.1. Introduction

We will survey and rationalize the structures of neutral aluminum compounds in the following. The intention has been to describe representative examples of all known structure types and to show how these may be rationalized, while including all structures that fail to conform to the patterns. We shall distinguish between normal and dative bonds.[1] When experimental information is lacking, normal bond distances may be estimated by using the modified Schomaker–

Stevenson rule.[2] Dative bonds are generally longer, weaker, and more sensitive to inductive effects than normal bonds between the same atom pair.

In many cases we shall consider bonds that are neither purely normal or purely dative. In such cases the normal and dative bonding contributions may be estimated by enumerating the number of canonical forms in which the given bond is normal or dative.[1] When rationalizing molecular shapes or discussing valence angles we shall employ the powerful Valence Shell Electron Pair Repulsion (VSEPR) model.[3,4] When observed structures deviate from those predicted by this model, we shall look for evidence of repulsion between geminal atoms or groups. In this connection we find the geminal repulsion radii of Bartell and Glidewell particularly useful.[5–7]

1.2 Normal and Dative Bonds

In order to delineate the difference between covalent and dative bonds it is useful to begin by considering the properties of ethane and amine–borane, H_3NBH_3. These two molecules are isoelectronic and, at least in principle, the covalently bonded ethane may be transformed into the datively bonded amine–borane by transferring a proton from one carbon nucleus to the other. Homolytic rupture of the C–C bond in ethane yields electrically neutral methyl radicals, heterolytic rupture yields two diamagnetic ions, CH_3^+ and CH_3^-, and the products obtained possess either net spin or net charge, but never both. Rupture of the dative bond in H_3NBH_3, on the other hand, yields products which have either both net charge and net spin or neither. Homolytic rupture yields the radical ions NH_3^+ and BH_3^-, whereas heterolytic rupture yields the neutral diamagnetic species H_3N and BH_3.

We therefore propose that a bond in a neutral molecule should be defined as *dative* if rupture yields either a pair of neutral diamagnetic species or a pair of radical ions, and as *normal* if rupture yields either a pair of uncharged radicals or a pair of diamagnetic ions.[1] In the following we shall use the formalism N→B when we wish to emphasize that a bond is dative.

According to this definition the gaseous monomeric LiF is held together by a normal bond. Homolytic rupture yields the paramagnetic Li and F atoms and heterolytic rupture yields the diamagnetic ions Li^+ and F^-. The term *normal bond* covers covalent bonds like the C–C bond in ethane, *ionic bonds* like the bond in a gaseous monomeric alkali halide, as well as bonds of intermediate polarity.

In the gas phase, minimum energy rupture of the dative B–N bond in the amine–borane, of the normal/covalent bond in ethane and of the normal/ionic bond in monomeric LiF all proceed to yield uncharged species:

$$H_3N{-}BH_{3(g)} \longrightarrow H_3N_{(g)} + BH_{3(g)} \qquad \Delta H^\circ = 31.1\pm1.0 \text{ kcal/mol,}$$ [1]

$$H_3C{-}CH_{3(g)} \longrightarrow H_3C\bullet_{(g)} + \bullet CH_{3(g)} \qquad \Delta H^\circ = 89.8\pm0.5 \text{ kcal/mol,}$$ [8]

$$Li{-}F_{(g)} \longrightarrow Li\bullet_{(g)} + F\bullet_{(g)} \qquad \Delta H^\circ = 138\pm2 \text{ kcal/mol,}$$ [9]

It is seen that the prototype normal/ionic bond between second period elements is 50% stronger than the prototype normal/covalent bond in ethane which in turn is three times stronger than the prototype dative bond in amine–borane.

The different strength of the prototype dative and normal/covalent bonds is reflected in the bond distances; C–C = 1.533 (2) Å and B–N = 1.658 Å.[10,11] (The bond distance in $LiF_{(g)}$ is 1.564 Å).[12]

Whereas the dipole moment of ethane is zero by symmetry, the dipole moment of H_3NBH_3 is 5.2 D in the gas phase.[11] The negative pole is undoubtedly at the acceptor end of the molecule (BH_3). Thus while the negative pole in the normally bonded $LiF_{(g)}$ is at the more electronegative F atom, in the datively bonded H_3NBH_3 it is at or near the more electropositive B atom.

The charge distribution in H_3NBH_3 has been investigated by molecular orbital calculations with a split valence basis.[13] The Mulliken population parameters thus obtained indicate that formation of the dative bond is accompanied by transfer of 0.20 e^- from the electron donor H_3N to the electron acceptor BH_3. Perfect sharing of the electron pair would, of course, correspond to a transfer of 1.00 e^-; the two electrons forming the dative bond are not equally shared, but remain more closely associated with the donor atom.

The discussion thus far may be summarized as follows: the normal/covalent bond in $H_3C{-}CH_3$ may be transformed into the dative bond in $H_3N{-}BH_3$ by transferring one proton from the C atom on the right to the C atom on the left. This introduces a large dipole moment which is diminished by electron relaxation as the electrons in the C–C bond are displaced in the direction of proton transfer. As the electrons are removed from the overlap region to a region closer to the N atom, the bond is weakened and elongated.

The electron population on the B atom does not appear to increase as the adduct is formed from the isolated donor and acceptor, and the electron density on the N atom does not appear to decrease. The acceptor atom B passes the acquired electron density to the H atoms bonded to it, and the N atom replaces the lost electron density by withdrawing negative charge from its H atoms.[13]

It is not surprising that dative bonds are very sensitive to inductive effects, particularly at the acceptor atom. The replacement of the three H atoms at B in H_3NBH_3 by more electron releasing methyl groups reduces the dissociation enthalpy by a factor of one half.[1]

1.3 Dative Bonds and the VSEPR Model

Figure 1.1 compares the molecular structures of the free donor Me_3P and the free acceptor $AlMe_3$ with the structure of the adduct Me_3P-$AlMe_3$.[14–16] In free $AlMe_3$ the metal atom is surrounded by three normal bond electron pairs, and the coordination is trigonal planar in agreement with the VSEPR model.[3,4]

As the adduct is formed and a fourth electron is introduced (accepted), the Al–C bonds are bent back, and the C–Al–C angles reduced from 120° to 113°.

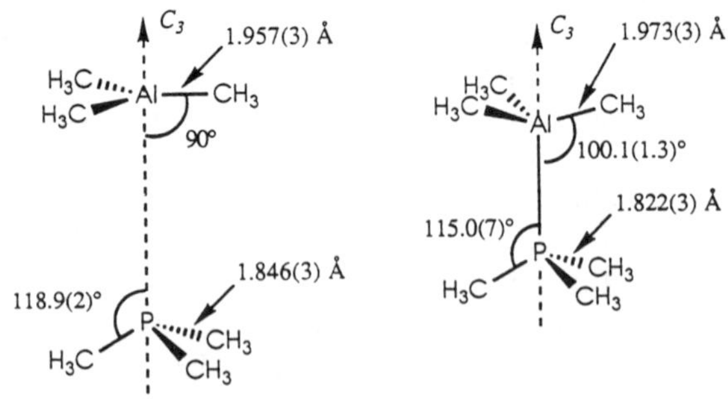

Figure 1.1. The gas phase structures of monomeric $AlMe_3$, PMe_3, and the adduct Me_3PAlMe_3.[13–15]

At the same time the Al–C bonds are slightly but significantly elongated from 1.957 (3) to 1.973 (3) Å. Both changes are in the direction predicted from the VSEPR model. It should be noted, however, that after the adduct has been formed, the C–Al–C angles remain greater than tetrahedral, suggesting that the spatial requirements of the three normal bond pairs (NBP) remain larger than the spatial requirements of the accepted bond pair (ABP):

$$\text{NBP} > \text{ABP}$$

The P atom in Me_3P is surrounded by three NBP and one electron lone pair (LP). The structure is pyramidal in accordance with the VSEPR model, and the C–P–C angles are less than tetrahedral in accordance with the assumption that the spacial requirements of an LP is greater than that of a NBP.

As the adduct is formed and the lone pair is partly removed (donated), the C–P–C angles increase from 98.6 (3) to 103.4 (8)°, and there is a slight but significant decrease of the P–C bond distances from 1.846 (3) to 1.822 (3) Å. Both changes are in the direction predicted by the VSEPR model. After the adduct has been formed, however, the angles C–P–C remain smaller than tetrahedreal, indicating that the spatial requirements of the donated bond pair (DBP) remain larger than those of a normal bond pair:

$$\text{LP} > \text{DBP} > \text{NBP}$$

1.4 Failures of the VSEPR Model and Geminal Repulsion Radii

In Figure 1.2 we show the valence angles found in some simple derivatives of C, N, and O.[17] While the valence angles in methane are tetrahedral, the valence angles in ammonia and water molecules are significantly smaller than tetradhedral, as expected from the VSEPR model. The C–N–C valence angle in dimethylamine and the C–O–C angle in dimethylether are, however, greater than tetrahedral in disagreement with the VSEPR model, and the angles in

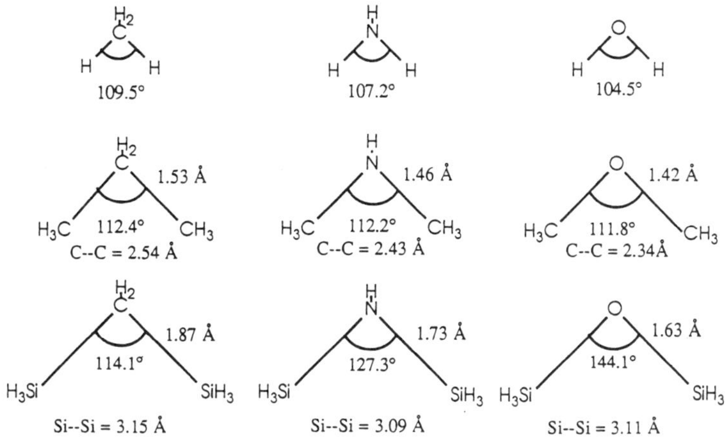

Figure 1.2. The gas phase structures of R_2CH_2, R_2NH, and R_2O where R = H, Me, or SiH_3.[17]

disilylamine and disilylether are 127° and 144°, respectively. It has often been assumed that these very large angles are due to dative pπ–dπ interactions between N or O and Si. This hypotheses, however, has received little support from high quality *ab initio* molecular orbital calculations.[18–20]

Bartell has suggested that when two ligating atoms are as large or larger than the central atom, repulsion between them may lead to a valence angle at the central atom that is significantly larger than the value predicted from the VSEPR model.[5] Bartell's ideas were extended by Glidewell who noted that the geminal Si—Si distances in disilylmethane, disilylamine, and disilylether are 3.15°, 3.09°, and 3.11 Å, respectively, and suggested that repulsion between the geminal Si atoms becomes prohibitively large when the Si—Si distance is less than 3.10 Å. On this basis Glidewell assigned an "intramolecular non-bonded atomic radius" of 1.55 Å to the Si atom.[6] Such radii, which we refer to as *geminal repulsion radii*, have been assigned for several main group elements. The accepted value for carbon is 1.25 Å.[7] Glidewell has assigned a radius of 1.85 Å to Al, while we have suggested a somewhat smaller value of 1.75 Å.[21]

In the following we shall use twice this value, 3.50 Å, when trying to assess whether steric repulsion between geminal Al atoms, and the sum of the geminal repulsion radii of C and Al, 3.00 Å, when trying to assess whether steric repulsion between geminal Al and C atoms may be large enough to determine the valence angle.

1.5 Some Mean Bond Enthalpies

Table 1.1 lists mean bond enthalpies in the gaseous compounds AlF, AlCl, Al_2O, $AlCl_2$, AlF_3, $AlCl_3$, Al_2F_6 and Al_2Cl_6 as a function of the coordination

Table 1.1. Mean bond enthalpies, D (in kcal/mol), as a function of coordination number (CN) of the Al atom. Uncertainties greater than 1 kcal/mol are indicated. Estimated values in square brackets, (see text).

CN	D(Al-C)	D(Al-N)	D(Al-O)	D(Al-F)	D(Al-Cl)
1	—	—	126±3	161±2	120±3
2	—	—	—	—	101±4
3	67	[84]	[111]	142	101±2
4	55	67	—	113	80
6	—	—	61	83	56

number (CN) of the metal. Also included are mean bond enthalpies in crystalline Al_2O_3, AlF_3, and $AlCl_3$ where the CN of Al is six, and the mean bond enthalpy in $AlN_{(cr)}$ where the CN is four. All these bond enthalpies are based on thermodynamic data from the JANAF tables.[8] Finally, we include the mean Al–C bond enthalpies in $AlMe_3$ and Al_2Me_6 where the CN is four.[22,23]

The mean Al–O bond enthalpy for CN = 3 was estimated assuming that the ratio between Al–O bond enthalpies for CN = 1 and 3 is equal to the ratio of Al–F bond enthalpies for the same coordination numbers. Similarly the Al–N bond enthalpy for CN = 3 was estimated by assuming that the ratio between Al–N bond enthalpies for CN = 3 and 4 is equal to the ratio of Al–F bond enthalpies for the same coordination numbers. The bond enthalpies for CN = 3 may be regarded as mean *normal* bond enthalpies.

For every bonding partner *X*, the mean Al–*X* bond enthalpy is seen to decrease with increasing coordination number. Keeping the coordination number constant, bond strength is seen to increase in the order Al–C < Al–N < Al–O < Al–F, while Al–Cl bonds are weaker than Al–F bonds.

We shall return to a discussion of the variation of mean Al–Cl bond enthalpies with coordination number later in Section 1.2.

1.6 Normal Bond Distances in Compounds With Three–coordinate Aluminum

The molecular structures of monomeric trimethylaluminum,[14] and monomeric aluminum-trifluoride[24] and –trichloride[25] have been determined by gas electron diffraction, and the structure of the monomeric trisamide $Al[N(SiMe_3)_2]_3$ by X-ray crystallography.[26] No structure of a monomeric trisalkoxide is known, but the structure of monomeric methylaluminum bis(aryloxide), $MeAl(OAr)_2$, (Ar = 2,6–di–*tert*–butyl–4–methylphenyl), has been reported by Ittel and co-workers.[27] The bond distances are listed in Table 1.2.

The Al–N distance in $[MeAlN(2,6\text{–di–}isopropylphenyl)]_3$, **22**, which contains a planar six-membered ring of alternating Al and N atoms, and where

Table 1.2. Mean normal bond enthalpies, D (in kcal/mol), and normal bond distances, R (in Å units), in compounds with three-coordinate Al.

	Al–C	Al–N	Al–O	AL–F	Al–Cl
D[a]	67	[84]	[111]	142	101
R(Exp)	1.957(3)	1.78(2)	1.686(2)	1.631(3)	2.068(4)
R(Calc)[b]	2.03	1.96	1.92	1.90	2.25
R(Calc)[c]	[1.957]	1.822	1.747	1.667	2.135

[a] From Table 1.1.
[b] Calculated sum of Paulings "tetrahedral convalent radii."[29]
[c] Calculated from the modified Schomaker–Stevensen rule[2] and R(Al)=1.256Å.

both Al and N are trigonal planar, is 1.782 (4) Å. This value is similar to that of the trisamide.[28]

As indicated in the third row of Table 1.2, the observed bond distances are up to 0.27 Å shorter than the estimates obtained by adding Pauling's *tetrahedral covalent radii.*[29] There are several possible reasons for this discrepancy: the estimates do not include any shortening due to bond polarity; to the contraction of the bonding radius of Al with an increasing number of electronegative substituents; or due to possible dative π-bonding, that is donation of pπ–electrons of the ligating atom into the empty valence shell pπ–orbital on aluminum.

The shortening expected from bond polarity is most easily included by using the modified Schomaker–Stevenson rule:

$$R(A{-}B) = R(A) + R(B) - c|\chi(A) - \chi(B)|^{1.4}$$

where R(A) and R(B) are the bonding radii of the two atoms, χ(A) and χ(B) their electronegativities, and the constant c = 0.085 Å. This relationship has been found to reproduce bond distances in simple molecules formed by H and the elements in periods 14 to 17 with an average deviation of 0.032 Å.[2] Using the bonding radius of Al, R(Al) = 1.256 Å, obtained from the Al–C bond distance in Me_3Al, in combination with the bonding radii of the other elements,[2] we calculate the Al–*X* bond distances given in the fourth row of Table 1.2. These improved estimates are found to be 0.04 Å, 0.06 Å, 0.04 Å, and 0.07 Å, too large for *X* = N, O, F, and Cl, respectively. Are these differences due to dative π-bonding?

Gropen and Wisløff Nilssen[30] have optimized the Al–O bond distance in the planar H_2AlOH molecule by *ab initio* molecular orbital calculations at the dz level and obtained a value 1.704 Å in reasonable agreement with the subsequent experimental value in Table 1.2. They then eliminated π-bonding by rotating the OH group into a plane perpendicular to the H_2AlO plane. The energy increased by 4.4 kcal/mol, not much more than the rotation barrier in ethane, and the Al–O bond distance increased by about 0.01 Å: It would seem that dative π-bonding has only a minor effect on the Al–O bond distance.

The two Al–O–C valence angles in $MeAl(OAr)_2$ are 141 and 147°, respectively.[26] The distances between Al and C atoms bonded to the same O atom are 2.88 Å and 2.93 Å (that is less than the sum of the geminal repulsion radii of Al and C, 3.00 Å). Steric repulsion between the geminal Al and C atoms thus provide a sufficient explanation for these unusually large valence angles. However, once established, the wide angles may in turn facilitate a certain amount of π-bonding between Al and O: *ab initio* MO calculations on the model compound H_2AlOH show that the opening of the Al–O–H angle raises the orbital energies of the O π-electrons and thus facilitates dative π-bonding to Al.[31] Lichtenberger, Barron and co-workers have found evidence for such interactions in the photoelectron spectra of the adduct of $MeAl(OAr)_2$ with PMe_3.[32]

1.7 The Dissociation Enthalpies of Simple Adducts of Alanes with Electron Donors

In Table 1.3 we compare the enthalpies of the reactions

$$\text{Donor} \rightarrow AlX_3 \longrightarrow \text{Donor} + AlX_3$$

in the gas phase or in hydrocarbon solution for X = Me, Cl and Br. When available, we list gas and solution values separately. When several studies have been made of the same adduct in the same phase, we list the highest and the lowest of the published values. The strength of the dative bonds in these adducts range from about 10 to about 50 kcal/mol. Even the strongest of the dative N→Al bonds (in Me_3NAlCl_3) is, however, only 60% of our estimate of the strength of a normal Al–N bond, and the strongest O→Al bond (in $Ph(H)C{=}OAlBr_3$) is only 45% of our estimate of the strength of a normal Al–O bond.

It might be objected that our estimates of normal bond strengths are too high, that an increase of the coordination number from three to four would be expected to weaken the normal bonds. This is certainly true; however, no estimates of normal bond strengths in adducts are available, since the dative bond will break first when the adduct is loaded with energy. The mean Al–N bond energy in $AlN_{(cr)}$ where Al is four-coordinate, constitutes an underestimate of a normal bond energy, since it represents the mean of three normal bonds and one dative bond. Even so, it is 30% higher than the highest observed dative N→Al bond energy.

Comparison of the dissociation enthalpies of adducts of the same electron donor, e.g. H_3NAlX_3; ΔH° = 41 kcal/mol for X = Cl or Br, and 28 kcal/mol for X = Me, or $pyrAlX_3$; ΔH° = 46 (X = Cl), 45(X = Br) and 28 kcal/mol (X = Me), or Et_2SAlX_3; ΔH° = 30(Cl), 31(Br), and 17 kcal/mol (Me), shows that dative bonds to $AlCl_3$ and $AlBr_3$ are equally strong, and about 50% stronger than the bonds to $AlMe_3$.

Comparison of the dissociation enthalpies of adducts with the same electron acceptor shows that while the strength of normal bonds increase in the order

Table 1.3. Dissociation enthalpies, ΔH° (in kcal/mol), of adducts of $AlMe_3$, $AlCl_3$ and $AlBr_3$ with electron donors.

Donor	Acceptor	ΔH°	Phase	Reference
Me_3N	$AlMe_3$	30.7±0.3	Gas	33, 34
Me_3N	$AlMe_3$	30.0±0.2	Solution	33
Me_2HN	$AlMe_3$	30.8±0.3	Solution	33
MeH_2N	$AlMe_3$	30.0±0.3	Solution	33
H_3N	$AlMe_3$	27.6±0.3	Solution	33
Et_3N	$AlMe_3$	26.5±0.2	Solution	33
pyr	$AlMe_3$	27.6±0.2	Solution	33
Me_3P	$AlMe_3$	21.0±0.3	Solution	33
Et_3P	$AlMe_3$	22.1±0.3	Solution	33
Ph_3P	$AlMe_3$	17.6±0.2	Solution	33
Me_2O	$AlMe_3$	21.9±0.2	Gas	33
Me_2O	$AlMe_3$	20.3±0.3	Solution	33
Et_2O	$AlMe_3$	20.2±0.2	Solution	33
thf	$AlMe_3$	22.9±0.2	Solution	33
Me_3PO	$AlMe_3$	32.0±0.2	Solution	35
Ph_3PO	$AlMe_3$	28.7±0.2	Solution	35
$(MeO)_3PO$	$AlMe_3$	25.6±0.2	Solution	35
Me_2SO	$AlMe_3$	28.6±0.3	Solution	36
Me_2S	$AlMe_3$	18.1±0.5	Gas	37
Me_2S	$AlMe_3$	16.7±0.2	Solution	37
Et_2S	$AlMe_3$	16.8±0.2	Solution	37
Me_3N	$AlCl_3$	48±2		34
H_3N	$AlCl_3$	41	Gas	38
pyr	$AlCl_3$	45.7/46.2	Gas	38
Cl_3PO	$AlCl_3$	28.4	Gas	38
Me_2S	$AlCl_3$	30.3	Gas	38
Et_2S	$AlCl_3$	29.9	Gas	38
Et_3N	$AlBr_3$	44.8	Solution	38
H_3N	$AlBr_3$	41.0	Gas	38
pyr	$AlBr_3$	39.6/45.3	Gas	38
pyr	$AlBr_3$	44.7	Solution	38
Bu_3P	$AlBr_3$	47.5	Solution	38
Ph_3P	$AlBr_3$	34.9	Solution	38
Br_3P	$AlBr_3$	8.8	Gas	38
Et_2O	$AlBr_3$	36.5/39.1	Solution	38
Bu_2O	$AlBr_3$	34.8/38.4	Solution	38
PhOMe	$AlBr_3$	27.9/31.3	Solution	38
PhOEt	$AlBr_3$	30.1/30.5	Solution	38

(Continued)

Table 1.3. *(Continued)*

Donor	Acceptor	$\Delta H°$	Phase	Reference
Ph_2O	$AlBr_3$	25.3/25.7	Solution	38
EtOH	$AlBr_3$	39.9	Solution	38
BuOH	$AlBr_3$	40.7	Solution	38
PhOH	$AlBr_3$	33.0	Solution	38
PhHC=O	$AlBr_3$	49.9	Solution	38
PhMeC=O	$AlBr_3$	33.7	Solution	38
Ph_2C=O	$AlBr_3$	33.2	Solution	38
Et_2S	$AlBr_3$	30.6	Gas	38
Bu_2S	$AlBr_3$	30.9	Solution	38

Al–N < Al–O < Al–F, amines form stronger dative bonds than ethers, which in turn form stronger dative bonds than alkyl flourides; the latter do not form dative bonds which are strong enough to compete with dimerization of the acceptor.

It should be noted, however, that electron donors such as ketones (e.g., Me_3PO or Me_2SO) which contain doubly bonded O atoms with a CN of one form stronger dative bonds than ethers.

We now turn our attention to molecular structures. In the following we shall omit estimated standard deviations to bond distances when they are less than or equal to 0.005 Å and to angles if they are less than 1°.

1.8 Aluminum–Nitrogen Compounds

1.8.1. Compounds with Four-coordinate Aluminum

1.8.1.1 Dative N→Al bonds

The Al atoms in **1**[39] and **2**[40] are normally bonded to a (three-coordinate) amide N and datively bonded to a (four-coordinate) amine N atom. The normal Al–N distance in **1** is similar to the reference bond distance in Table 1.2, the corresponding bond in **2** appears to be slightly elongated as expected from the VSEPR model. The dative N→Al bonds are seen to be 0.19 and 0.16 Å longer than the normal bonds.

The three crystallographically independent Al–N bonds in **3**[41] are 2.07, 1.91 and 2.03 Å, repsectively. Al(1) forms three normal bonds to the three ethyl groups, the fourth bond, N(1)→Al(1) = 2.03 Å, must therefore be dative. Al(2) bridges two four-coordinate N atoms at very different distances. However, since N(2) is normally bonded to three C atoms, the N(2)→Al(2) = 2.07 Å bond must be dative. Finally, if Al(2) and N(1) are to form three normal bonds each, Al(2)–N(1) = 1.91 Å must be a normal bond. This normal bond is 0.13 Å

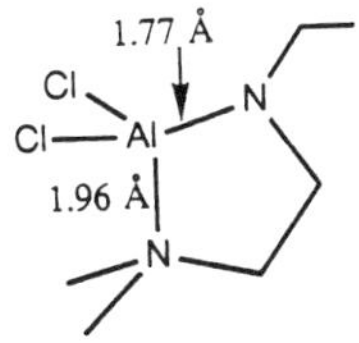

1

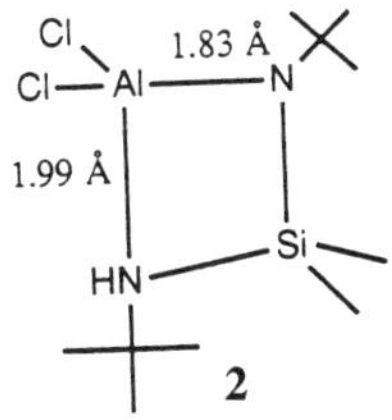

2

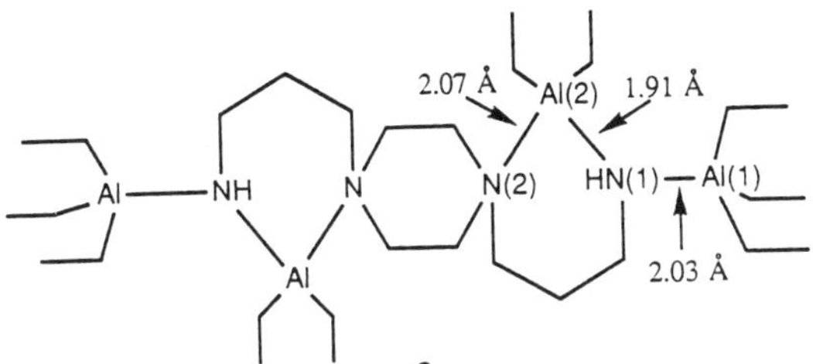

3

longer than the normal Al–N bond distance in Table 1.2, the two dative bonds are 0.25 and 0.27 Å longer.

Table 1.4 lists N→Al bond distances in representative simple adducts of alanes with N-centered electron donors. The coordination polyhedron of the Al atom in $Me_3NAl(BH_4)_3$ is a pentagonal bipyramid: two BH_4 groups are bonded to Al through two bridging H atoms in the equatorial plane; the two bridging H atoms in the third group occupy one equatorial and one axial position; the second axial position is taken by the N atom. The Al atoms in the other complexes are four-coordinate.

The dative N→Al bond distances are in the range 1.94 to 2.10 Å. The shortest bonds are found in t-BuH_2NAlCl_3 and Me_3NAlCl_3. The latter has the highest known dissociation enthalpy (Table 1.3). The longest bond is found in

Table 1.4. N→Al bond distances, R(N→Al) (in Å units), in simple adducts of alanes with N-centered electron donors.

Donor	Acceptor	R(N→Al)	Reference
Me_3N	$AlMe_3$	2.10	34
N-tetramethylcyclam	$AlMe_3$ (4X)	2.10 (4X)	42
TMEDA	$Al(SiMe_3)_3$ (2X)	2.07 (2X)	43
Me_3N	AlH_3	2.06	44
Quinoclidine	AlM_3	2.06	45
MeCN	$AlMe_3$	2.06	46
Me_3N	$AlIMe_2$	2.02	47
H_3N	$AlCl_3$	2.00(2)	48
H_3N	$AlBr_3$	2.00(2)	49
Me_3N	$Al(BH_4)_3$	1.99	50
Me_3N	$AlCl_3$	1.96	51
t-BuH_2N	$AlCl_3$	1.94	52

Me_3AlMe_3, which has the lowest known dissociation enthalpy. The dative bond distances in H_3NAlCl_3 and H_3NAlBr_3, which have intermediate dissociation enthalpies, fall near the middle of the range.

The dissociation enthalpy of Me_3NAlH_3 is unknown, the N→Al bond distance suggests that it is slightly greater than that of Me_3NAlMe_3. In the following sections we shall adopt Al–N = 2.08 Å as a reference value for a dative bond from an amine N atom to an Al atom carrying only H atoms or organyl substituents.

In Figure 1.3 we compare the structures of free $AlMe_3$,[15] free $AlCl_3$,[25] and free Me_3N[53,54] with the structures of the adducts Me_3NAlMe_3[34] and Me_3NAlCl_3.[51,55] The structure of the acceptor $AlMe_3$ in the adduct with Me_3N is very similar to the structure in the adduct with Me_3P (Figure 1.1), and may be rationalized in the same way. Monomeric $AlCl_3$ is trigonal planar in the gas phase with Al–Cl = 2.06 Å. On formation of the adduct, the anlge Cl–Al–Cl is reduced from 120 to 114°, and the Al–Cl bonds are elongated to 2.12 Å; again the changes are as predicted by the VSEPR model.

We now turn our attention to the structure of the donor: according to the VSEPR model, free Me_3N should be pyramidal and the C–N–C angle less than tetrahedral. However, this angle is found to be slightly greater than tetrahedral, 110.6 (6)° by gas phase electron diffraction[53] and 110.9 (6)° by microwave spectroscopy.[54] Formation of the adduct leaves the C–N–C angle essentially

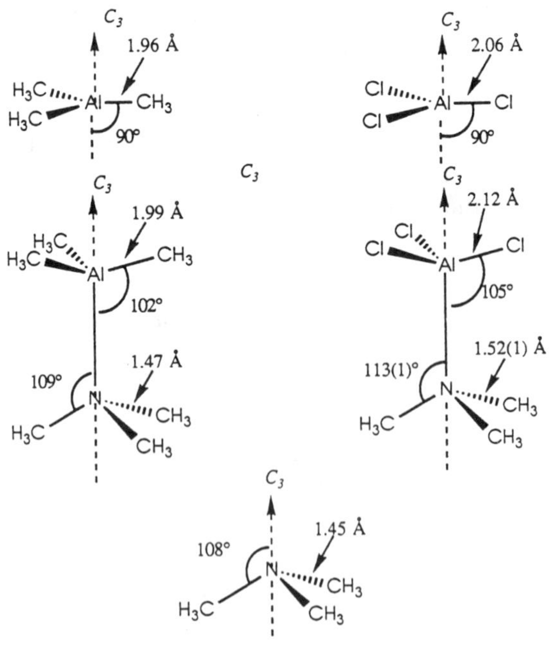

Figure 1.3. The gas phase structures of monomeric $AlMe_3$,[15] monomeric $AlCl_3$,[25] Me_3N[53,54] and the adduct Me_3NAlMe_3.[34] The structure of the adduct Me_3NAlCl_3 as determined by X-ray diffraction[51] and gas phase electron diffraction.[34]

unchanged, while the N–C bond distance increases from 1.452 (2) Å in free Me_3N to 1.473 (3) Å in the adduct with $AlMe_3$ and to 1.516 (12) Å in the still stronger adduct with $AlCl_3$. These changes are in the opposite direction to those expected from the VSEPR model, and we suggest that they are due to repulsion between the Al atom and the *N*-methyl groups; the non-bonded Al—C distances in the two compounds are both about 0.10 Å less than the sum of the geminal repulsion radii.

1.8.1.2. Aminoalanes; Al–N bonds with Equal Normal and Dative Bonding Contributions

With the exception of the trisamide $Al[N(SiMe_3)_2]_3$ (described earlier), all known aluminum amides are oligomeric.[56]

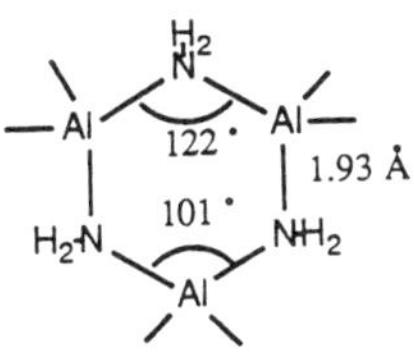

4

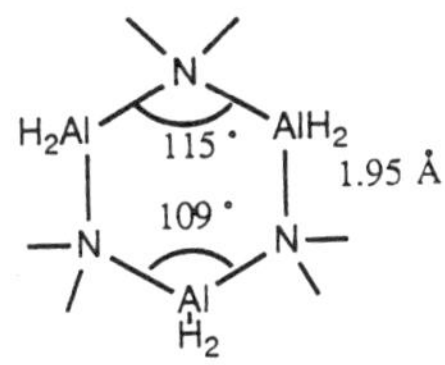

5

Dimethyl(amino)alane[57] **4** and dimethylaminoalane[58] **5** are both trimeric in the crystalline phase with non-planar six-membered rings with alternating Al and N atoms. The conformation of **4** is that of a skewed boat, that of **5** a chair. Individual Al–N bond distances in each are indistinguishable, in **4** they average 1.93 Å, in **5**, 1.95 Å. Just as it is convenient to describe the structure of benzene in terms of two canonical forms with alternating single and double C–C bonds, so it is convenient to describe the structure of these compounds in terms of two canonical forms with alternating normal and dative N–Al bonds. Such a description is consistent with our definition of normal and dative bonds: three bonds in the rings are expected to rupture heterolytically to yield three monomeric neutral aminoalanes, whereas the bonds in the monomers are expected to rupture homolytically to yield R_2N and AlR_2 radicals.

The observed bond distances are indeed close to the average of our reference value for a dative bond from N to an Al atom carrying only H and organyl substituents, and the reference value for a normal N–Al bond to four-coordinate Al, 1.81 Å, which we shall justify on p. 15:

$$(1/2)2.08\ \text{Å} + (1/2)1.81\ \text{Å} = 1.945\ \text{Å}$$

We have noted that the steric requirements of a dative bond electron pair are greater than the steric requirements of a normal bond electron pair at the donor atom, and less than the steric requirements of a nominal bond electron pair at the acceptor atom. To the extent that these properties are transferred to bonds of partial dative character, one would expect the endocyclic valence angles at

N to be greater than tetrahedral, and the endocyclic valence angle at Al to be less than tetrahedral; in **4** the average angles are Al–N–Al = 122° and N–Al–N = 101°. It should also be noted, however, that the Al—Al distances in **4** average 3.38 Å, slightly less than twice the geminal repulsion radius of Al, 3.50 Å. The large endocyclic valence angle at nitrogen may, at least in part, be due to Al—Al repulsion.

NMR spectra of **4** in toluene solution contain peaks attributable to both trimeric and dimeric species, and the enthalpy of the reaction

$$2(\text{trimer}) \longrightarrow 3(\text{dimer})$$

has been determined as 36.0±0.7 kcal/mol at 100°C.[59] There can be little doubt the dimer is destabilized by angle strain, since the conversion into a dimer necessitates the reduction of the Al–N–Al angle by about 30° to a value near 90°. The angle must, however, be relatively flexible; even if it is assumed that the entire enthalpy difference is due to the deformation of Al–N–Al angles, it amounts to no more than 6 kcal/mol for each.

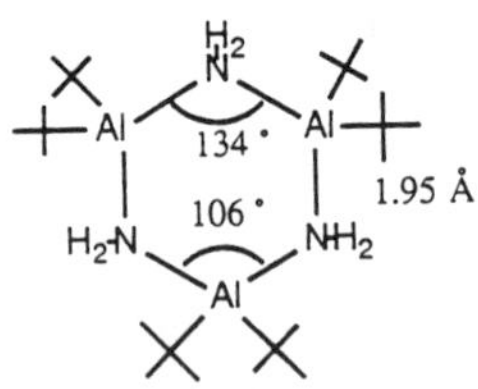

6

Structure **6**, obtained from **4** by replacing the methyl groups on Al by *tert*-butyl groups, is unique among trimeric aluminum amides in having a planar six-membered ring.[57] The planarity must be due to repulsion between *tert*-butyl groups at different Al atoms; such repulsion would not be reduced by conversion into a dimer. The endocyclic valence angle at N has been increased by 12 to 134°, again suggesting great flexibility.

The N atom in **5** is bonded to two Al atoms and two methyl groups, and must be very congested; the angle Al–N–Al is 7° smaller than in **4**, the Al···Al distance has been reduced to 3.29 Å, and the shortest geminal Al–C distances are 2.75 Å while the sum of the geminal repulsion radii of Al and C is 3.00 Å. In consequence the C–N–C angles are only 106°. Morevover, the Al–N–C angles are 6° larger for axial than for equatorial C atoms, indicating that there is significant repulsion between axial methyl groups.

Indeed, introduction of two methyl groups on each N atom in **4** to form Me_2AlNMe_2 leads to breakdown of the trimer: Me_2NAlMe_2, **7**, and all other known dialkyl(dialkylamino)alanes are dimeric.[56] Since both **4** and **5** are trimeric, it must be steric repulsion between methyl groups on neighboring atoms in the ring that forces Me_2AlNMe_2 to become dimeric.

Dimethyl(dimethylamino)alane **7** contains a planar, nearly square Al_2N_2 ring with bond distances Al–N = 1.96 Å.[60] The bonding in this compound must

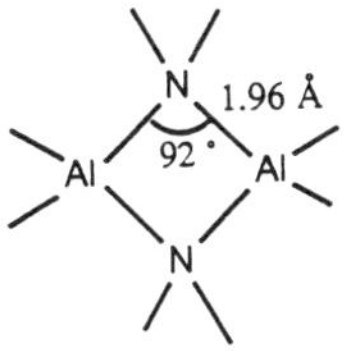

7

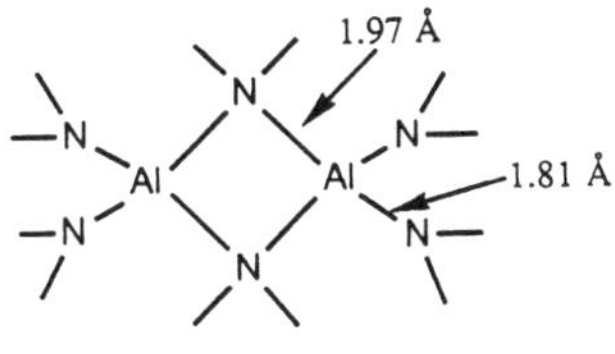

8

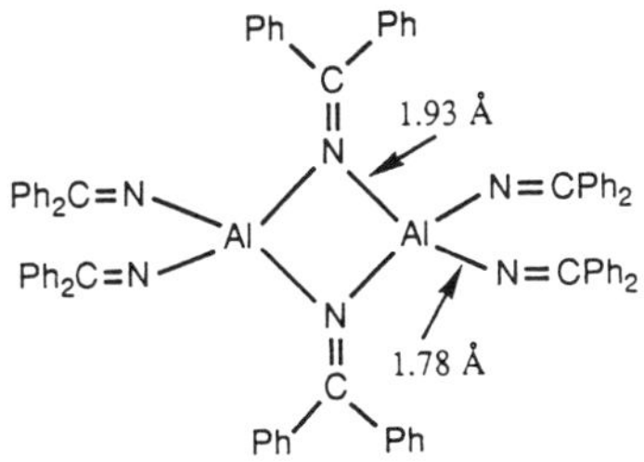

9

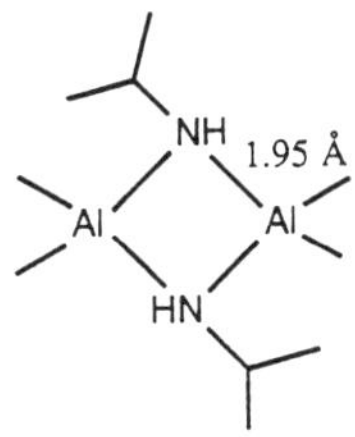

10

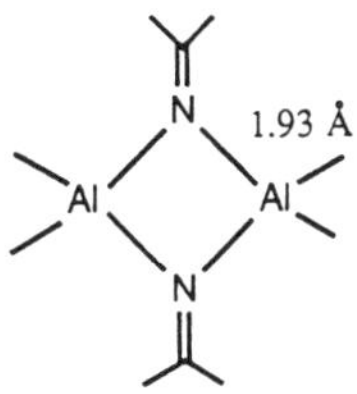

11

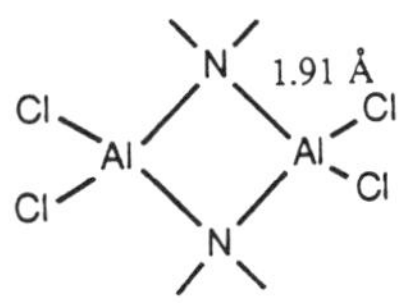

12

be described in terms of two canonical forms each with two normal and two dative Al–N bonds. The Al–N bond distance is in agreement with this description.

The structure of tris(dimethylamino)alane **8**[58] is also found to be dimeric with the central Al_2N_2 ring which must be described in the same manner, and with endocyclic Al–N bond distances of the same magnitude. The terminal Al–N bonds must, however, be described as normal, and are found to be 0.02 to 0.04 Å longer than the reference value for three-coordinate Al. Such an elongation is expected when the coordination number is increased from three to four through the formation of the (partly dative) bridge bonds. We have already adopted (on page 13) Al–N = 1.81 Å as a reference value for a normal bond distance from N to a four-coordinate Al atom.

It might be objected that this rationalization of the difference between bridging and terminal Al–N bond distances in **8** is unnecessarily cumbersome: since the bridging N atoms are four-coordinate and the terminal N atoms are three-coordinate, would it not be simpler to suggest that the difference is due to the different coordination numbers?

In the complex **9**[61] which, like **8**, contains two bridging and four terminal N atoms, the coordination number of the bridging N atoms has been reduced from four to three and the coordination number of the terminal N atoms from three to two. The $Al{-}N_{(bridge)}$ distances are reduced by 0.04 Å, the $Al{-}N_{(terminal)}$

distances by 0.03 Å. The difference between bridging and terminal Al–N bond distances remain essentially unaltered at 0.15 Å. Similarly, the reduction of the coordination number of the bridging N atom from four in complex **10**[62] to three in complex **11**[63] leads to a reduction of the Al–N bridge bond distance by only about 0.02 Å. We submit, therefore, that the coordination number is a less-important variable than the normal and dative bonding contributions.

The shortest endocyclic Al–N bond is found in **12**[64] where the bridging N is four-coordinate: the shortening relative to **7** is presumably an inductive effect of the electronegative Cl atoms on the acceptor atom.

In conclusion, the endocyclic Al–N bond distances in dimeric and trimeric Al amides, which we describe as having equal normal and dative bonding contributions, all fall in the range 1.94±0.3 Å. Repulsion between geminal Al atoms favor Al–N–Al angles greater than 115°, which in turn favors the formation of six-membered rings. However, the Al–N–Al angles appear to be very flexible, and when the substituents on neighboring Al and N atoms are all as bulky as or bulkier than Methyl groups, repulsion between them leads to the formation of four-membered rather than six-membered rings.

1.8.1.3. Other Species with Bonds with Equal Normal and Dative Bonding Contributions

The mixed-bridge complex **13**[65] must be described in terms of normal bond/no-bond Al–C canonical forms and normal bond/dative bond Al–N canonical forms, **13a** and **13b**, respectively. The Al–N bonds are about 0.05 Å longer, and the Al–N–Al angle somewhat sharper, than in the dinuclear complexes described in the preceding section.

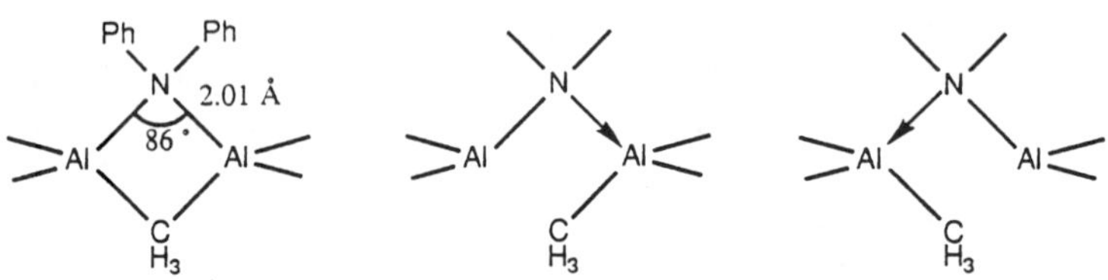

13 **13a** **13b**

The Al–N bond distances in each of the complexes **14**[66] and **15**[67] are indistinguishable and average 1.98 and 1.93 Å, respectively. Both may be described in terms of two canonical forms with bonding sequences

normal Al–N/ normal N–P/ double P=N/ dative N→Al

and the reverse, implying equal normal and dative bonding contributions to each Al–N bond.

The Al–N bond distances in the complexes **16**[68] and **17**[69] are also indistinguishable, and average 1.93 and 1.90 Å, respectively. Each may be described in terms of two canonical forms:

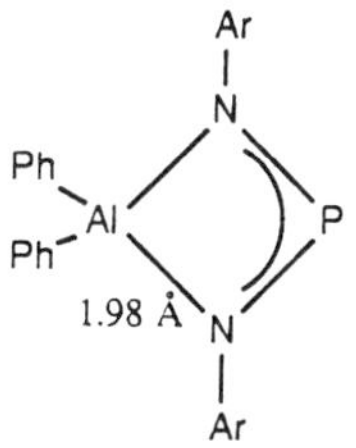

Ar = 2,4,6 -*t*-$Bu_3C_6H_2$
14

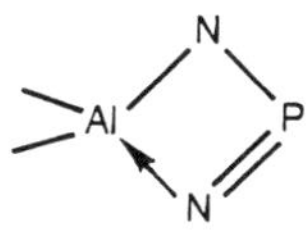

14a

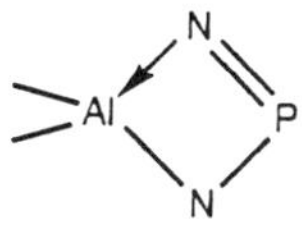

14b

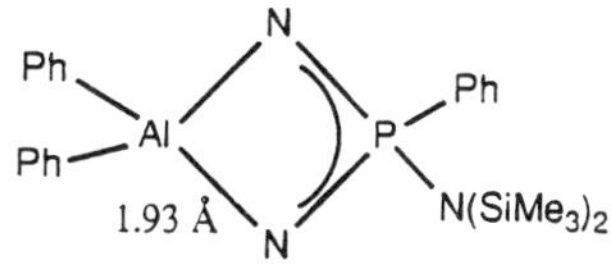

15

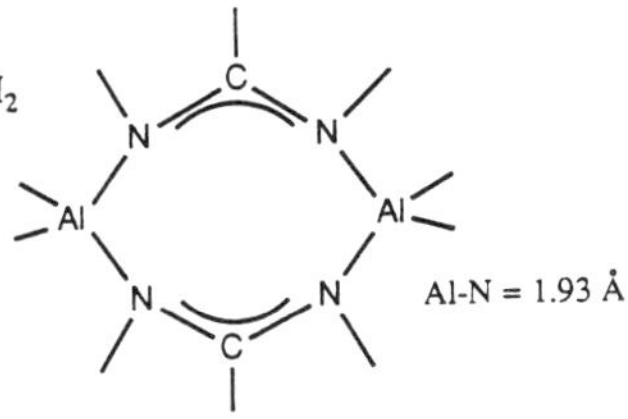

16

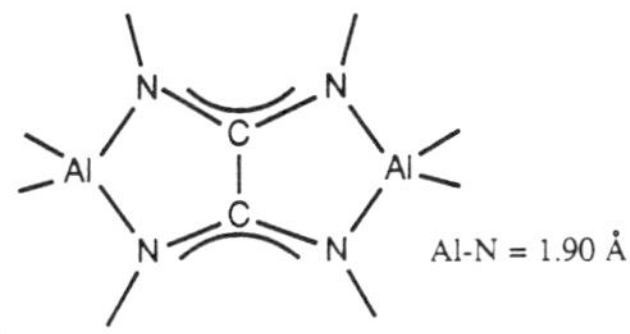

17

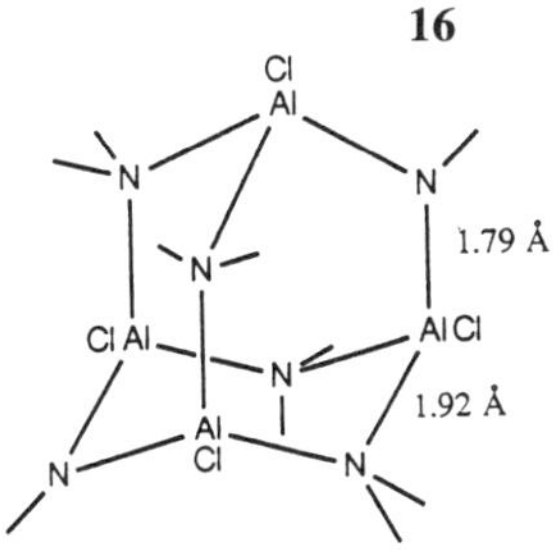

18

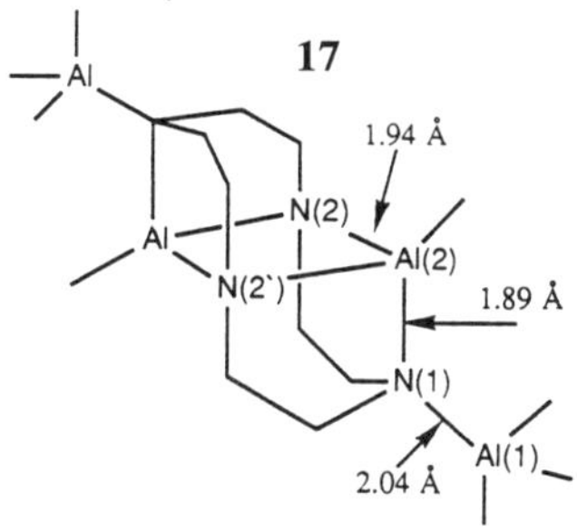

19

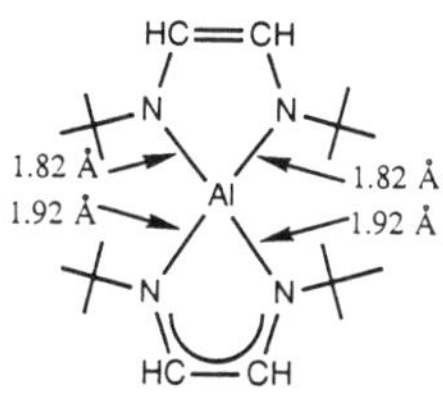

20

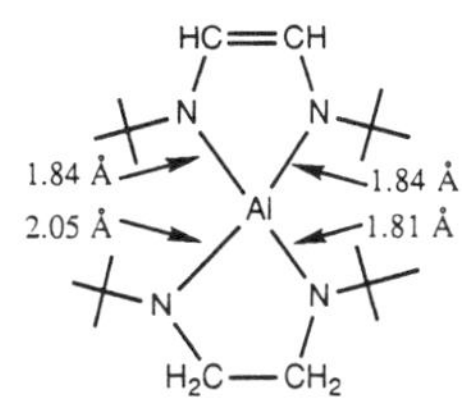

21

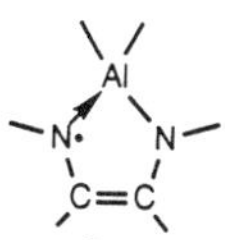

20a

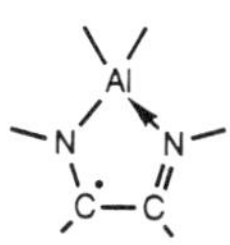

20b

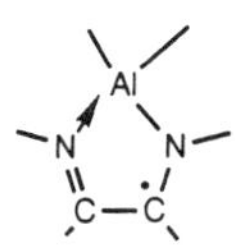

20c

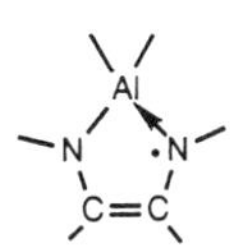

20d

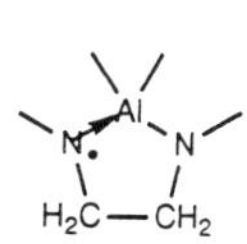

21a

normal Al–N/ normal N–C/ double C=N/ dative N→Al

and the reverse, bridging the two aluminum atoms rather than chelating them as in **14** and **15**.

Similarly,

normal/{normal/double}$_2$/dative

and reverse sequences are possible with five-atomic chelating and bridging ligands, but not, as we shall see below, with four-atomic ligands unless they carry an unpaired electron.

We now turn our attention to three complexes which contain more than one type of Al–N bond. In the cage-compound $Al_4Cl_4(\mu\text{-NMe})_2(\mu\text{-NMe}_2)_4$,[70] **18**, the bonds between Al and three-coordinate N which must be normal bonds, average 1.79 Å, while the bonds to the four-coordinate N, which must have equal normal and dative bonding contributions, average 1.92 Å.

The structure of the centrosymmetric tetranuclear complex **19**[71] is related to that of **3**: The N(1)–Al(1) bond at 2.04 Å must be dative, the N(1)–Al(2) bond at 1.89 Å normal, and the Al–N(2) and –N(2') bonds at 1.94 Å half normal and half dative.

The complex **20** is formed in a reaction of aluminum vapor with 1,4-di-*tert*-butyl-1,4-diazabutadiene (dbdab), and complex **21** in a reaction of Me_3NAlH_3 with dbdab in hexane.[72] Both **20** and **21** are paramagnetic; ESR spectra indicate that the unpaired electron in each is ligand centered. In both compounds the bond distances from the Al atom to the N atoms in the upper ring and the C=C double bond distance suggest that the ligand is bonded to the metal through two normal Al–N bonds. The magnitude of the Al–N bond distances to the lower ring in **20** as well as the C–C and C–N distances in the ligand, suggests that each Al–N bond is half normal and half dative.

The structure of **21**, on the other hand, suggests that addition of the two H atoms to the two C atoms in the lower ring freezes the complex in the canonical form **21a**.

1.8.1.4. Iminoalanes, Al–N bonds with 2:1 Normal:Dative Bonding Contributions

As mentioned earlier, the iminoalane MeAlN(2,6-di-*iso*propylphenyl), **22**,[28] is unique in forming a trimer with three-coordinate Al and N atoms, all other known iminoalanes form tetramers and higher aggregates with four-coordinate Al and N atoms. In the case of **22** formation of such aggregates is presumably prevented by the bulk of the aryl ligand. The trimer consists of a planar six-membered ring with alternating Al and N atoms. The average Al–N bond distance 1.78 Å, is consistent with a normal Al–N bond possibly shortened by 0.03 Å or 0.04 Å due to π-bonding. With three pairs of π-electrons on the N atoms, the complex is pseudo-isoelectronic with benzene. Homodesmotic reaction energies obtained by *ab initio* molecular orbital calculations suggest, however, that the delocalization energy relative to three *single* and three *double* Al–N bonds is about 2 kcal/mol,[73] about one tenth of the delocalization energy of benzene. This difference suggests that Al–N π-bonding energy in a monomeric aminoalane R_2NAlR_2 is about one tenth of the C=C π-bond energy in $H_2C{=}CH_2$, that is about 6 kcal/mol.

All other iminoalanes which have been investigated thus far, have been found to form tetramers or higher oligomers. Therefore, MeAlN(*iso*propyl)

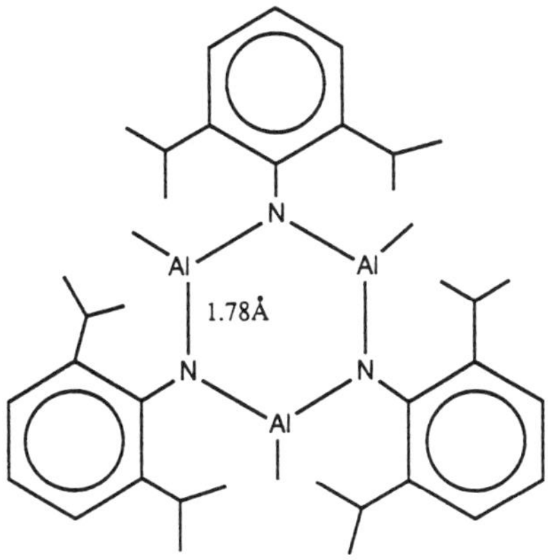

22

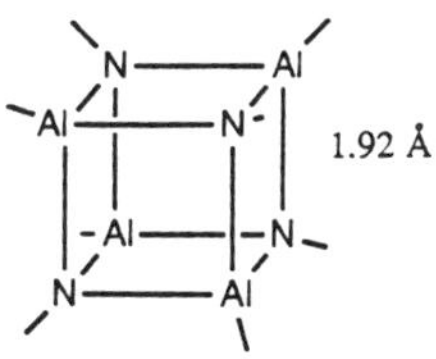

$(MeAlN\text{-}i\text{-}Pr)_4$
23

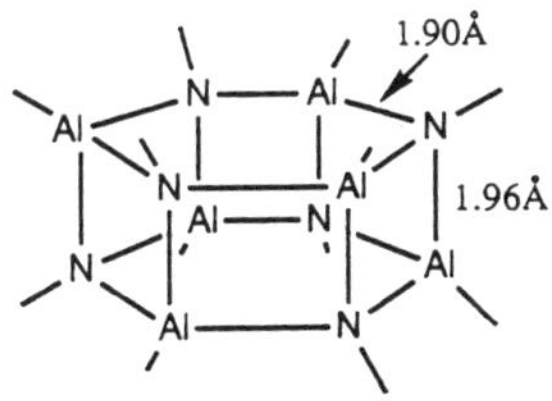

$(HAlN\text{-}i\text{-}Pr)_6$
24

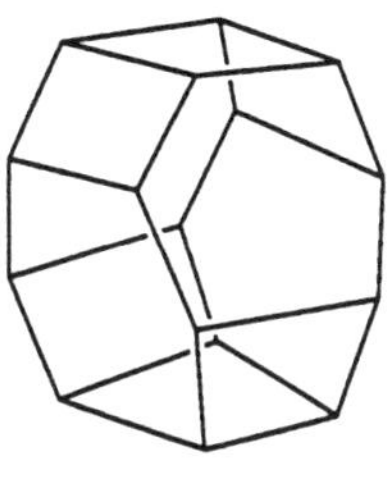

$(HAlN\text{-}i\text{-}Pr)_8$
25

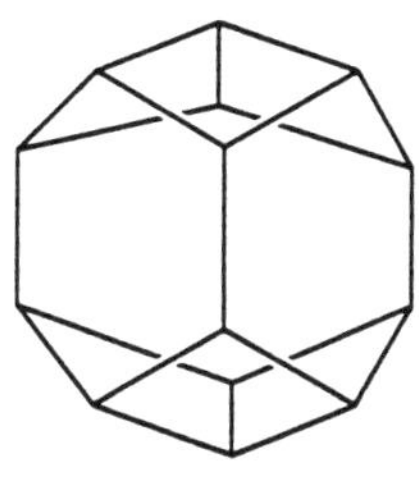

$(MeAlNMe)_7$
26

forms a cubic tetramer **23**,[74] HAlN(*iso*propyl) forms a tetramer and a hexamer **24**,[75] HAlN(*n*-propyl) forms a hexamer and an octamer **25**,[76] and MeAlNMe forms a heptamer **26** and an octamer.[77]

The Al–N bond distances in each of the tetramers are indistinguishable: in **23** the average is Al–N = 1.92 Å; in $[HAlN(\textit{iso}propyl)]_4$ the average is 0.01 Å smaller. The terminal Al–C or Al–H bonds, as well as the terminal N–C bonds, must be normal. Since each Al or N atom can form two more normal bonds, each Al–N bond in the cube must be two-thirds normal and one-third dative. This rationalization is consistent with the observed bond distances which are close to the appropriate mean of the reference value for normal and dative bond distances to Al carrying only H or organyl substituents:

$$(2/3)1.81\ \text{Å} + (1/3)2.08\ \text{Å} = 1.90\ \text{Å}$$

The average Al–N bond distances in the hexamers **24** and [HAlN(*n*-propyl)]$_6$[17] are 1.92 and 1.91 Å, respectively,[76,77] and may also be described as having two-thirds normal and one-third dative bonding contributions. It should be noted, however, that the bonding in six-membered rings, average values 1.90 and 1.89 Å, respectively, are significantly different from the length of the transverse bonds between hexagons, 1.96 and 1.96 Å, respectively. The latter bond distances are very close to the bond distances in simple trimeric or dimeric aminoalanes (e.g. **7** or **9**) which we have described as having equal normal and dative bonding contributions. If the transverse bonds are assumed to have equal normal and dative bonding contributions, it follows that each of the two bonds in the hexagon formed by each Al or N atom, must be three-quarters normal and one-quarter dative. The corresponding mean of the reference values for pure normal and dative bond distances in good agreement with the observed values are:

$$(3/4)1.81\ \text{Å} + (1/4)2.08\ \text{Å} = 1.88\ \text{Å}$$

It may be of interest to note that the mean bond distance in crystalline aluminum nitride in which each Al or N atom is four-coordinate and where each Al–N bond must be similarly described as three-quarters normal and one-quarter dative, is 1.89 Å.[78]

In **25** and **26** we indicate the shapes of the polyhedra formed by the skeletal Al and N atoms in an octamer[76] and a heptamer.[77] Al and N atoms occupy alternating positions in such a way that each Al(N) is bonded to three N(Al) atoms. In addition, each skeletal atom carries one terminal H atom or organyl group. The surface of the octamer **25** is formed by four hexagons and six squares. The longest Al–N bonds are those shared by two squares, followed by bonds shared between a square and a hexagon: the shortest bonds are those shared by two hexagons. The mean Al–N bond distance is 1.92 Å, indistinguishable from that found in the other oligomeric iminoalanes.

The surface of the heptamer **26** is formed by three hexagons and six squares. The structure has not been determined with sufficient accuracy to allow discussion of individual bond distances; the average Al–N bond distance is 1.91 Å.

1.8.2. Compounds with Five-coordinate Aluminum

1.8.2.1. Aluminum Atoms with Three Normal and Two Dative bonds

We have seen that the N→Al bond in Me_3NAlCl_3 is significantly stronger than the bond in Me_3NAlMe_3. No thermochemical information on the bond strength

Table 1.5. N–Al bond distances (in Å units) in adducts with three normal and two dative bonds.

Donor	Acceptor	R(N→Al)	Reference
TMEDA[a]	AlH_3	2.21 (2X)	79
Me_3N (2X)	AlH_3	2.18(1)[c] (2X)	80
Me_3N (2X)	AlH_3	2.19(2)[d] (2X)	81
Morph[b] (2X)	$AlCl_3$	2.09 (2X)	82
Me_2HN (2X)	$AlCl_3$	2.06 (2X)	83

[a] TMEDA=*N,N,N',N'*-tetramethylethylenediamine.
[b] Morph=morpholine=*cyclo*-$NH(CH_2)_4O$.
[c] by X-ray crystallography.
[d] by gas phase electron diffraction.

in Me_3NAlH_3 is available, but the N→Al bond distance indicates that it is marginally stronger than in the $AlMe_3$ analogue. AlH_3 and $AlCl_3$ are found to form adducts where two N atoms are joined to Al through dative N→Al bonds; no such adducts are known for $AlMe_3$.

The coordination polyhedron of Al in such adducts is trigonal bipyramidal with the donor atoms in axial positions in agreement with the hypothesis that the spatial requirements of accepted electron pairs are smaller than the spatial requirements of the normal bond pairs.

The N→Al bond distances in these adducts are listed in Table 1.5.

The crystalline adduct of AlH_3 with *N,N,N',N'*-tetramethylethylenediamine (TMEDA) consists of polymeric chains of alternating AlH_3 and TMEDA units. The mean N→Al bond distance is 2.21 Å, that is about 0.15 Å longer than in Me_3NAlH_3. An investigation of the 1:2 adduct $(Me_3N)_2AlH_3$ in the gas phase has yielded a N→Al bond distance of 2.19 (2) Å and an Al–H bond distance of 1.53 (3) Å; the latter, though inaccurate, is not significantly longer than in Me_3NAlH_3.

The morpholine rings are coordinated to $AlCl_3$ through the N rather than through the O atoms. This is hardly surprising since we have seen that dative N→Al bonds are generally stronger than dative O→Al bonds. The N→Al bond distance in this adduct and in $(Me_2HN)_2AlCl_3$ are 0.13 and 0.09 Å longer than in the 1:1 adduct with Me_3N. The Al–Cl bonds are 0.06 Å longer than in the 1:1 adduct; the dative bond distances appear to be more sensitive to the increase of coordination number than the normal bond distances.

1.8.2.2. Aluminum Atoms with Two Normal, One Dative Bond and Two Bonds of Intermediate Type

The five-coordinate Al(1) atom in **27**[84] is trigonal bipyramidal, the four Al(1)–N distances range from 1.83 to 2.14 Å. The Al(1)–C and the Al(1)–N(1)

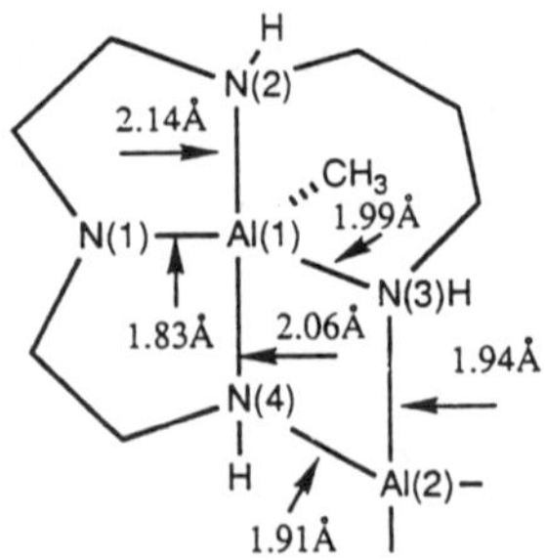

27

bonds must be normal; the observed Al(1)–N(1) bond distance is 1.83 Å. The Al(1)–N(2) bond must be dative; the observed bond distance is 2.14 Å. As expected the normally bonded C and N(1) occupy equatorial positions, while the datively bonded N(2) occupies an axial position.

Al(1) is bonded to the four-coordinate Al(2) through two amido N atoms, N(3) and N(4). We have seen that when two four-coordinate Al atoms are connected through two amido bridges as in $(Me_2AlNMe_2)_2$, **7**, normal and dative bonding is delocalized to yield four equal bonds at about 1.95 Å. In **27** the bond distances in the four-membered Al_2N_2 ring ranged from 1.91 to 2.06 Å.

There are probably several reasons for this variation. First, bond distances to the acceptor atom are expected to increase with increasing coordination number. The mean of the bond distances from N(3) and N(4) to the five-coordinate Al(1) is about 0.10 Å greater than the mean of the bond distances to the four-coordinate Al(2).

Second, in compounds of main group elements with trigonal bipyramidal coordination, the bonds to axial ligands tend to be a few hundreths of an Å unit longer than bonds to equatorial ligands. Finally, we interpret the small, but significant, differences between the bond distances to the four-coordinate Al(2), 1.935(2) Å and 1.911(3) Å, as an indication that normal and dative bonding are incompletely delocalized over the Al_2N_2 ring: the shortest of the two, Al(2)–N(4) and consequently also Al(1)–N(3) bond appear to be more than half normal, while the longer bonds Al(2)–N(3) and Al(1)–N(4) appear to be more than half dative. We further suggest that the reason for the incomplete delocalization is that it would require energy to increase the normal bonding contribution to the axial Al(1)–N(4) bond at the expense of the equatorial Al(1)–N(3) bond.

1.8.2.3. Aluminum Atoms with One Normal Bond and Four Bonds of Intermediate Type

The coordination geometries of the five-coordinate Al atoms in chloro(phthalocyaninato)aluminum, **28**,[85] and **29**,[86] are square pyramidal. The Al–Cl bond distance in **28** is 2.19 Å, the four Al–N bond distances range from 1.96 to 2.02 Å and average 1.98 Å, slightly greater than 1:1 normal:dative bond distances to

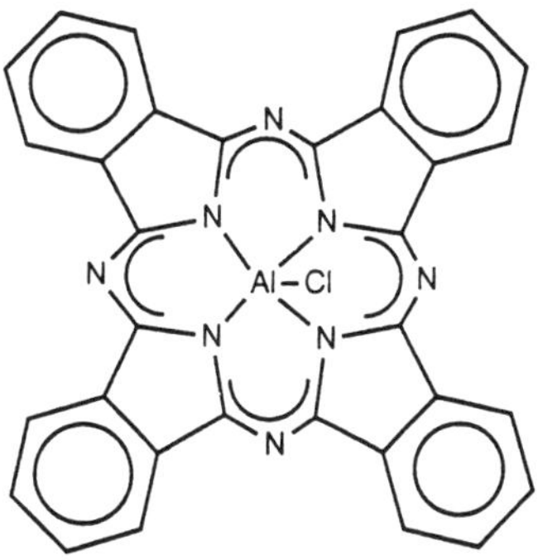

28

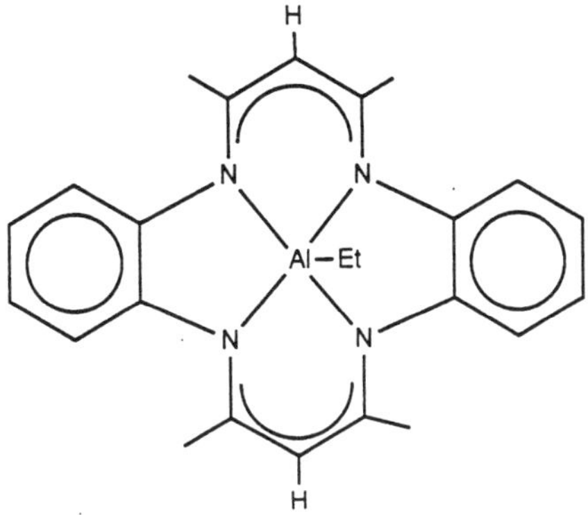

29

four-coordinate Al. The four *cis* N-Al-N angles range from 85 to 90°, the *trans* N-Al-N angles are 155° and 157°.

The Al–C bond distance at 1.98 Å in **29** is unexceptional, the four basal Al–N bonds average 1.97 Å. The *trans* N–Al–N angles average 145°, about 10° smaller than in **28**.

The square pyramidal configuration in these compounds may be imposed by the macrocyclic ligands. On the other hand, the symmetry of the configuration does reflect a situation where the metal atom forms one strong normal bond and four equally strong bonds of intermediate type.

The trinuclear complexes **30**[87] and **31**[88] are closely related. The two terminal Al atoms are four-coordinate, the central Al atom, Al(1), is five-coordinate. The N–Al bonds to the four-coordinate Al range from 1.91 to 1.94 Å and may be described as 1:1 normal:dative bonds.

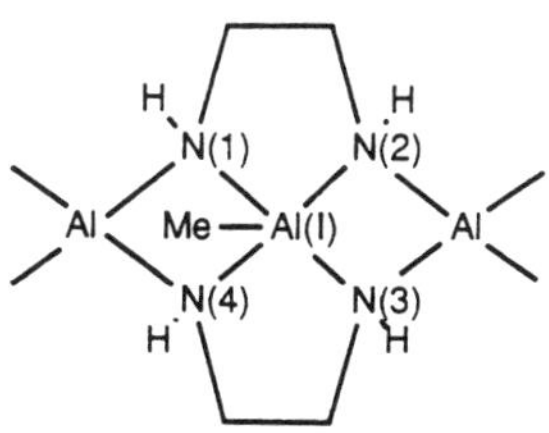

30

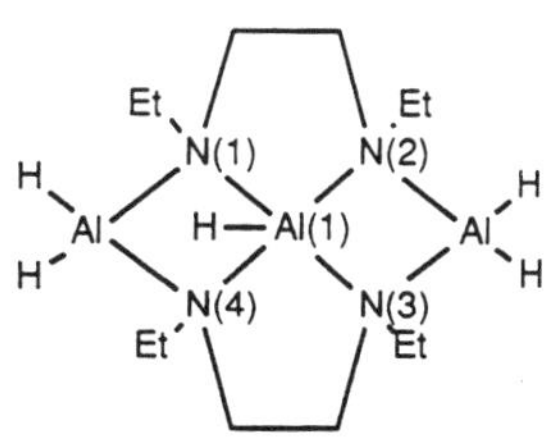

31

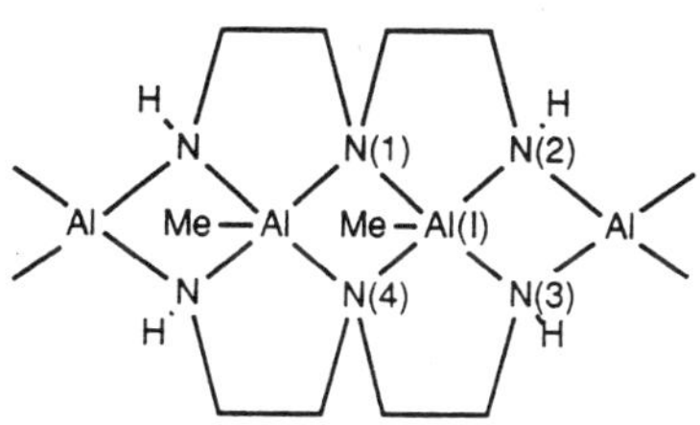

32

In **30** the largest N-Al-N angles at the central atom are N(1)–Al(1)–N(3) = 149° and N(2)–Al(1)–N(4) = 125°, a difference of 24°. For perfect trigonal bipyramidal configuration this difference should be 60°, for a square pyramidal configuration 0°. In **31** the corresponding angles are 154 and 127°, a difference of 27°. The observed geometries are thus very near the transition point between the two symmetric configurations. The bond distances defining the larger N–Al–N angle, which may be regarded as being closer to axial position in a trigonal bipyramid, are perceptibly longer than the bond distances defining the lesser N–Al–N angle.

The composition of **30** and **31** may be written schematically as $[R_2AlNR_2][RAl(NR_2)_2][R_2NAlR_2]$, and regarded as being formed by insertion of $[RAl(NR_2)_2]$ into a cyclic $[R_2AlNR_2]$ dimer.

Structure **32**[89] is closely related to **30** and **31** and may be regarded as formed by insertion of another $[RAl(NR_2)_2]$ unit. The structure is centrosymmetric; N(1)–Al(1)–N(3) = 151° and N(2)-Al(1)–N(3) = 124°.

Thus, of the five species containing four-coordinate Al atoms with one normal and four 1:1 normal:dative Al–N bonds, two are near perfect square pyramids, and the other three are at the transition from square pyramidal to trigonal bipyramidal structure.

1.8.3. Compounds with Six-coordinate Aluminum

The Al atom in **33**[90] contains one normal Al–C bond and one dative N→Al bond from the pyridine ligand. The coordination of the metal is distorted octahedral.

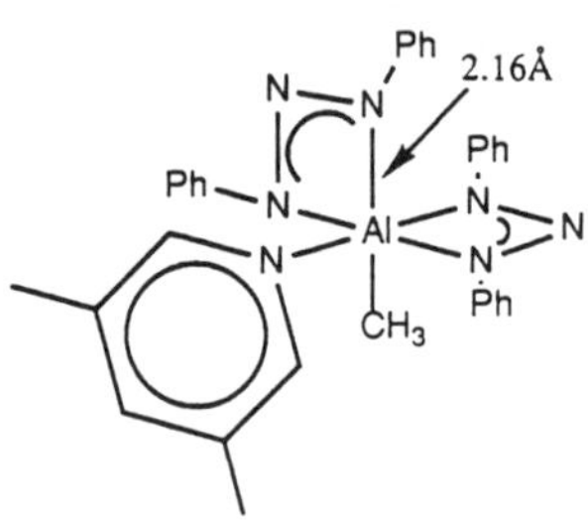

33

The N–N bond distances in the triazene ligands suggest that N–N–N π-bonding is delocalized, and that the four $Al–N_{(triazene)}$ bonds consequently are 1:1 normal:dative bonds. One would therefore expect the four Al–N(triazene) bond distances to be essentially equal, while the bond to the pyridine N-atom should be significantly longer. Unexpectedly, the pyridine N→Al bond and the three other Al–N bonds that are *cis* to the methyl group are found to be comparable, ranging from 2.01 to 2.05 Å, whereas the Al–N(triazene) bond which is *trans* to the methyl group is 2.16 Å. Leman and Barron [90] describe the lengthening of this bond as a *trans* effect, which is a reasonable proposal. What is

Table 1.6. P→Al bond distances (in Å units) in simple adducts of alanes P-centred donors.

Donor	Acceptor	R(P Al)	Reference
Me_3P	$AlMe_3$	2.53 (4)	16
Ph_3P	$AlMe_3$	2.54	91
$(o\text{-Tolyl})_3P$	$AlMe_3$	2.58	91
$Ph_2PCH_2PPh_2$	$AlMe_3$ (2X)	2.58, 2.52	92
$Ph_2P(CH_2)_2\text{-}PPh_2$	$AlMe_3$ (3X)	2.54 (2X)	93
$[Ph_2P(CH_2)_2]_2\text{-}$	$AlMe_3$ (3X)	2.54 (2X)	94
PPh		2.52 (1X)	
Me_3P	$AlMe_2OAr$[a]	2.50	95

[a] Ar=4-methyl,2,6-di-*tert*-butylphenyl.

surprising is that the *trans* position is not occupied by the datively bonded pyridine. We are left to suggest that this might have occurred if the bite of the triazene ligands had been large enough to give a reasonably square coordination if all four N atoms had occupied positions in the same plane.

1.9. Aluminum–Phosphorus Compounds

1.9.1. Compounds with Four-coordinate Aluminum

1.9.1.1. Dative P→Al bonds

The dissociation enthalpy of the adduct Me_3PAlMe_3 in hydrocarbon solution is 21.0±0.3 kcal/mol (Table 1.3). The bond distances and valence angles in this adduct have already been discussed in Section 1.3. The P→Al bond distance has been determined with large error limits, but comparison with the other entries of Table 1.6 shows that the P→Al bond distances in several adducts of $AlMe_3$ with P-centered donors are close to 2.54 Å.

Only one adduct where the Al atom carries more electronegative substituents than methyl groups has been characterized structurally: in an adduct of a dimethylaluminum-aryloxide with Me_3P, P→Al = 2.50 Å.[95] The bond distance in an adduct of $AlCl_3$ would probably be even shorter.

To the best of our knowledge, no structure of a complex containing a normal Al–P bond has been determined. The modified Schomaker–Stevensen rule yields an estimate of 2.31 Å.

1.9.1.2. Compounds Containing Al–P bonds of Intermediate Type

We now turn to the description of complexes where the bonding must be described in terms of more than one canonical form.

Dimethylaluminum dimethylphosphamide Me_2AlPMe_2, **34**, is trimeric in the gas phase.[96] The Al–P bonds must be described as having equal normal and dative bonding contributions. The Al–P bond distance is equal to the mean of the standard value of a P→Al bond (2.54 Å), and the Al–P normal bond distance estimated from the modified Schomaker–Stevensen rule (2.31 Å); 2.43 Å. The bond angles are similar to those found in the sterically least encumbered trimeric dimethylaluminum amide, **4**: in agreement with the VSEPR model, the endocyclic angles at P are greater than tetrahedral, whereas the angles at Al are less than tetrahedral.

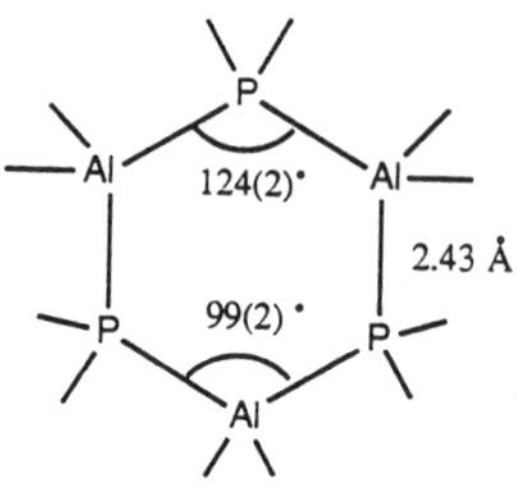

34

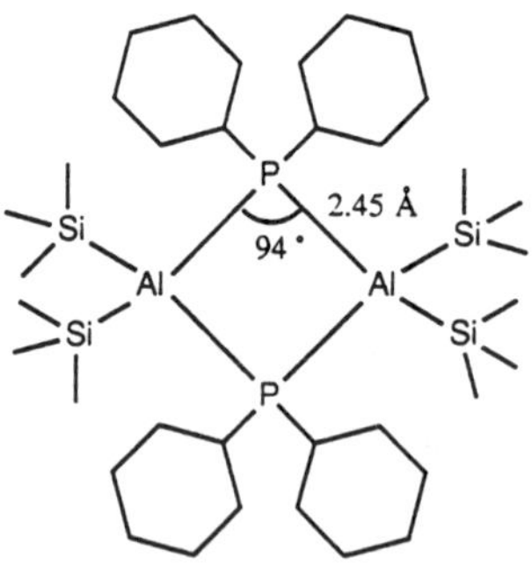

36

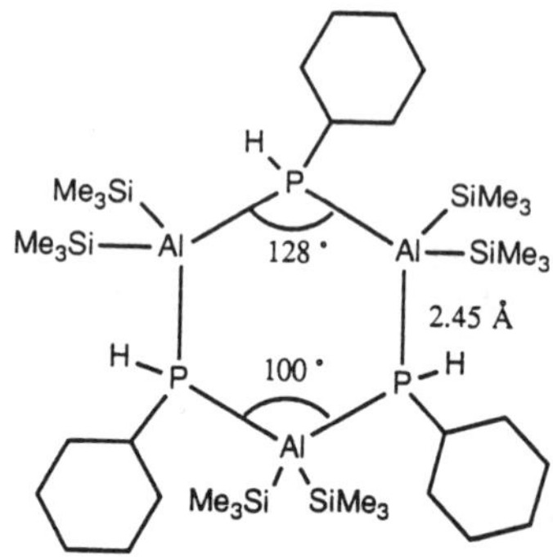

35

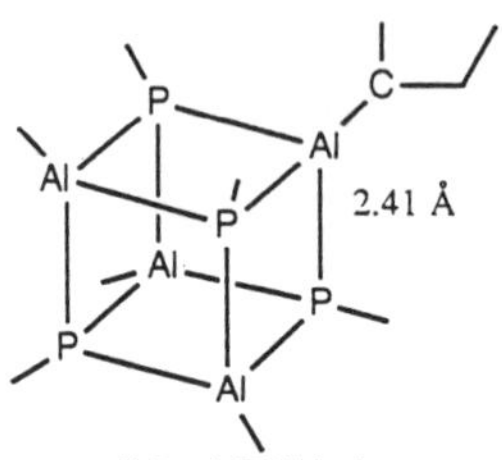

$[i\text{-Bu-AlPSiMe}_3]_4$

37

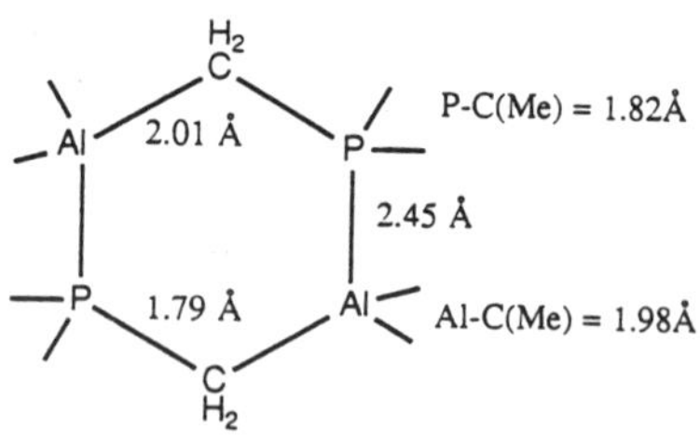

38

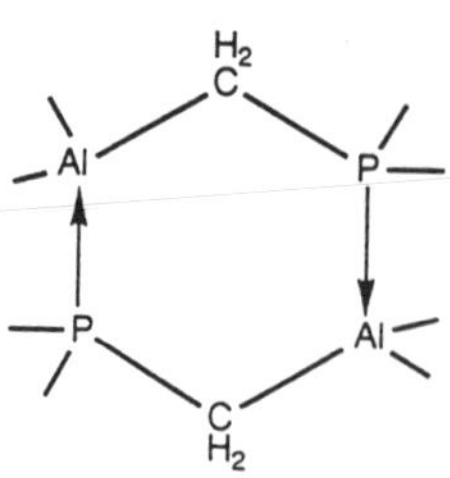

38a

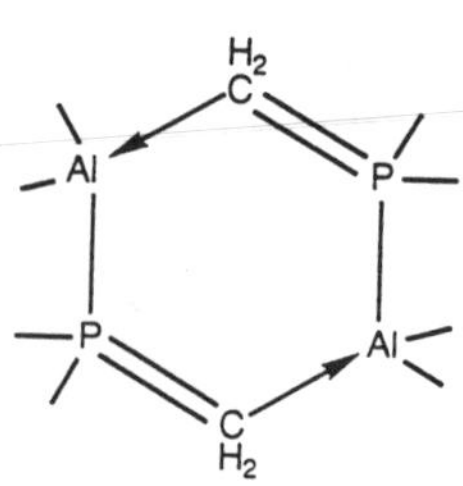

38b

Another trimeric phosphamide is $[(Me_3Si)_2AlPH(cyclo\text{-}C_6H_5)]_3$, **35**.[97] The Al–P bond distance and the endocyclic valence angles are similar to those of **34**. The six-membered Al_3P_3 ring is in a twist boat conformation. Introduction of a second cyclohexyl group at the phosphorus atoms leads to a dimeric compound **36**,[97] presumably because the trimer is destabilized by steric interactions between the large ring substituents. The four-membered ring is planar, and the Al–P bond distance equal to that in **35**.

Recently the structure of the first alkylaluminum phosphimide, $[i\text{-}BuAl\text{-}PSiMe_3]_4$, **37**, has been published.[98] As with some N analogs, this is tetramer with cubane structure. The mean Al–P–Al angle is slightly less than 90°. The Al–P bonds, which must be described in terms of 2:1 normal:dative contributions, average 2.41 Å, versus 2.39 Å calculated from the reference values for normal and dative bond distances. The mean Al–P bond distance in crystalline aluminum phosphide, where the coordination number of the atoms are four, and the bonds must be described as one-quarter dative and three-quarters normal, is 2.37 Å, which is again equal to the value estimated from the reference values.[99]

The complex **38**[100] which contains a non-planar $C_2P_2Al_2$ six-membered ring in the chair conformation looks deceptively simple; at first glance one might expect the structure to be adequately described by the canonical form **38a**. The Al–P bond distance is, however, significantly shorter than the P→Al bond in Me_3PAlMe_3, which in turn suggests that it is significantly stronger. One possible explanation for the increased strength would be through space electrostatic interaction between the antiparallel Al→P bonds which may stabilize a greater charge transfer from P to Al and hence a strengthening of the dative bond.

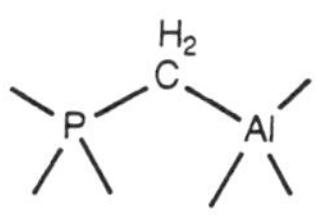

39

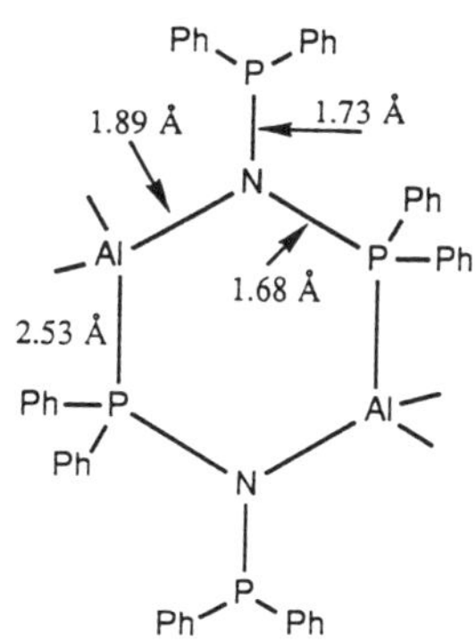

40

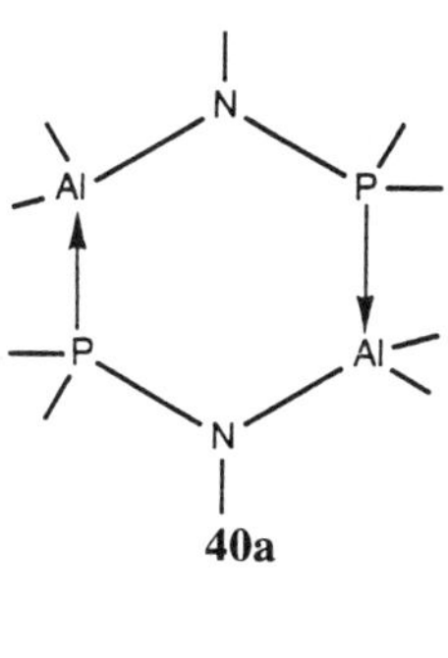

40a

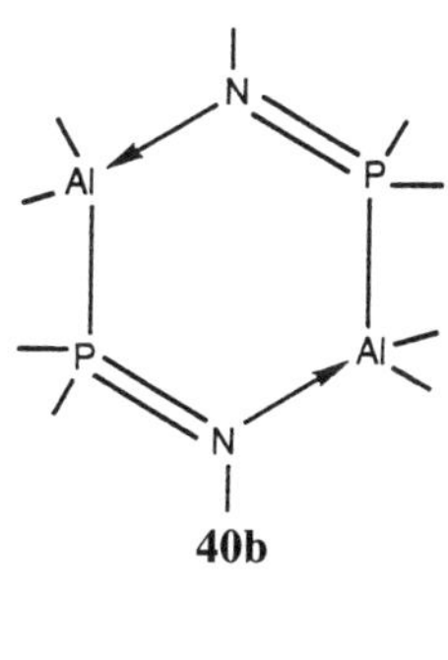

40b

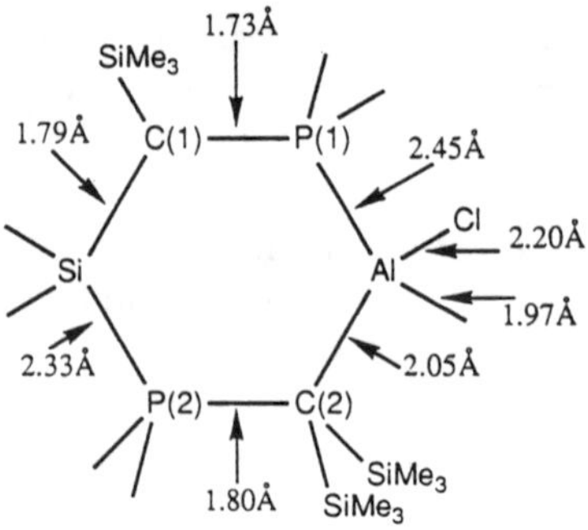

41

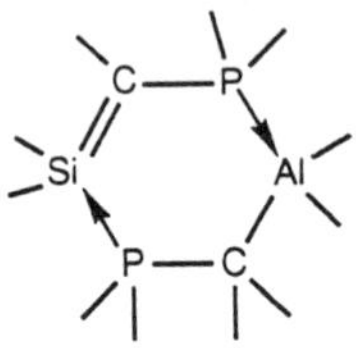

41a

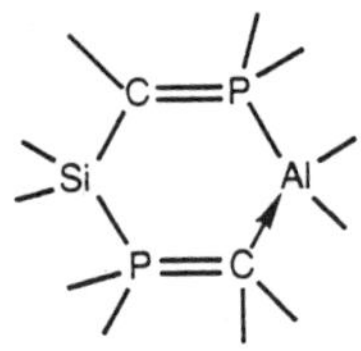

41b

Karsch and co-workers noted than the endocyclic P–C bond is 0.03 Å shorter than the average of the exocyclic P–$C_{(Me)}$ bonds, and that the endocyclic Al–C bond is 0.03 Å longer than the exocyclic Al–$C_{(Me)}$ bonds, and suggest that the canonical form **38b** contributes.[100] Such a suggestion seems reasonable if it is recalled that the adduct **39** can be synthesized by direct combination of Me_3PCH_2 and $AlMe_3$.[101] This adduct is monomeric in benzene and may be sublimed without decomposition. The structure has been established by spectroscopic techniques, but has not been studied by diffraction methods.

The bonding in **40**[102] is related to that in **38** in so far as it is possible to draw two canonical forms, one **40a** with three-valent P and a dative P→Al bond, and the other **40b** with five-valent P and a dative N→Al bond. However, the average Al–P, Al–N, and N–P bond distances all indicate that **40a** contributes more than **40b**.

The structure of the cyclic complex **41**[103] is more difficult to unravel. Again two canonical forms are possible, **41a** with three-valent P and **41b** with five-valent P. The bonding contributions from both are in agreement with the observed Al–P bond distance, with the observation that the endocyclic Al–C bond is shorter than the exocyclic, with the shortness of the P(1)–C(1) bond and C(1)–Si bonds and with the length of the Si–P bond (about 0.09 Å shorter than in $P(SiH_3)_3$). However, it is not clear why the P(1)–C(1) bond is significantly shorter than the P(2)–C(2) bond.

1.9.2. Compounds with Six-coordinate Aluminum

We are not aware of any structure determination of a complex where phosphorus is bonded to a five-coordinate aluminum atom and only one example where

P is bonded to a six-coordinate aluminum. The bonding in the four-membered AlP_2C rings in **42**[104] must be described in terms of the canonical forms

normal Al–P / normal P–C / double C=P / dative P→Al

and reverse. The Al–P bonds are then half dative and half normal. The mean Al–P bond distance is 2.50 Å, only 0.05 Å longer than 1:1 bonds to four-coordinate Al.

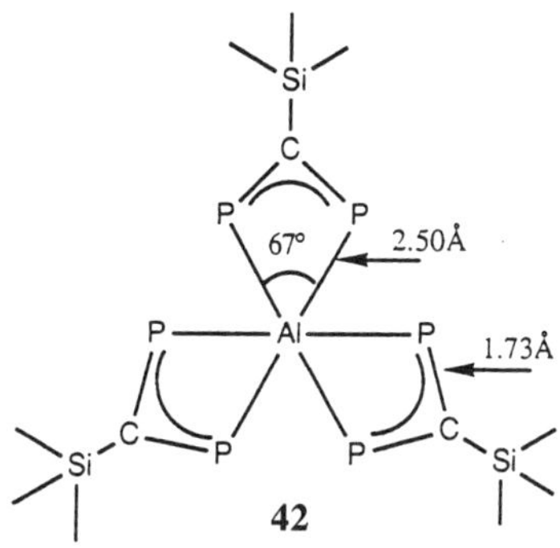

42

1.10. Aluminum–Oxygen Compounds

1.10.1. Compounds with Four-coordinate Aluminum

1.10.1.1. Dative O→Al bonds

The compound $MeAl(OAr)_2$, (Ar = 2,6,4-*tert*-butyl-4-methylphenyl), where aluminum is three-coordinate and forms two normal Al–O bonds at 1.69 Å, has been mentioned previously. The adduct with isopentaldehyde, **43**, and the adduct of Me_2AlOAr with benzophenone, **44**, have been studied by X-ray crystallography.[105] These adducts contain both normal and dative Al–O bonds, the

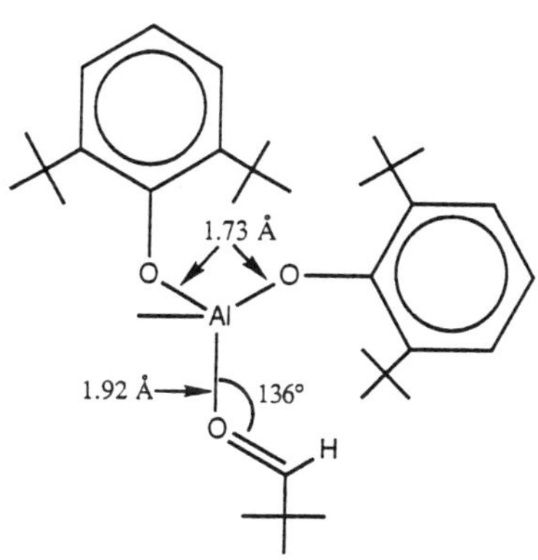

43

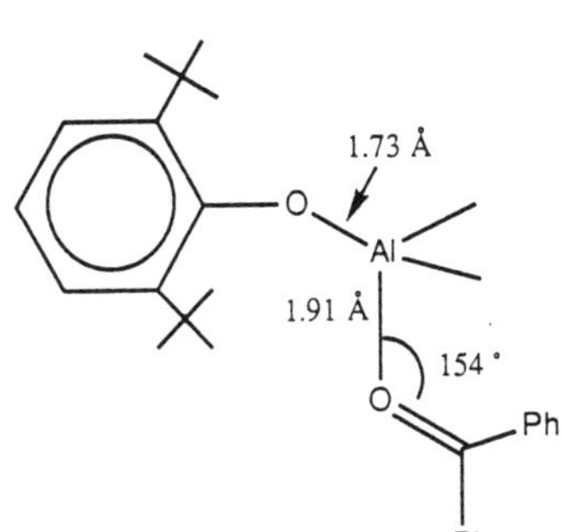

44

dative bonds are nearly 0.20 Å longer than the normal bonds. Since both the normally and the datively bonded O atoms are two-coordinate, these complexes also demonstrate that the bonding mode, rather than the coordination number of the atoms, is the dominant factor determining bond distances. The valence angles and conformations of the two adducts will be discussed later.

The dissociation enthalpies in Table 1.3 show that data are available for a greater variety of O-centered than N-centered donors. In particular, the data for adducts of $AlBr_3$ suggest a number of points:

1. The strongest complex is formed by the aldehyde PhHC=O; ΔH = 50 kcal/mol which amounts to less than 50% of the mean bond dissociation enthalpy of a normal Al–O bond. Since the donor atom is mono-coordinate and the β-carbon carries a H atom, PhHC=O is probably the sterically least encumbered of the electron donors investigated.
2. The dissociation enthalpies of the ketones PhMeC=O and Ph_2C=O are 30% lower than the dissociation enthalpy of PhHC=O and similar to the dissociation enthalpy of the related alcohol PhOH. The dissociation enthalpies of the adducts of the sterically less encumbered alcohols EtOH and BuOH are higher than for PhOH.
3. Comparison of the dissociation enthalpies of the adducts of EtOH and Et_2O, of BuOH and Bu_2O, and of PhOH and Ph_2O shows that the alcohols form stronger adducts than the ethers.
4. The dissociation enthalpies of adducts with ethers decrease in the sequence $Et_2O > Bu_2O >$ PhOMe = PhOEt $> Ph_2O$, that is with increasing bulk of the organyl groups bonded to O.

To conclude, we suggest that dissociation enthalpies decrease in the order:

aldehyde > ketone = alcohol > ether

and that this order primarily is due to increasing steric encumberance of the donor atom. Similarly, when donors within each of these classes are compared, the strongest dative bonds are formed when the donor atom is bonded to the least bulky substituents.

The dative O→Al bond distances listed in Table 1.7 fall into four distinct groups:

1. The O→Al bond distances from ether donors range from 2.01 Å in the simplest form, Me_2OAlMe_3, to 1.90 Å as the methyl groups on the metal atom are replaced by more electron-withdrawing silyl, aryl or benzyl groups. We assume that this decrease reflects an increasing strength of the dative bonds. In the following, 2.00 Å is used as the reference value for a dative bond from an ether O to an Al atom bonded to three alkyl groups.
2. As has been shown, aldehydes and ketones tend to form stronger bonds to alanes than ethers. The O→Al bond distances in adducts of aldehydes and ketones with methylaluminum aryloxides fall in the narrow range of 1.90 – 1.92 Å, that is, in the lower end of the range of ether adducts.

Table 1.7. O→Al bond distances (in Å units) in simple adducts of alanes with O-C centred donors.

Donor	Acceptor	O→Al	Reference
Me_2O	$AlMe_3$	2.01 (1)	106
Dioxane	$AlMe_3$ (2X)	2.02 (2)	107
15-crown-5-ether	$AlMe_3$ (4X)	2.01 (4X)	108
18-crown-6-ether	$AlMe_3$ (4X)	1.98 (4X)	109
THF	Al(mesityl)$_3$	1.97	110
Et_2O	$Al(SiMe_3)_3$	1.95	111
THF	$AlPh_2Si(SiMe_3)_3$	1.93	112
Et_2O	Al(*o*-tolyl)$_3$	1.93	113
Et_2O	Al(benzyl)$_3$	1.90	114
$Me_3CHC{=}O$	$AlMe(OAr)_2$	1.92	105
$Ph_2C{=}O$	$AlMe(OAr)_2$	1.90	115
$Ph_2C{=}O$	$AlMe_2(OAr)$	1.91	105
EtClC=O	$AlCl_3$	1.85	116
PhClCl=O	$AlCl_3$	1.82	117
(*o*-Tolyl)ClC=O	$AlCl_3$	1.82	118
(*p*-Tolyl)ClC=O	$AlCl_3$	1.83	118
$Ph_3P{=}O$	$AlBr_3$	1.74	119
$Ph_3P{=}O$	$AlCl_3$	1.73	119

3. The O→Al distances in the third group, adducts of acid chlorides with $AlCl_3$, fall in the range 1.82 – 1.85 Å, and are thus only 0.10 Å longer than normal bonds from oxygen to four-coordinate aluminum. We do not believe that acid chlorides are better electron donors than aldehydes or ketones. In particular the replacement of the aldehyde H atom in PhHC=O by Cl should both decrease the electron density at the donor atom and increase the steric encumberance. We believe, rather, that the shortening of the O→Al bond relative to the second group is due to the greater acceptor ability of $AlCl_3$.
4. Finally the O→Al distances in Ph_3POAlX_3 (X = Cl or Br) are more than 0.25 Å shorter than in Me_2OAlMe_3, and fall in the range of normal bond distances from oxygen to four-coordinate aluminum. Does this mean that these dative bonds are as strong as normal bonds?

Not necessarily: The dissociation enthalpy of $Me_3POAlMe_3$ is similar to that of Me_3NAlMe_3, suggesting that Me_3PO and Me_3N form equally strong dative bonds to alanes. Since the dissociation enthalpy of Me_3NAlCl_3 is 49 kcal/mol, the dissociation enthalpies of $Me_3POAlCl_3$ or $Ph_3POAlCl_3$ are expected to be about 50 kcal/mol, less than 50% of our estimate of the strength of a normal Al–O bond. The valence angles P-O→Al are linear in both the $AlCl_3$ and $AlBr_3$ adducts. Burford and co-workers[119] suggest that the two π lone pairs of electrons on oxygen are engaged in π-bonding to phosphorus, and that the σ lone pair, which presumably has considerable s-character, forms a dative bond to aluminum.

We now turn our attention to the simplest ether adduct, Me_2OAlMe_3, **45**. The dissociation enthalpy is 21.9±0.2 kcal/mol, one-fifth of our estimated mean Al–O normal bond energy. The molecular structure has been determined by gas phase electron diffraction: the O→Al bond distance is 2.01(1) Å, about 0.30 Å longer than the reference value for normal bonds from O to four-coordinate Al which we shall justify below.

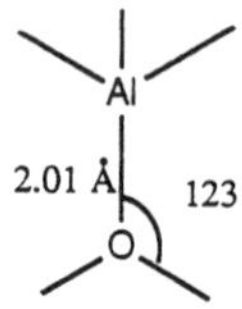

45

We have seen that when trimethylamine is coordinated to $AlMe_3$, the N–C bonds were slightly but significantly elongated, whereas the angle between them remained unaltered. The VSEPR model predicts that N–C bonds should become shorter and that the angle between them should increase on formation of the adduct. The elongation of the bonds was therefore rationalized as due to repulsion between the Al atom and the three C atoms bonded to N. The O–C bond distance in free dimethylether, 1.416 (1) Å, is slightly, but significantly elongated to 1.436 (3) Å on coordination to $AlMe_3$. Again we ascribe the elongation to repulsion between the atoms and groups bonded to the donor atom.

The C—C distance in free Me_2O is 2.34 Å, less than twice the geminal repulsion radius of C (2.50 Å), and the C–O–C angle is 111.5 (5)°, slightly greater than tetrahedral. On formation of the adduct the angle increases slightly, but significantly to 114.5 (7)°. This is somewhat surprising since the elongation of the C–O bonds have been ascribed to increased steric repulsion.

The most striking aspect of the structure of Me_2OAlMe_3 is perhaps the configuration of the oxygen atom which is nearly planar: the sum of the valence angles at oxygen is 359.7°. The value obtained for the angle between the OC_2 plane of the ether and the O→Al bond is 5(4)°, that is, not significantly different from zero. The non-bonded Al—C(O) distance is 3.04 (2) Å, compared to the sum of the geminal repulsion radii of aluminum and oxygen, 3.00 Å. We therefore suggest that the planarity, or near planarity, of the O atom is due to repulsion between Al and C(O) atoms.

Me_2OAlMe_3 is closely related to the *iso*-electronic molecule Me_2NSiMe_3,[120] the latter can be obtained from the former by transferring a proton from the O to the Al atom, at the same time the dative O→Al bond is transformed to a normal N–Si bond and shortened by 0.30 to 1.71 Å. The nitrogen atom in Me_2NSiMe_3 is found to be planar in the gas phase, in the solid phase the sum of the valence angles at nitrogen is 357.8 (2)°.[120] The non-bonded Si—C(N) distance is 2.77 Å, compared to the sum of the relevant geminal repulsion radii, 2.80 Å.

The bonding in the model adduct H_2OAlH_3 has been investigated by *ab initio* MO calculations.[121] The O→Al bond distance was calculated as 2.02 Å and the dissociation enthalpy as 17 kcal/mol, both of which are in reasonable agreement with the experimental values for Me_2OAlMe_3, of 2.01 (1) Å and 22 kcal/mol, respectively.

The electronic energy as a function of the angle ϕ between the plane of the H_2O molecule and the O→Al bond showed a minimum at ϕ = 27°, that is equally far from the planar ϕ = O° and tetrahedral ϕ = 55° configurations. The potential curve $V(\phi)$ is however very soft: a more pyramidal configuration with ϕ = 40°, a planar configuration or a lightly pyramidal model with eclipsed Al–H and O–H bonds; ϕ = −20°, are all less than 0.25 kcal/mol above the equilibrium conformation. The calculations therefore give credence to the proposal that the planarity of the oxygen atom in Me_2OAlMe_3 is due to geminal repulsion between the Al and the C(O) atoms.

Incorporation of the oxygen atom into the four-membered ring in oxetane, Me_3O, reduces the valence angle C–O–C to a value close to 90°. If the configuration at the oxygen atom in ether adducts is determined by geminal Al—C(O) repulsion, the oxygen atom in the adduct of oxetane with $AlMe_3$ should be non-planar, since the reduction of C–O–C would increase the non-bonded Al—C(O) distances. And indeed, a gas phase electron diffraction investigation yielded the angle ϕ = 37(7)° and the distance Al—C(O) = 3.07 (3) Å, that is indistinguishable from the corresponding distance in the dimethylether adduct.[122]

The adducts of aldehydes, ketones and acid chlorides all have similar structures. The structure of an adduct of *iso*pentylaldehyde is shown in **43**. According to the VSEPR model, the two electron lone pairs on oxygen should lie in the H–C–O plane and along vectors forming angles of about 120° with the C=O bond. The Al atom is indeed lying in the H–C=O plane of the donor and *cis* to the H atom. The valence angle C–O–Al is, however, 136°.[105] The distance from Al to the carbonyl carbon atom is 2.87 Å, about 0.13 Å smaller than the sum of the geminal repulsion radii of carbon and aluminum; again the angle at the oxygen donor atom appears to be determined by repulsion between geminal aluminum and carbon atoms.

The ketone adduct **44** and the acid chloride adduct **46** both contain planar C–C=O→Al fragments. In **44** the valence angle C=O→Al = 154° corresponding to a geminal Al—C(O) distance of 3.00 Å.[105] We suggest that the increase of the angle relative to **43** is primarily due to the replacement of the sterically undemanding aldehyde hydrogen atom by a bulky Ph group. In **46** the

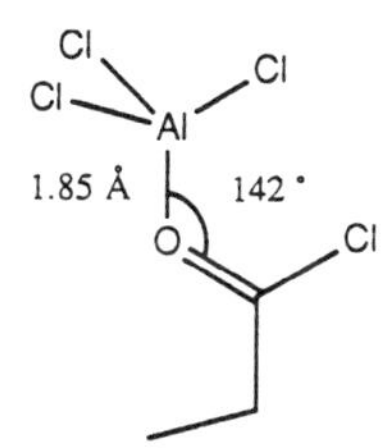

46

C=O→Al angle is 136° corresponding to a geminal Al—C(O) distance of 2.91 Å.

In **46**, the Al–Cl distance of 2.09 Å, falls in the middle of the range of observed Al–Cl distances in adducts of $AlCl_3$, and serves as our reference value for normal Al–Cl bond distances to four-coordinate aluminum atoms.

1.10.1.2. Alkoxyalanes; Al–O Bonds with Equal Normal and Dative Bonding Contributions

Dimethylaluminum methoxide is trimeric in solution and in the gas phase, with a non-planar six-membered ring of alternating Al and O atoms, **47**.[123] The gas electron diffraction data are consistent with a model of C_{3v} symmetry, that is with a chair conformation of the ring, but less symmetric conformations were not ruled out. The Al–O bond distance is 1.85 Å. It is convenient to describe the structure in terms of two canonical forms, each with alternating normal and dative O–Al bonds, and we note that the observed bond distance is close to the mean of the reference values for normal and dative bonds from oxygen to four-coordinate aluminum:

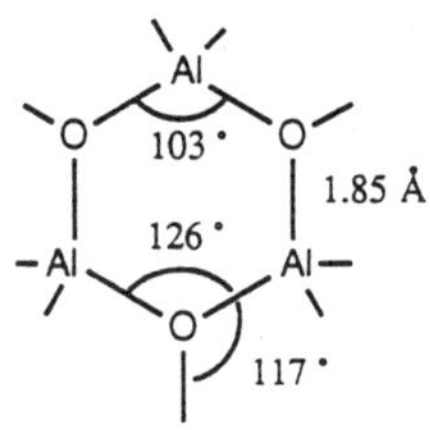

47

$$(½)(1.70 \text{ Å} + 2.00 \text{ Å}) = 1.85 \text{ Å}$$

The most striking aspects of the structure are the large Al–O–Al angles and the planar, or near planar, configuration of the oxygen atoms. Both the large angle and the planarity at oxygen may be rationalized as due to geminal repulsion between Al atoms and the C(O) atom; the observed Al—Al distance, 3.30 Å and the Al—C(O) distance, 2.82 Å, are both considerably smaller than the sums of the relevant geminal repulsion radii, 3.50 and 3.00 Å, respectively.

The chair conformation of the ring and the planarity at the O atoms lead to six very close Me(Al) —— Me(O) contacts of 3.20 Å compared to a van der Waals diameter of the methyl group of 4.00 Å. These Me——Me distances would increase if the ring adopts another, less symmetric, conformation, but there will always be six such contacts at 3.50 Å or less.[123] It is hardly surprising, then, that the degree of association of dialkylaluminum alkoxides decreases with increasing bulk of both aluminum alkyl and oxygen alkyl groups, just as we have seen for the dialkylaluminum amides. Thus Me_2AlOPh

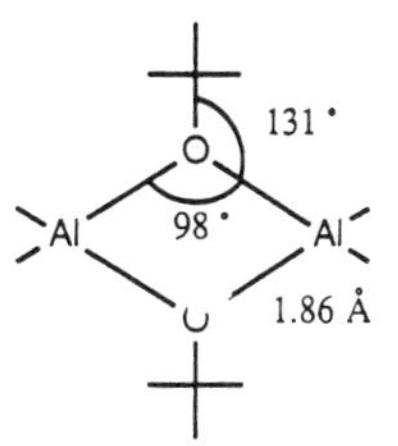

48

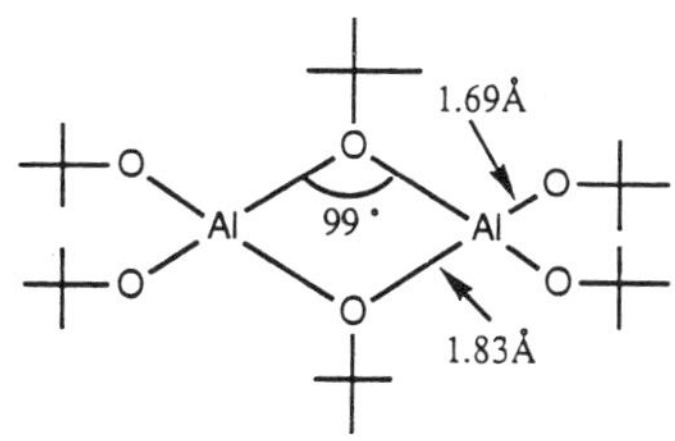

49

equilibrates to a mixture of trimers and dimers,[124] whereas dimethylaluminum *tert*-butoxide is dimeric, **48**.[125]

The central Al_2O_2 ring in the dimer is planar or near-planar. The Al–O distance is hardly different from the bond distance in the trimer, but the Al–O–Al angle has been reduced by 28° to 98°, and the O–Al–O angle by 21° to 82°. The forced reduction of Al–O–Al reduces the Al—Al distance to 2.82 Å, significantly less than twice the geminal repulsion radius. The configuration of the oxygen atom is planar, the Al—C(O) distance is 2.99 Å, equal to the sum of the geminal repulsion radii.

The structure of dimeric $Al(O\text{-}t\text{-}Bu)_3$, **49**, is similar to that of **48**; the central ring is planar and the bridging oxygen atoms adopt a planar configuration.[126] The $Al{-}O_{(bridge)}$ bonds at 1.83 Å are 0.14 Å longer than the pure, normal $Al{-}O_{(terminal)}$ bonds. Our reference value for a normal bond distance from oxygen to four-coordinate aluminum is taken from the terminal Al–O distance in this complex.

The structure of $Br_2Al(\mu\text{-}OSiMe_3)_2AlBr_2$ is similar to those of **48** and **49**. The $Al{-}O_{(bridge)}$ distance has, however, decreased to 1.80 Å, presumably due to the electronegativity of the Br atoms.[127]

The dinuclear complex **50**[128] is related to **16** where delocalization of bonding in the three-atom bridge resulted in four nearly equal Al–N bond distances of the length expected for bonds with equal normal and dative bonding contributions, 1.93 Å. The observed Al–O and Al–N bond distances in **50**, as well as the bond distances in the bridge suggest that both are adequately desribed as 1:1 normal:dative bond types.[128]

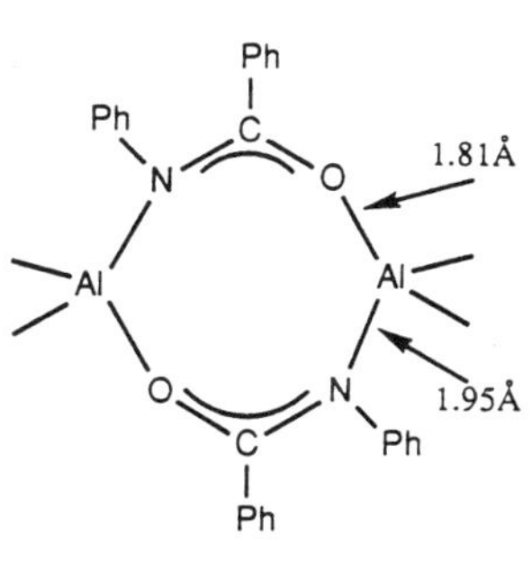

50

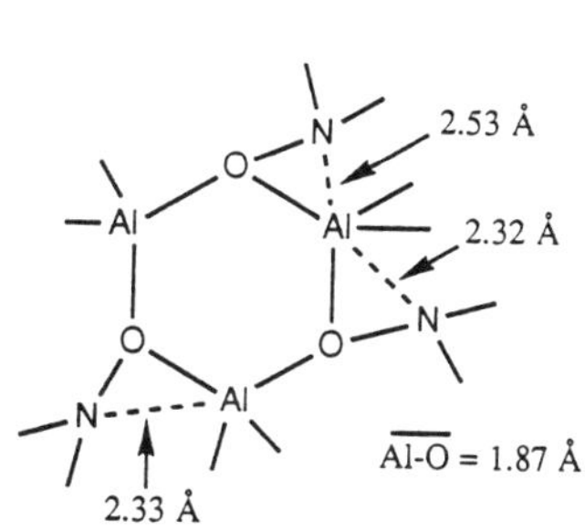

51

Structure **51** consists of a six-membered ring of alternating Al and O atoms with a mean of Al–O bond distance slightly longer than those of **47**.[129] The valence angles at the oxygen atoms are significantly distorted to allow the formation of three very long, and presumably very weak, dative N→Al interactions. The endocyclic Al–O bonds to the aluminum atoms with insipient five- and six-coordination are 0.02 – 0.04 Å longer than the bond distances to the aluminum atom which remain strictly four coordinate.

1.10.1.3. Acyclic Chains of Alternating Dative and Normal Bonds

We have noted that normal bonds from oxygen to four-coordinate aluminum usually fall in the range 1.73±0.04 Å, while dative O→Al distances range from 1.73 Å to 2.02 Å. In **43** and **44**, where the same aluminum atom is bonded normally to one oxygen atom and datively to another, the dative bonds are more than 0.15 Å longer than the normal bond.

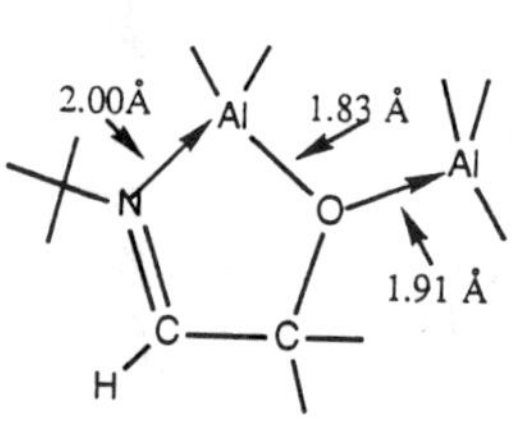

52

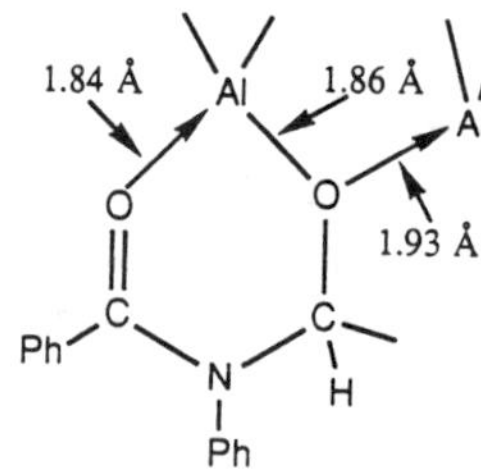

53

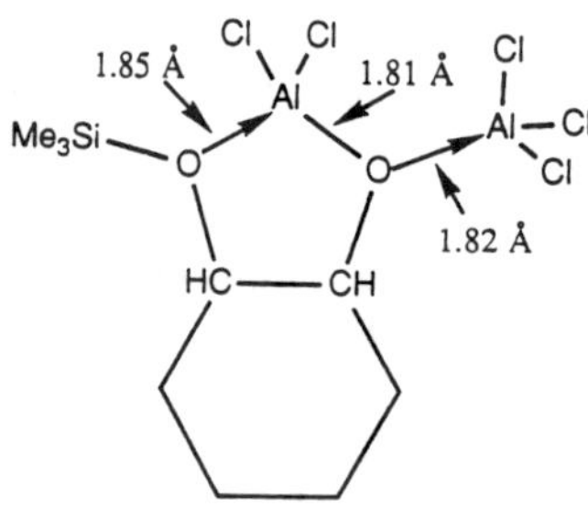

54

Molelcules **52**, **53**, and **54** contain chains of three alternating bonds: D→Al–O→Al.[130–132] In **52** D=N, in **53** and **54** D=O. In all three species the normal Al–O bond is longer than the usual range. The normal and dative Al–O bonds in **52** differ by only 0.08 Å; in **53** one dative O→Al bond is 0.02 Å shorter than the normal Al–O bond. In **54** the three bonds are very nearly equal. We suggest that the two dative bonds are strengthened and the normal bond is weakened through a synergetic effect. Formation of a dative bond is accompanied by elongation and increased polarization of normal bonds to the acceptor atom. Elongation and increased polarization of the normal Al–O bond in **52**, **53**, or **54** leads to the enhancement of two dative bonds rather than one.

1.10.2. Compounds with Five-coordinate Aluminum

1.10.2.1. Aluminum Atoms with Three Normal and Two Dative Bonds

The coordination polyhedron of the aluminum atoms in **55** is close to perfect trigonal bipyramidal: the angle N–Al–N = 179°, mean O–Al–$N_{(pyr)}$ = 95°, and mean O–Al–O = 119.4°.[133] The three normal, equatorial bonds to oxygen average 1.76 Å, 0.03 Å longer than our reference value for a normal bond from oxygen to four-coordinate aluminum bond. The length of the dative bond from the pyridine nitrogen atom, 2.16 Å, significantly greater than the length of the

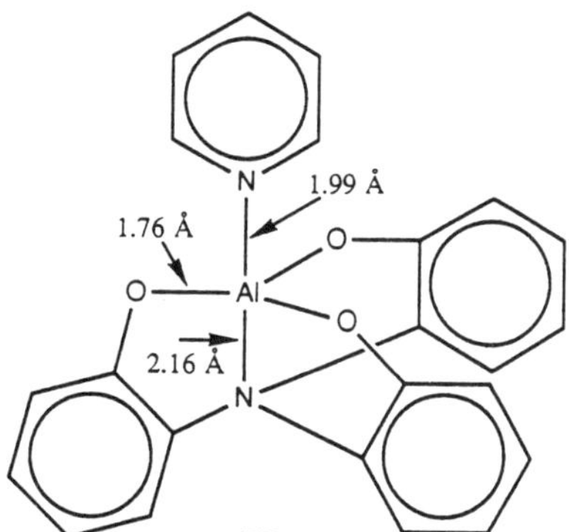

55

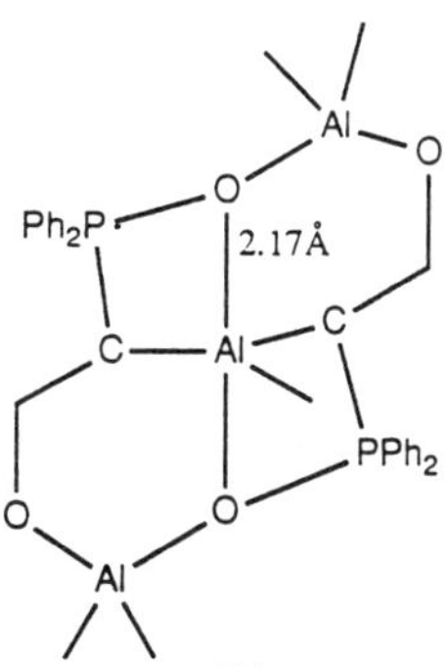

56

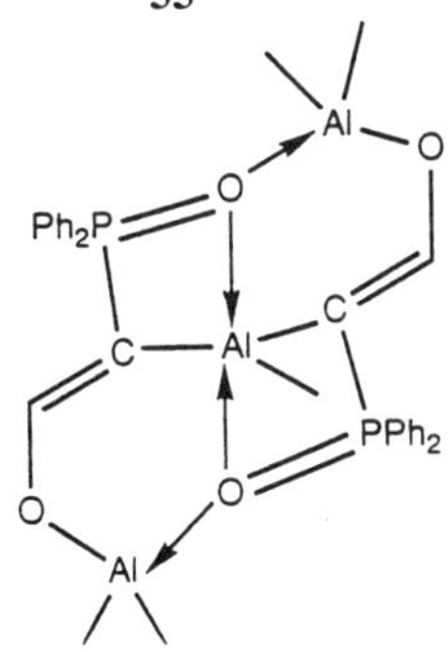

56a

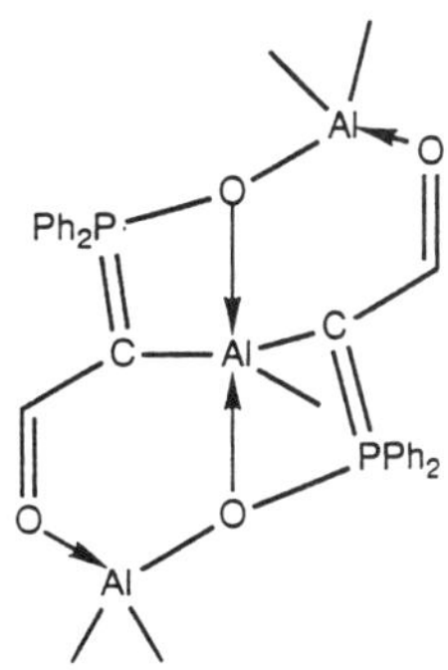

56b

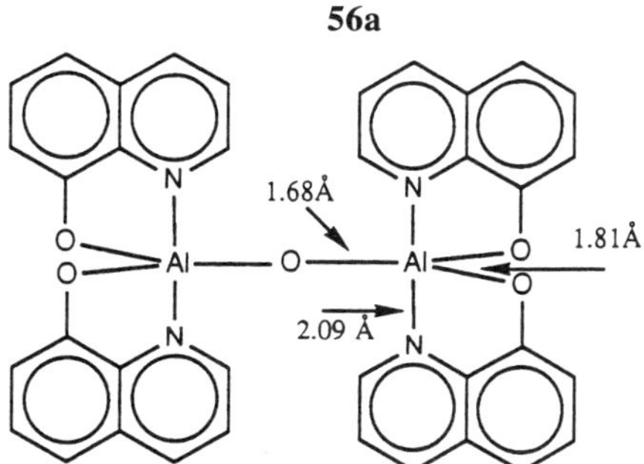

57

dative bond from the amine N, 1.99 Å. The average of the two is close to the dative bonds in the 1:2 adduct of $AlCl_3$ with $NHMe_2$ or morpholine, 2.06 Å or 2.09 Å (Table 1.5).

The coordination polyhedron of the central, five-coordinate aluminum atom in the trinuclear complex **56** is somewhat distorted from perfect trigonal bipyramidal geometry. The aluminum atom and the three ligating carbon atoms are coplanar and the three C–Al–C angles close to 120°, but the O–Al–O angle is only 163°.[134] It appears likely that this angle is imposed by the cyclic ligand system. The complex may be described in terms of the canonical forms **56a** and **56b**; no canonical forms with normal bonds from oxygen to the central aluminum atom are possible. The three equatorial normal Al−C bonds average 1.99 Å, not much longer than normal Al−C bonds in four-coordinate complexes. The dative O→Al bonds at 2.17 Å are 0.15 Å longer than the longest dative bonds from oxygen to four-coordinate aluminum.

The dinuclear complex **57** contains two five-coordinate aluminum atoms.[135] The reader may be convinced that there is only one possible canonical form in which all Al−O bonds are normal and all Al−N bonds dative. The coordination polyhedron is distorted trigonal bipyramidal. The AlO_3 fragment is trigonal planar, but N–Al–N = 163°. The N→Al bond distances are of the expected magnitude. The Al−O bonds may be compared to the reference value of a normal bond distance from oxygen to four-coordinate aluminum, 1.73 Å; the bonds to the bridging oxygen are unexpectedly short. The central Al–O–Al fragment is linear. Though this is not indicated in the sketch, the two ends of the molecule are oriented in such a way that the two N–Al–N axes are perpendicular.

1.10.2.2. Aluminum Atoms with Two Normal Bonds, One Dative Bond and Two Bonds of Intermediate Type

The dinuclear, centrosymmetric complexes **58**, **59**, **60**, and **61** all contain aluminum atoms in a distorted trigonal bipyramidal configuration.[136–139]

In **58** and **60** each aluminum forms normal bonds to two equatorial methyl groups and a purely dative bond to an apical O atom. In **59** and **61** each aluminum forms normal bonds to two equatorial methyl groups and a purely dative bond to an apical nitrogen atom. Each of the four molecules contains a central four-membered Al_2O_2 ring where each O atom is in the equatorial position of one aluminum atom and in an axial position to the other. The equatorial Al−O bond length is approximately constant through the series, but the axial Al−O bond grows monotonically from 1.89 Å in **58** to 1.95 Å in **59** to 1.99 Å in **60** and to 2.06 Å in **61**. In **58** axial and equatorial Al−O distances are reasonably close to the Al−O distance in **47** which we have described as having equal dative and normal bonding contributions. The difference between the contributions is similar to that normally found between axial and equatorial normal bonds in trigonal bipyramidal compounds of main-group elements like PF_5.[140]

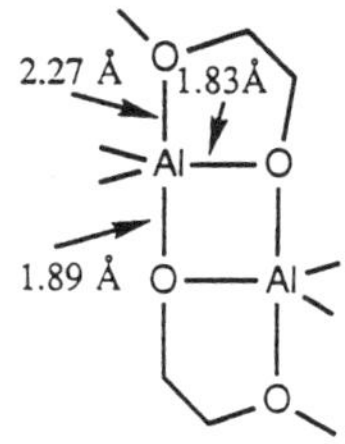

58

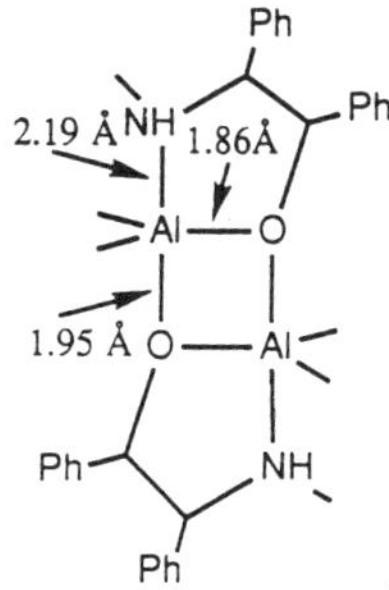

59

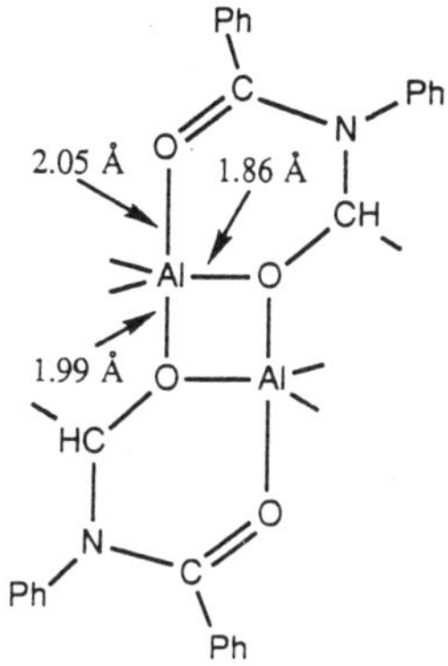

60

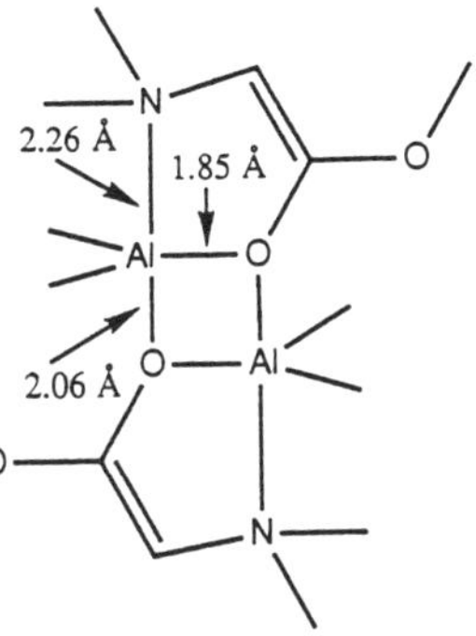

61

In **61** the difference is so large that it seems reasonable to assume that the axial bond is predominantly dative and the equatorial bond predominantly normal. The series of complexes from **58** to **61** thus appears to form a gradual transition from delocalized to localized dative and normal bonding in the Al_2O_2 rings.

1.10.2.3. *Aluminum Atoms with One Normal Bond and Four Bonds of Intermediate Type*

The coordination polyhedron of the central Al atom in the trinuclear complex **62** is trigonal bipyramidal with O_{ax}–Al–O_{ax} = 171° and O_{eq}-Al-O_{eq} = 111°.[141]

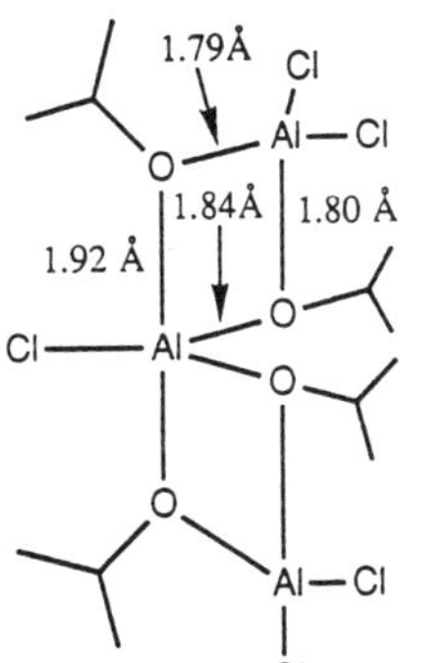

62

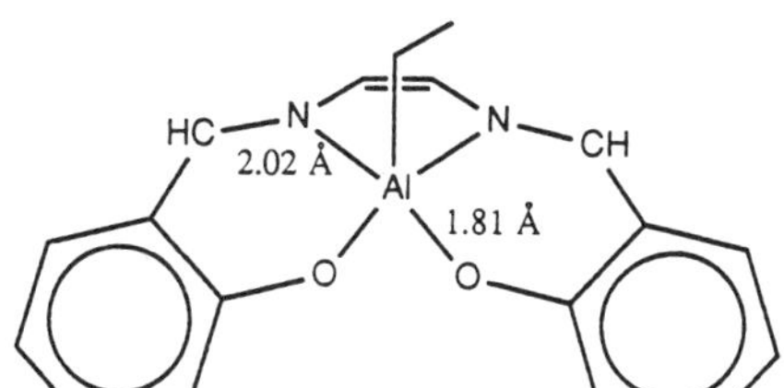

63

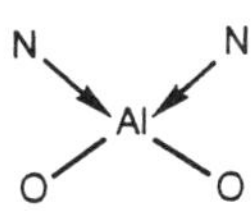

63a

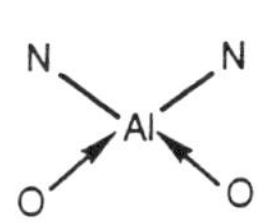

63b

The similarity of the Al–O bond distances in the four-membered rings indicate that dative and normal bonding is delocalized. The difference between equatorial and axial bond distances to the five-coordinate aluminum is of the magnitude expected for such bonds in compounds of main-group elements.

The coordination geometry of **63**, on the other hand, is close to square pyramidal with N–Al–O angles of 142° and 151°.[142] As indicated in the canonical forms **63a** and **63b**, each half of the chelating ligand system may be described in terms of two canonical forms: one with a normal Al–N bond and a dative O→Al bond and the other with a dative N→Al bond and a normal Al–O bond. The observed Al–O and Al–N bond distances suggest that both canonical forms contribute, though **63a** may contribute more than **63b**.

1.10.3. Compounds with Six-coordinate Aluminum

Aluminum tris(*iso*propoxide), **64**, is tetrameric in the solid phase.[143] The coordination polyhedron of the central six-coordinate Al atom is distorted octahedral, that of the four-coordinate Al atoms is distorted tetrahedral. The terminal, normal Al–O bonds are not significantly different from the terminal Al–O distances in dimeric aluminum tris(*tert*-butoxide), **49**; the bonds from the four-coordinate Al to bridging oxygen atoms are about 0.04 Å shorter than in **49**. The bond distances from the bridging oxygen atoms to the six-coordinate aluminum atom are indistinguishable and average 1.94 Å; 0.15 Å longer than the average bond distances to the four-coordinate Al. The structure is consistent with a description in terms of a model where each Al–O_{bridge} bond is half normal and half dative.

Tris(2,4-pentanedionato)aluminum, **65**, has been studied by X-ray crystallography in two crystallographic modifications, monoclinic and

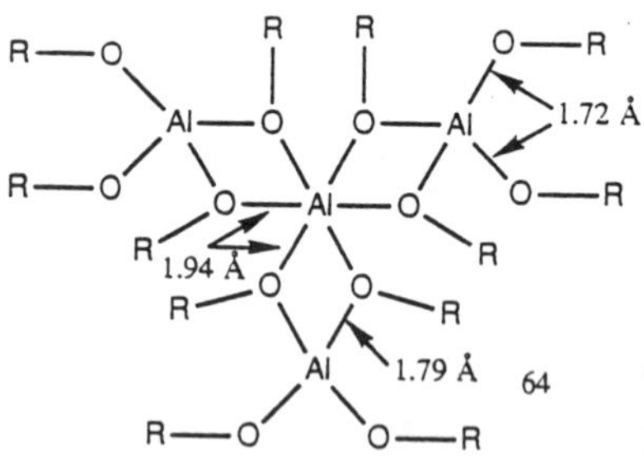

64

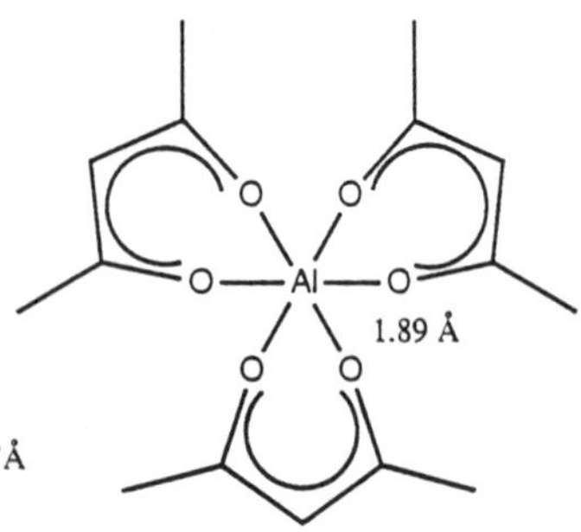

65

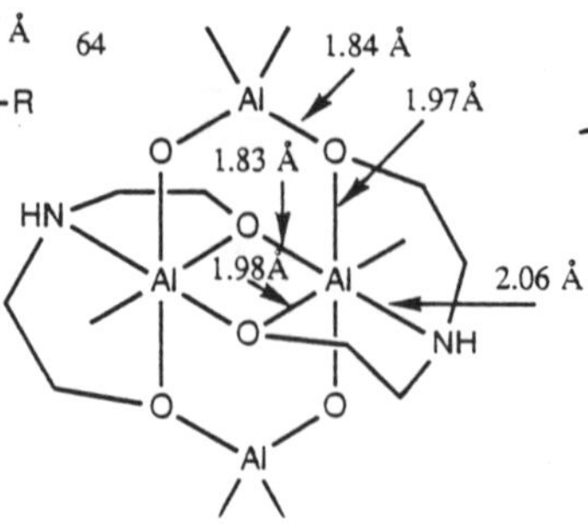

66

orthorhombic.[143,145] The molecular structures obtained in the two studies are not significantly different. The coordination sphere of the aluminum atoms is nearly perfect octahedral. The three *trans* O–Al–O angles are all greater than 178°, the *cis* O–Al–O angles range from 88° to 91°. The Al–O distances range from 1.87 to 1.90 Å. The bond distances in the chelating ligands show that the π-bonding is delocalized corresponding to the canonical forms

$$\text{normal} / \{\text{normal} / \text{double}\}_2 / \text{dative}$$

and the reverse. The six Al–O bonds must therefore be described as 1:1 normal Al:dative bonds. The average Al–O bond distance, 1.89 Å, is 0.05 Å smaller than 1:1 bonds from oxygen to the six-coordinate aluminum in **64**. It can be seen in Table 1.7 that carbonyl oxygen atoms form shorter dative bonds than ether oxygen atoms.

The six Al–O bond distances in *tris*(troponolato)aluminum, $Al(O_2C_7H_5)_3$, are indistinguishable from those in **65**.[146] The coordination polyhedron of the metal atom is intermediate between octahedral and trigonal prismatic, presumably because the ligand bite is too small to allow octahedral coordination.

The centrosymmetric complex **66** contains two four-coordinate and two six-coordinate aluminum atoms.[147] The four-coordinate metal atoms form two normal bonds to terminal methyl groups and two 1:1 normal:dative bonds to bridging alkoxide groups. Each of the six-coordinate aluminum atoms forms one normal bond to a methyl group, one dative bond to an amine nitrogen atom, and four bonds to bridging O atoms. The two Al–O bonds that are perpendicular to the Al–C–N plane are of equal length, and may be described as having equal normal and dative bonding contributions. The Al–O bonds in the central four-membered ring are asymmetric; the bond which is *trans* to the dative N→Al bond is short at 1.83 Å, whereas the bond which is *trans* to the normal Al–C bond is long at 1.98 Å. As in the six-coordinate complex **33** there is a pronounced *trans*-effect. But in **66** the aluminum atom can adjust to the situation by concentrating dative bonding in the Al–O bond which is *trans* to the normally bonded methyl group.

1.11. Aluminum-Sulfur Compounds

1.11.1. Compounds with Four-coordinate Aluminum

1.11.1.1. Dative S→Al bonds

The dissociation enthalpy of the adduct Me_2SAlMe_3 in the gas phase is 18.1±0.5 kcal/mol (Table 1.3). The S→Al bond distance in the gas phase, 2.55(2) Å,[122] is more than 0.30 Å longer than the Al–S normal bond distance of 2.19 Å found in the monomeric thiolate $Al(SAr)_3$, (Ar = 2,4,6-tri-*tert*-butylphenyl),[148] or estimated from the modified Schomaker–Stevensen rule, 2.21 Å. Robinson and co-workers[149] have determined the solid-state structure of a

Table 1.8. S→Al bond distances (in Å units) in simple adducts of alanes with S-centered donors.

Donor	Acceptor	S→Al	Reference
Me_2S	$AlMe_3$	2.55 (2)	122
14-Crown-4-thioether	$AlMe_3$	2.52 (4X)	149
H_2S	$AlBr_3$	2.40	150
Ph_3PS	$AlCl_3$	2.30	151
$(i\text{-}Pr)_2ClPS$	$AlCl_3$	2.29	152

tetrathia-crown-ether with four moles of $AlMe_3$. The resulting S→Al bond distance, 2.52 Å, is more accurately determined than the gas phase value, and in the following we adopt this as our reference value for a dative bond from a thioether to a four-coordinate Al atom carrying only H or organyl substituents. Table 1.8 gives S→Al bond distances in adducts with S centered donors.

The configuration at the sulfur atom in Me_2SAlMe_3 is non-planar, but the Al–S–C angles at 124 (2)° are greater than tetrahedral. The Al—C(S) distance is 3.9 Å, so the large angle cannot be rationalized by geminal repulsion. The structure has not been determined with sufficient accuracy to allow a meaningful comparison with the S–C bond distance or C–S–C valence angle of the free donor.

The dissociation enthalpy of Et_2SAlBr_3 at 30.6 kcal/mol is 50% higher than that of Me_2SAlMe_3; presumably an inductive effect of the electronegative Br atoms. The structure of this adduct is unknown, but the S→Al distance in solid H_2SAlBr_3 is 2.40 Å.[180]

It has been shown earlier that the O→Al distances in adducts of $AlCl_3$ or $AlBr_3$ with Ph_3PO are unusually short and fall in the range of normal Al–O distances. Similarly the S→Al bond distance in $Ph_3PSAlCl_3$ is 0.22 Å shorter than the bond distance from thioether sulfur to $AlMe_3$, and only 0.11 Å longer than the normal Al–S bond in $Al(SAr)_3$.[151] Unlike the O analogues, $Ph_3PSAlCl_3$ is non-linear at the donor atom; the P–S–Al angle being 112°.

The crystal structure of an adduct of 12-Crown-4-thioether with one mole $AlMe_3$ shows that the aluminum atom interacts with two S atoms in different rings. The configuration is trigonal bipyramidal as expected and the axial S→Al bond distances of 2.72 Å and 3.905 Å.[153]

1.11.1.2. Compounds Containing Al–S Bonds of Intermediate Type

Dimethylaluminum methylthiolate has been investigated by both X-ray crystallography and gas diffraction. In the gas phase the compound forms centrosymmetric dimers, **67**.[154] The S-atoms are pyramidal with all valence angles less than tetrahedral, and the Me(S) groups are in *transoid* positions. The endocyclic valence angle at Al is 85°. The Al–S bond distance of 2.37 Å is

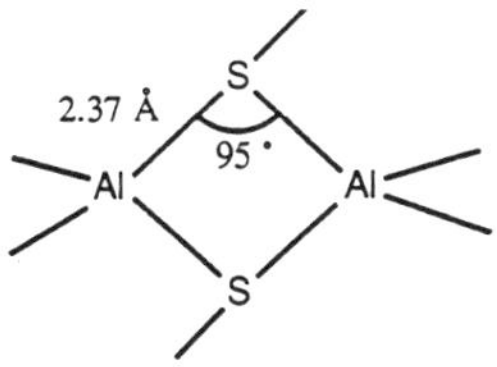

67

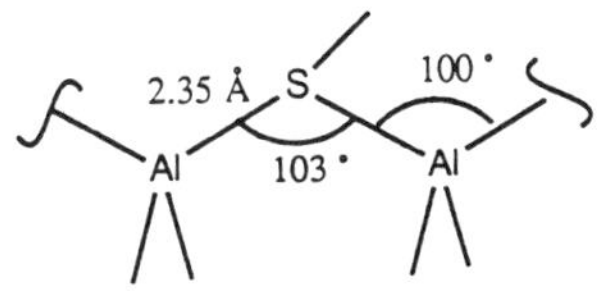

68

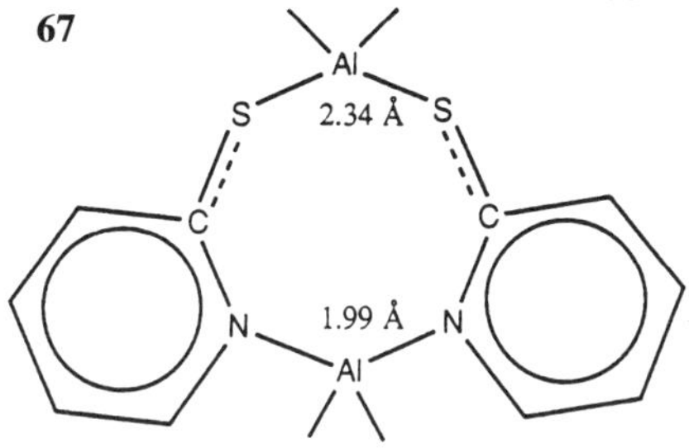

69

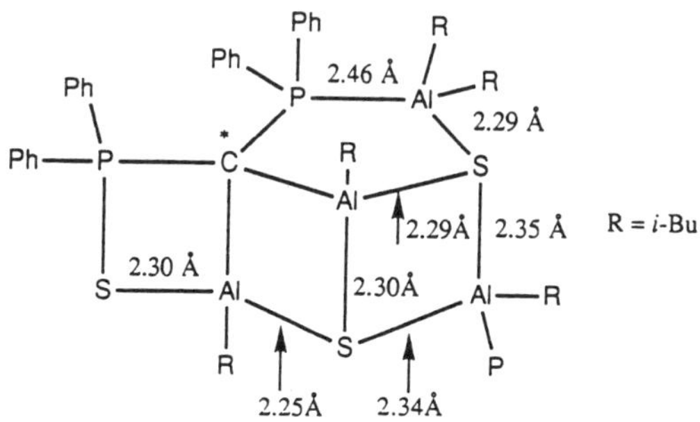

70

close to the average of the reference value for a dative bond from a thioether S to Al carrying organyl substituents and the normal bond distance in $Al(SAr)_3$; 2.355 Å. In the solid state $MeSAlMe_2$ forms infinite polymers with non-planar zigzag chains of alternating Al and S atoms, **68**.[155] The bond distances are unaltered, but the valence angle Al-S-Al is increased by 8°, and S–Al–S by 15°.

Me_2AlSAr (Ar = 2,6-dimethyl-phenyl) forms cyclic tetramers in the solid phase. The average Al–S distance in the Al_4S_4 ring is 2.36 Å.[156]

The cubic crystal modification of Al_2S_3 contains both four- and six-coordinate Al atoms. The bonds from S to the four-coordinate Al atoms average 2.29 Å.[157] These bonds may be described as three-quarters normal and one-quarter dative. The appropriate mean of the reference bond distances is 2.27 Å.

Oliver and co-workers[158] prepared and studied the dinuclear complex **69**. The Al–S and Al–N bond distances both correspond with 50% dative contributions and suggest that the three-atom bridge may be described in terms of the canonical forms dative/double/single/normal and reverse.

Finally we turn our attention to the tetranuclear complex **70** which contains seven topologically different Al–S bonds.[159] This complex is so large and the symmetry so low, that it is possible to draw at least six canonical formulas, each containing one or two ylidic P = S or P = $C^{*}_{(bridgehead)}$ bonds. Each Al–S or Al–P bond in the cluster is normal in at least one canonical form and dative in at least one other. Under such circumstances our model obviously loses its predictive power.

1.12. Aluminum–Halogen Compounds

1.12.1. Compounds with Four-, Five-, or Six-coordinate Aluminum

1.12.1.1. Dative X→Al Bonds

As noted above, alkyl flourides do not form adducts with alkyl alanes, presumably because the dative F→Al bonds in such adducts would be too weak to compensate for the enthalpy of dissociation of the dimeric alkyl alanes. Schmidbaur and Kline[160] have, however, synthesized the adduct $Me_3SiFAlEt_3$ by combining equimolar quantities of donor and acceptor units at 0°C. The adduct which was characterized by NMR spectroscopy, freezes at −79 to −81°C, but has unfortunately not been characterized structurally. Its mere existence demonstrates, however, that the strength of a F→Al bond may be sufficient to compensate for the enthalpy required to form one mole of $AlEt_3$ from the dimer, 16.9±2 kcal/mol.[161]

We are only aware of two cases where the structures of complexes containing dative bonds from halogen to aluminum have been determined. The aluminum atom in **71** is normally bonded to one chlorine atom and two amide nitrogen atoms and datively bonded to two F atoms.[52] The valence angles F–Al–F and N–Al–N are 152 and 141°, respectively. We suggest that the distortion from trigonal bipyramidal coordination is due to the limited bite of the three-atom N–Si–F bridge. The F→Al bonds are 0.46 Å longer than the normal Al–F bonds in monomeric AlF_3, and the normal Si–F bonds are 0.11 Å longer than in Me_3SiF.[162]

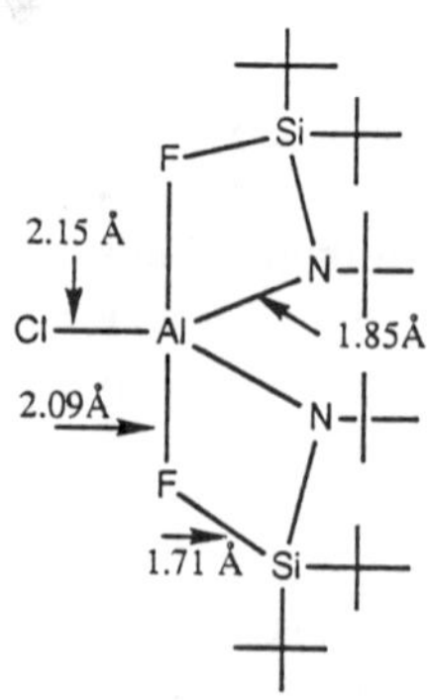

71

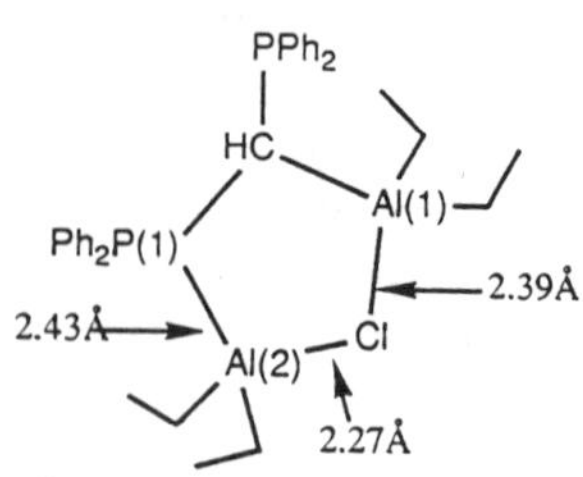

72

The complex **72** contains both a normal and a dative Al–Cl bond. Since Al(1) is normal bonded to three C atoms, the bond to Cl must be dative. If this bond is dative Al(2)–Cl must be a normal bond and P→Al(2) must be dative.[163] The difference between the dative and normal Al–Cl bonds is, however, only 0.12 Å. We have seen in Section 1.10.1.3. that in molecules which contain a normal Al–O bond between two dative bonds in a sequence D→Al–O→Al, the difference between the dative and normal Al–O bonds is unusually small. It is proposed that this is due to a synergetic effect where the elongation and polarization of the normal Al–O bond allows the strengthening of both dative bonds. It appears that a similar effect may be operative in **72**.

While $AlCl_3$ dissolves as a *dimer* in methyl or ethyl chloride, molecular-weight data indicate that $AlBr_3$ and AlF_3 are monomeric in MeBr and Mel, respectively.[164] The monomers are presumably stabilized through electron donor acceptor interactions with the solvents, but structural information is lacking.

1.12.1.2. Compounds Containing Al–X bonds of Intermediate Type

Dimethylaluminum flouride is tetrameric in freezing benzene and in the gas phase, **73**.[165] The tetramer contains a non-planar eight-membered ring of alternating aluminum and flourine atoms joined by bonds which must be half dative and half normal. The Al–F bond distance at 1.81 Å is 0.05 Å shorter than the mean of the normal Al–F bond distance in gaseous monomeric AlF_3 and the dative F→Al bond distance in **71** where the aluminum atom is five-coordinate and the F→Al bond distances presumably are longer than in a four-coordinate adduct.

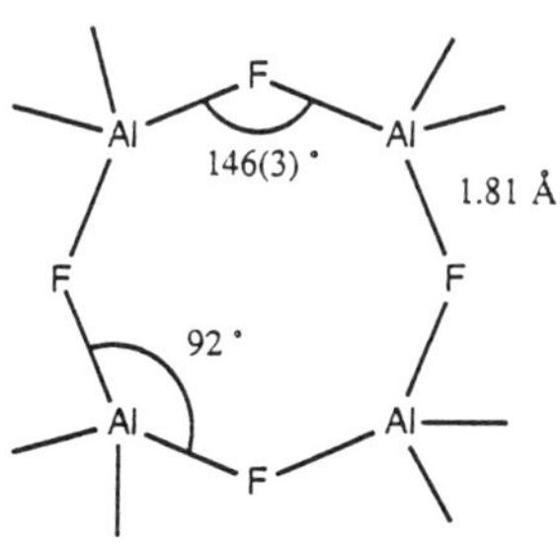

73

As in the trimeric dimethylaluminum amides, for example **4**, and in trimeric dimethylaluminum methoxide, **47**, the endocyclic angle at the bridging atom is significantly greater than tetrahedral. Since the fluorine atom is surrounded by two bond electron pairs and two lone pairs, the VSEPR model predicts a valence angle Al–F–Al less than tetrahedral. The Al—Al distance of 3.46 Å suggests that the angle is determined by repulsion between geminal aluminum atoms.

As we have seen earlier, introduction of more bulky ligands on the ring atoms tends to reduce the degree of association. Et_2AlF is tetrameric in benzene,[166] but Pr_2AlF and *iso*-Bu_2AlF are trimeric.[167] However, *tert*-Bu_2AlF is tetrameric.[168]

Dimethylaluminum chloride is dimeric in the gas phase **74**.[169] The $Al-Cl_{(bridge)}$ bonds, which must be half normal and half dative, are 0.27 Å longer than the normal bonds in gaseous $AlCl_3$ where the coordination number, CN, is three and 0.18 Å longer than the normal Al–Cl bonds in Me_3NAlCl_3 where the CN is four.

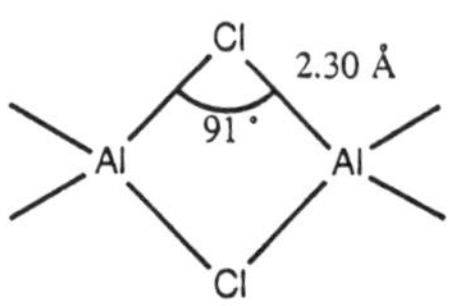

74

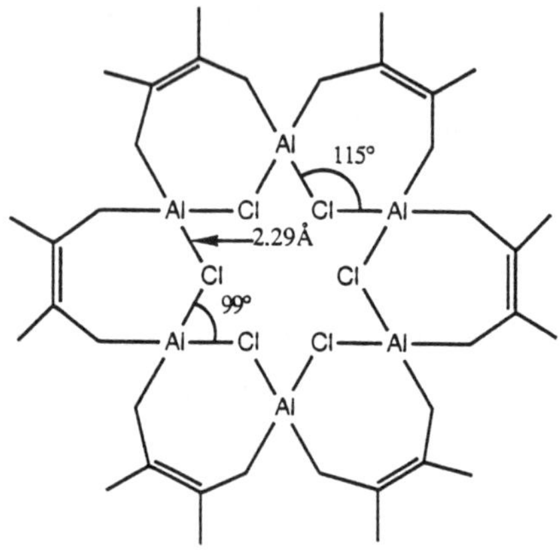

75

The $Al-Cl_{(bridge)}$ bond distances in the hexanuclear complex **75**, 2.29 Å, are indistinguishable from those in **74**.[170] The valence angles in the hexamer are Cl–Al–Cl = 99° and Al–Cl–Al = 115°.

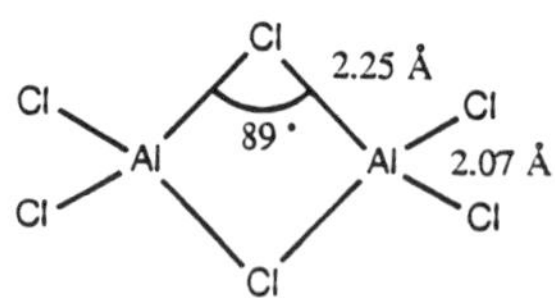

76

Aluminum trichloride is dimeric in non-coordinating solvents and as saturated vapor at 150°, **76**.[171] The terminal Al–Cl bond distance is indistinguishable from the distance in the monomer; the $Al-Cl_{(bridge)}$ bond, which must be half dative, is 0.18 Å longer. The $Al-Cl_{(bridge)}$ bond is 0.05 Å shorter than in **74**, presumably due to an inductive effect of the terminal chlorine atoms.

In Table 1.1 we listed the mean bond Al–Cl bond enthalpies, D, in Al–Cl(g, CN = 1), $AlCl_2$(g, CN = 2), $AlCl_3$(g, CN = 3), Al_2Cl_6(g, CN = 4) and $AlCl_3$(cr, CN = 6). Figure 1.4a shows D as a function of CN; mean bond enthalpies are

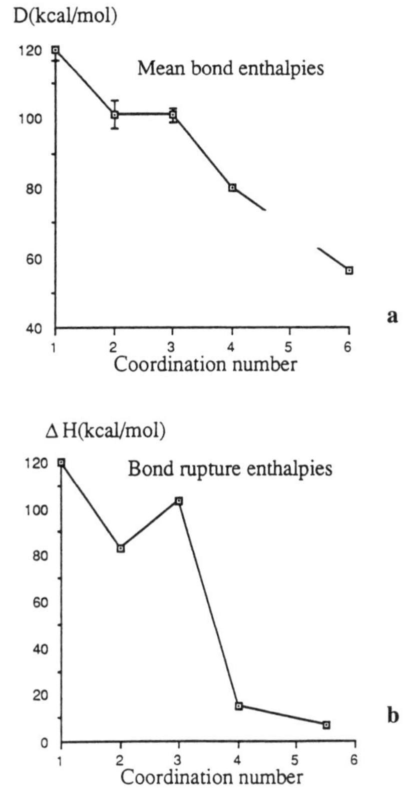

Figure 1.4. (a) Mean Al–Cl bond enthalpies for Al coordination numbers 1, 2, 3, 4, and 6. (b) Al–Cl bond rupture enthalpies for Al coordination numbers 1, 2, 3, and 4. The mean of the bond rupture enthalpies for coordination number 5 and 6 are plotted at CN = 5.5.

seen to decrease in a fairly regular manner with increasing CN.

Bond rupture enthalpies $\Delta H_r(CN)$ for CN one, two, and three may be defined as the enthalpies of the reaction

$$AlCl_{CN(g)} \longrightarrow AlCl_{CN-1(g)} + Cl_{(g)} \quad \Delta H° = \Delta H_r(CN)$$

The values thus obtained are $\Delta H_r(1) = 120\pm3$, $\Delta H_r(2) = 83\pm5$, and $\Delta H_r(3) = 103\pm4$ kcal/mol.[8]

The bond rupture enthalpy for CN = 4 is defined as half the dissociation enthalpy of Al_2Cl_6:

$$Al_2Cl_{6(g,\ CN=4)} \longrightarrow 2AlCl_{3(g,\ CN=3)}$$

$$\Delta H° = 30\pm2 \text{ kcal/mol}^8$$

$$\Delta H_r(4) = 15\pm1 \text{ kcal/mol}$$

$\Delta H_r(4)$ may be regarded as the enthalpy required to break a dative Cl→Al bond (neglecting the delocalization energy in the Al(μ-Cl)$_2$Al ring).

The crystal structure of $AlCl_3$ may be described as consisting of edge-sharing $AlCl_6$ octahedra.[172] We are aware that the solid state may be described in terms of an ionic binding model, even if a charge of +3 on a metal ion seems very high. It should be noted, however, that an alternative description in terms of Al(μCl)$_2$Al fragments is possible: the mean Al–Cl distance, 2.31 Å, is reasonable when compared to the $AlCl_{(bridge)}$ distance in the gaseous dimer where the Al atom is four coordinate. The sum of the bond rupture enthalpies $H(5)$ and $H(6)$ may be defined as half the sublimation enthalpy of one mole of the dimer.

$$2AlCl_{3(cr,\ CN=6)} \longrightarrow Al_2Cl_{6(g,CN=4)}$$

$$\Delta H° = 28 \pm 2 \text{ kcal/mol}^{8},$$

$$\Delta H_r(5) + \Delta H_r(6) = 14 \pm 1 \text{ kcal/mol}$$

This sum may be regarded as the enthalpy required to break two dative Cl→Al bonds (again neglecting the delocalization energy of the Al(μ-Cl)$_2$Al ring); the experimental value, 14±1 kcal/mol, suggests that the sum of the dissociation enthalpies of the second and third dative bond to each Al atom is a little less than the dissociation enthalpy of the first.

Figure 1.4b plots bond rupture enthalpies as a function of the CN. We see that the first three normal bonds formed between Al and Cl are much stronger than additional dative bonds.

1.13. Conclusions

1.13.1. General

Herein we have attempted to rationalize and interpret the structures of neutral aluminum complexes in terms of the model cluster described in the introduction; that is the distinction between normal and dative bonds, the modified Schomaker–Stevensen rule for prediction of normal bond distances, the VSEPR model, and the geminal repulsion model. In the cases when a complex may be described in terms of more than one canonical formula with pure normal and dative bonds between aluminum and donor atoms, we have estimated the normal and dative bonding contributions to a particular bond by enumerating the fraction of the canonical forms in which the bond is normal or dative respectively. Finally, we have assumed that the bond strength decreases and the bond length increases with increasing dative contributions.

Only in three cases have experimentally determined structures been found to exhibit features directly opposite to those predicted by the model cluster. In the crystalline adducts Ph_3POAlX_3, X = Cl or Br, the dative O→Al bonds are as short as (but probably not as strong as) normal Al–O bonds. In the six-coordinate complex **33** the purely dative N→Al from the pyridine ligand is significantly shorter than the Al–N bond *trans* to the methyl group, which appears to be half dative and half normal. Finally, it must be recognized that when a polynuclear complex is large enough, and the symmetry low enough, the number of canonical forms may become so large as to deprive the model cluster of any predictive power. This is, for instance, the case with the tetranuclear complex **70** which contains seven topologically different Al–S bonds, and for which at least six different canonical forms are possible.

1.13.2. Compounds with Four-coordinate Aluminum

Table 1.9 lists reference values for *normal* Al–N, Al–P, Al–O, Al–S, Al–F and Al-Cl bond distances to four-coordinate aluminum, and reference values for dative N→Al bonds (in amine adducts of AlR_3), P→Al bonds (from triorganyl phosphines to AlR_3), O→Al bonds (in ether adducts of AlR_3), and for S→Al bonds (in thioether adducts of AlR_3 where R is a hydrogen atom or an alkyl group). Since dative bond distances are very sensitive to inductive effects introduction of more electronegative substituents on the aluminum atom may lead to dative bond distances which are up to 0.20 Å shorter than the reference values. In the case of oxygen-centered donors further shortening is observed when the oxygen atom is doubly bonded (mono-coordinate) in the isolated donor.

Table 1.9 also contains reference values for 1:1 normal:dative bonds taken from observed bond distances in cyclic species of type $(R_2AlNR'_2)_n$, $(R_3AlPR'_2)_n$, $(R_2AlOR')_n$, $(R_2AlSR')_n$, $(R_2AlF)_n$, or $(R_2AlCl)_n$ where R and R' are H atoms or alkyl groups. If the substituents on the Al atoms are replaced by

Table 1.9. Reference values for normal and dative bond distances (in Å units) to four-coordinate Al atoms, and for bond distances with equal normal and dative (1:1) bonding contributions.

	Normal	Dative	1:1
Al–N	1.82	2.08	1.95
Al–O	1.69	2.00	1.85
Al–F	1.67[a]	2.09	1.81
Al–P	2.31[a]	2.54	2.43
Al–S	2.19	2.52	2.37
Al–Cl	2.09	2.39	2.30

[a] Estimated from the modified Schomaker-Stevensen rule.

more electronegative groups, observed bond distances may be up to 0.10 Å shorter.

1.13.3. Compounds with Five-coordinate Aluminum

The available evidence suggests that those complexes containing five-coordinate aluminum for which only one canonical form (with three normal and two dative bonds) is possible, invariably adopts a trigonal bipyramidal conformation with the datively bonded donor atoms in axial positions. The aluminum atom and the three equatorial normally bonded atoms generally form a system of close to trigonal symmetry, but with bi- or poly-dentate ligands of limited bite the angle D_{ax}–Al–D_{ax} may be significantly less than 180°.

For some complexes containing five-coordinate aluminum it is possible to draw two canonical forms differing in the bonding mode of two ligating atoms, each canonical form with one atom normally bonded and the other datively bonded. In some cases the observed bond distances indicate that bonding is adequately described by *one* of the canonical forms, and the configuration is trigonal bipyramidal with the datively bonded atoms in axial positions.

More often the observed bond distances indicate that bonding is delocalized, that is both canonical forms contribute. The aluminum atom may then be described as forming two normal, one dative and two bonds of intermediate type. All such complexes which have so far been investigated, appear to contain aluminum atoms in a trigonal bipyramidal configuration with the purely dative bond and one of the bonds of intermediate type in the axial directions.

For a third group of complexes containing five-coordinate aluminum it is possible to draw two or more canonical forms in such a way that two atom pairs differ in their bonding mode; one member of each pair being normally, the other datively bonded in each canonical form. In all such complexes that have so far been investigated, observed bond distances indicate that bonding is delocalized, and that the aluminum atom forms one normal bond plus four bonds of intermediate type. Of the seven complexes of this type described here, one is close to trigonal bipyramidal configuration, three are close to square pyramidal configuration, and three are close to the transition point between trigonal bipyramidal and square pyramidal configurations.

Thus it seems clear that complexes containing two or one purely dative bonds prefer the trigonal bipyramidal configuration with the datively bonded atoms in axial positions. For complexes with no purely dative bonds, that is with one normal bond and four bonds of intermediate type, the energy difference between the two symmetric configurations seems to be so small that the configuration adopted may be determined by the nature of the ligands.

Acknowledgments

We are grateful to the Norwegian Research Council for Science and the Humanities (NAVF) and the VISTA program of The Norwegian Academy of

Science and STATOIL for financial support, to Dr. Kristin Rypdal for critical readings of this manuscript and to Dr. Paul Kiprof for assistance with the drawings.

References

1. Haaland, A. *Angew. Chem. Int. Ed. Engl.* **1989**, *28*, 992.
2. Blom, R.; Haaland, A. *J. Mol. Struct.* **1985**, *128*, 21.
3. Gillespie, R. J. *J. Chem. Ed.* **1970**, *47*, 18.
4. Gillespie, R. J.; Hargittai, I. *The VSEPR Model of Molecular Geometry*; Allyn and Bacon: Boston, 1991.
5. Bartell, L. S. *J. Chem. Ed.* **1968**, *45*, 754.
6. Glidewell, C. *Inorg. Chim. Acta* **1975**, *12*, 219.
7. Bartell, L. S. *J. Chem. Phys.* **1969**, *32*, 827.
8. *CRC Handbook of Chemistry and Physics*, 67th ed.; Weast, R. C., Ed., CRC Press: Boca Raton, FL, 1987.
9. Chase, M. W. Jr; Davies, C. A.; Downey, J. R., Jr; Frurip, D. J.; McDonald, R. A.; Syverud, A. N. *JANAF Thermochemical Tables*, 3rd ed.; *J. Phys. Chem. Ref. Data*, **1985** *14*, Suppl. 1.
10. Kuchitsu, K. *J. Chem. Phys.* **1968**, *49*, 4456.
11. Thorne, L. R.; Suenram, R. D.; Lovas, F. J. *J. Chem. Phys.* **1983**, *78*, 167.
12. Huber, K. P.; Herzberg, G. *Molecular Spectra and Molecular Structure*, Vol. IV, *Constants of Diatomic Molecules*, Van Nostrand: New York, 1979.
13. Umeyama, H.; Morokuma, K. *J. Am. Chem. Soc.* **1976**, *98*, 7208.
14. Bartell, L. S.; Brockway, L. O. *J. Chem. Phys.* **1960**, *32*, 512.
15. Almenningen, A.; Halvorsen, S.; Haaland, A. *Acta Chem. Scand.* **1972**, *25*, 1937.
16. Almenningen, A.: Fernholt, L.; Haaland, A.; Weidlein, J. *J. Organomet. Chem.* **1978**, *145*, 109.
17. Landolt-Börnstein, L. *Numerical Data and Functional Relationships in Science and Technology*, New Series, Group II: Atomic and Molecular Physics, Vol. 7 and Vol. 15: Structure Data of Free Polyatomic Molecules, Springer Verlag, Berlin, **1976** and **1987**.
18. Oberhammer, H.; Boggs, J. E. *J. Am. Chem. Soc.* **1980**,*102*, 7241.
19. Glidewell, C.; Thomson, C. *J. Comput. Chem.* **1982**, *3*, 495.
20. Glidewell, C.; Thomson, C. *J. Comput. Chem.* **1983**, *4*, 9.
21. Haaland, A.: Samdal, S.; Stokkeland, O.; Weidein, J. *J. Organomet. Chem.* **1977**, *134*, 165.
22. Smith, M. B. *J. Organomet. Chem.* **1974**, *76*, 171.
23. Smith, M. B. *J. Organomet. Chem.* **1972**, *46*, 31.
24. Hargittai, M.; Kolonits, M.; Tremmel, J.; Fourquet, J.-L.; Ferey, G. *Struct. Chem.* **1990**, *1*, 75.
25. Spiridonov, V. P.; Gershikov, A. G.; Zasorin, E. Z.; Popenko, N. I.; Ivanov, A. A.; Ermolayea, L. I. *High Temp. Sci.* **1981**, *14*, 285.
26. Sheldrick, G. M.; Sheldrick, W. S. *J. Chem. Soc. A.* **1969**, 2279.

27. Shreve, A. P.; Mulhaupt, R.; Fultz, W.; Calabrese, J.; Robbins, W.; Ittel, S.D. *Organometallics* **1988**, *7*, 409.

28. Waggoner, K. M.; Power, P. P. *J. Am. Chem. Soc.* **1991**, *113*, 3385.

29. Pauling, L. *The Nature of the Chemical Bond*, 3rd ed.; Cornell University Press: Ithaca, NY, 1960.

30. Gropen, O.; Nilssen, E. W. *J. Organomet. Chem.* **1976**, *111*, 257.

31. Barron, A. R.; Dobbs, K. D.; Frankl, M. M. *J. Am. Chem. Soc.* **1991**, *113*, 39.

32. Lichtenburger, D. L.; Hogan, R. H.; Healy, M. D.; Barron, A. R. *J. Am. Chem. Soc.* **1990**, *112*, 3369.

33. Henrickson, C. H.; Duffy, D.; Eyman, D. P. *Inorg. Chem.* **1968**, *7*, 1047.

34. Anderson, G. A,; Forgaard, F. R.; Haaland, A. *Acta Chem. Scand.* **1972**, *26*, 1947.

35. Nykerk, K. M.; Eyman, D. P. *Inorg. Nucl. Chem. Letters.* **1968**, *4*, 253

36. Henrickson, C. H.; Nykerk, K. M.; Eyman, D.P. *Inorg. Chem.* **1968**, *7*, 1028.

37. Henrickson, C. H.; Eyman., D. P. *Inorg. Chem.* **1967**, *6*, 1461.

38. Guryonova, E. N.; Goldstein, I. P.; Romm, I. P. *Donor-Acceptor Bond*, Wiley: New York, **1975**.

39. Zaworotko, M.J.; Atwood, J. L. *Inorg. Chem.* **1980**, *19*, 268.

40. Veith, M; Lange, H.; Recktenwald, O.; Frank, W. *J. Organomet. Chem.* **1985**, *294*, 273.

41. Moise, F.; Pennington, W. T.; Robinson, G. H.; Sangokoya, S. A. *Acta Crystallogr. Sect. C*, **1990**, *46*, 1110.

42. Robinson, G. H.; Zhang, H.; Atwood, J. L. *J. Organomet. Chem.* **1987**, *331*, 153.

43. Goebel, D. W.; Hencher, J.L.; Oliver, J.P. *Organometallics* **1983**, *2*, 746.

44. Almenningen, A.; Gundersen, G.; Haugen, T.; Haaland, A. *Acta Chem. Scand.* **1972**, *26*, 3928.

45. Whitt, C.D.; Parker, L.M.; Atwood, J. L. *J. Organomet. Chem.* **1971**, *32*, 291.

46. Atwood, J. L.; Seale, S. K.; Roberts, D. H. *J. Organomet. Chem.* **1973**, *51*, 105.

47. Atwood, J. L.; Milton, P.A. *J. Organomet. Chem.* **1973**, *52*, 275.

48. Hargittai, M.; Hargattai, I.; Spiridonov, V. P.; Pelissier, M.; Labarre, J. F. *J. Mol. Struct.* **1975**, *24*, 27.

49. Hargittai, M.; Hargittai, I.; Spiridonov, V. P.; Ivanov, A.A. *J. Mol. Struct.* **1977**. *39*, 225.

50. Bailey, N. A.; Bird, P. H.; Wallbridge, M. G. H. *Inorg. Chem.* **1968**, *7*, 1575.

51. Grant, D. F.; Killean, R. C. G.; Lawrence, J.L. *Acta Crystallogr.*, Sect. B **1969**, *25*, 377.

52. Clegg, W.; Klingebiel, U.; Neemann, J.; Sheldrick, G. *J. Organomet. Chem.* **1983**, *249*, 47.

53. Beagley, B.; Hewitt, T. G. *Trans. Far. Soc.* **1968**, *64*, 2561.

54. Wollrab, J. E.; Laurie, V. W. *J. Chem. Phys.* **1969**, *51*, 1580.

55. Almenningen, A,; Haaland, A.; Haugen, T.; Novak, D. P. *Acta Chem. Scand.* **1973**, *27*, 1821.

56. Mole, T.; Jeffrey, E. A. *Organoaluminium Compounds*, Elsevier: Amsterdam, **1972**.

57. Interrante, L. V.; Sigel, G. A.; Garbauskas, M.; Hejna, C.; Slack, G. A. *Inorg. Chem.* **1989**, *28*, 252.

58. Ouzounis, K.; Riffel, H.; Hess, H.; Kohler, U.; Weidlein, J. Z. *Anorg. Allg. Chem.* **1983**, *504*, 67.

59. Sauls, F, C.; Czekaj, C. L.; Interrante, L. V. *Inorg. Chem.* **1990**, *29*, 4688.

60. Hess, H.; Hinderer, A.; Steinhauser, S. *Z. Anorg. Allg. Chem.* **1970**, *377*, 1.

61. Bryan, S. J.; Clegg, W.; Snaith, R.; Wade, K.; Wong, E. H. *J. Chem. Soc., Chem. Commun.* **1989**, 1223.

62. Amirkhalili, S.; Hitchcock, P. B.; Jenkins, A. D.; Nyathi, J. Z.; Smith, J. D. *J. Chem. Soc., Dalton Trans.* **1981**, 377.

63. Seale, S. K.; Atwood, J. L. *J. Organomet. Chem.* **1974**, *73*, 27.

64. Ahmed, A.; Schwarz, W.; Hess, H. *Acta Crystallogr., Sect. B* **1977** *33*, 3574.

65. Magnussen, V. R.; Stucky, G. D. *J. Am. Chem. Soc.* **1969**, *91*, 2544.

66. Hitchcock, P. B.; Jasim, H. A.; Lappert, M. F.; Williams, H. D. *J. Chem. Soc., Chem. Commun.* **1986**, 1634.

67. Chernega, A. N.; Antipin, M. Yu.; Struchkov, Yu. T.; Boldeskul, I. E. Shulgin, V. F.; Romanenko, V. D. *Zh. Stukt. Khim.* **1985**, *26*, 120.

68. Hausen, H. D.; Gerstner, F.; Weidlein, J. *J. Organomet. Chem.* **1978**, *145*, 277.

69. Gerstner, F.; Schwarz, W.; Hausen, H. D.; Weidlein, J. *J. Organomet. Chem.* **1979**, *175*, 33.

70. Thewalt, U.; Kawada, I. *Chem. Ber.* **1979**, *103*, 2754.

71. Robinson, G. H.; Rae, A. D.; Campana, C. F.; Byram, S. K. *Organometallics* **1987**, *6*, 1227.

72. Geoffrey, F.; Cloke, N.; Dalby, C. I.; Henderson, M. J.; Hitchcock, P. B.; Kennard, C. H. L.; Lamb, R. N.; Raston, C. *J. Chem . Soc., Chem. Commun.* **1990**, 1394.

73. Fink, W. H.; Richards, J. C. *J. Am. Chem. Soc.* **1990**, *113*, 3393.

74. Del Piero, G.; Cesari, M.; Dozzi, G.; Mazzei, A. *J. Organomet. Chem.* **1977**, *129*, 281.

75. Cesari, M.; Perego, G.; Del Piero, G.; Cusinella, S.; Cernia, E. *J. Organomet. Chem.* **1974**, *78*, 203.

76. Del Piero, G.; Cesari, M.; Perego, G.; Cusinella, S.; Cernia, E. *J. Organomet. Chem.* **1977**, *129*, 129

77. Hitchcock, P. B.; Smith, J. D.; Thomas, K. M. *J. Chem. Soc., Dalton Trans.* **1976**, 1433.

78. Jeffrey, G. A.; Parry, Q. S.; Mozzi, R. L. *J. Chem. Phys.* **1956**, *25*, 1024.

79. Palenik, G. *Acta Crystallogr.* **1964**, *17*, 1573.

80. Heitsch, C. W.; Nordman, C. E.; Parry, R. W. *Inorg. Chem.* **1963**, *2*, 508.

81. Mastryukov, V. S.; Golubiskii, A. V.; Vilkov, L. V. *J. Struct. Chem.* **1979**, *20*, 921.

82. Müller, G.; Krüger, C. *Acta Crystallogr., Sect. C* **1984**, *40*, 628

83. Ahmed, A.; Schwarz, W.; Weidlein, J.; Hess, H. *Z. Anorg. Allg. Chem.* **1977**, *434*, 207

84. Robinson, G. H.; Sangokoya, S. A.; Moise, F.; Pennington, W. T. *Organometallics* **1988**, *7*, 1887.

85. Wynne, K. J. *Inorg. Chem.* **1984**, *23*, 4658.

86. Goedken, V. L.; Ito, H.; Ito, T. *J. Chem. Soc., Chem. Commun.* **1984**, 1453.

87. Jiang, Z.; Interrante, L. V.; Kwon, D.; Tham, F. S.; Kullnig, R. *Inorg. Chem.* **1991**, *30*, 995.

88. Perego, G.; Del Peiro, G.; Corbellini, M.; Bruzzone, M. *J. Organomet. Chem.* **1977**, *136*, 301.

89. Robinson, G. H.; Sangokoya, S. A. *J. Am. Chem. Soc.* **1987**, *109*, 6852.

90. Leman, J. T.; Barron, A. R. *Organometallics* **1989**, *8*, 1828.

91. Wierda, D. A.; Barron, A. R. *Polyhedron* **1989**, *8*, 831.

92. Robinson. G. H.; Self, M. F.; Sangokoya, S. A.; Pennington, W. T. *J. Cryst. Spectr. Res.* **1988**, *18*, 285.

93. Bradley, D. C.; Chudzynska, H.; Faktor, M. M.; Frigo, D. M.; Hursthouse, M. B.; Hussain, B.; Smith, L. M. *Polyhedron* **1988**, *7*, 1289.

94. Henrickson, C. H.; Duffy, D.; Eyman, D. P. *Inorg. Chem.* **1968**, *7*, 1047.

95. Healey, M. D.; Wierda, D. A.; Barron, A. R. *Organometallics* **1988**, *7*, 2543.

96. Haaland, A.; Hougen, J.; Volden, H. V.; Hanika, G. H.; Karsh, H. H. *J. Organomet. Chem.* **1987**, *322*, C24.

97. Janik, J. F.; Duesler, E. N.; McNamara, W. F.; Westerhausen, M.; Paine, R. *Organometallics* **1989**, *8*, 506.

98. Cowley, A. H.; Jones, R. H.; Mardones, M. A.; Atwood, J. L.; Bott, S. G. *Angew. Chem. Int. Ed. Engl.* **1990**, *29*, 1409.

99. Wang. C. C.; Zaheeruddin, M.; Spinar, L. H. *J. Inorg. Nucl. Chem.* **1963**, 326.

100. Karsch, H. H.; Appelt, A,; Köhler, F.; Müller, G. *Organometallics* **1985**, *4*, 231.

101. Schmidbaur, H.; Tronich, W. *Chem. Ber.* **1968**, *101*, 595.

102. Schmidbaur, H.; Lauteschläger, S.; Milewski-Mahrla, B. *Chem. Ber.* **1983**, *116*, 1403

103. Karsch, H. H.; Zellner, K.; Muller, G. *J. Chem. Soc., Chem. Commun.* **1990**, 1621.

104. Karsch, H. H.; Appelt, A.; Riede, J.; Müller, G. *Orgamometallics* **1987**, *6*, 316.

105. Power, M. B.; Bott, S. G.; Clark, D. L.; Atwood, J. L.; Barron, A. R. *Organometallics* **1990**, *9*, 3086.

106. Haaland, A.; Samdal, S.; Stokkeland, O.; Weidlein, J. *J. Organomet. Chem.* **1977**, *134*, 165.

107. Atwood, J. L.; Stucky, G. D. *J. Am. Chem. Soc.* **1967**, *89*, 5362.

108. Atwood, J. L.; Hrncir, D. C.; Shakir, R.; Dalton, M. S.; Priester, R. D.; Rogers, R. D. *Organometallics* **1982**, *1*, 1021.

109. Atwood. J. L.; Priester, R. D.; Rogers, R. D.; Canada, L. G. *J. Inclusion Phenom.* **1983**, *1*, 61.

110. Srini, V.; de Mel, J.; Oliver, J. P. *Organometallics* **1989**, *8*, 827.

111. Rösch, L.; Altnau, G.; Krüger, C.; Tsay, Y. -H. *Z. Naturforsch.* **1983**, *38b*, 34.

112. Siera, M. L.; Srini, V.; de Mel, J.; Oliver, J. P. *Organometallics* **1989**, *8*, 2312.

113. Barber, M.; Liptak, D.; Oliver, J. P. *Organometallics* **1982**, *1*, 1307.

114. Rahman, A. F. M. M.; Siddiqui, K. F.; Oliver, J. P. *J. Organomet. Chem.* **1987**, *319*, 161.

115. Power, M. B.; Bott, S. G.; Atwood, J. L.; Barron, A. R. *J. Am. Chem. Soc.* **1990**, *112*, 3446.

116. Le Carpentier, J. -M.; Weiss, R. *Acta Crystallogr., Sect. B,* **1972**, *28*, 1437.

117. Rasmussen, S. E.; Broch, N. C. *Acta Chem. Scand.* **1966**, *20*, 1351.

118. Chevrier, B.; Le Carpentier, J. M.; Weiss R. *Acta Crystallogr., Sect. B* **1972**, *28*, 2659.

119. Burford, N.; Royan, B. W.; Spence, R. E. v. H.; Cameron, T. S.; Linden, A.; Rogers, R. D. *J. Chem. Soc., Dalton Trans.* **1990**, 1521.

120. Blake, A. J.; Ebsworth, E. A. V.; Rankin, D. W. H.; Robertson, H. E.; Smith, D. E.; Welch, A. J. *J. Chem. Soc., Dalton Trans.* **1986**, 91.

121. Gropen, O.; Johansen, R.; Haaland, A.; Stokkeland, O. *J. Organomet. Chem.* **1975**, *92*, 147.

122. Fernholt, L.; Haaland, A.; Hargittiai. M.; Seip, R.; Weidlein, J. *Acta Chem. Scand.* **1981**, *A35*, 529.

123. Drew, D. A.; Haaland, A.; Weidlein, J. *Z. Anorg. Allg. Chem.* **1973**, *398*, 241.

124. Jeffrey, E. A.; Mole, T. *Aust. J. Chem.* **1968**, *21*, 2683.

125. Haaland, A.; Stokkeland, O. *J. Organomet. Chem.* **1975**, *94*, 345.

126. Cayton, R. H.; Chisholm, M H.; Davidson, E. R.; DiStaś, V. F.; Du, P.; Huffmann, J. C. *Inorg. Chem.* **1991**, *30*, 1020.

127. Bonamico, M.; Dessy, G. *J. Chem. Soc. A.* **1967**, 1786.

128. Kai, Y.; Yasuoka, N.; Kasai, N.; Kakudo, M. *J. Organomet. Chem.* **1971**, *32,* 165.

129. Hausen, H. D.; Schmöger, G.;Weidlein, J. *J. Organomet. Chem.* **1978**, *153*, 271.

130. van Vliet, M. R. P.; van Koten, G.; Rotteveel, M. A. Schrap, M.; Vrieze, K.; Koji'c, B.; Spek, A. L.; Duisenberg, A. J. M. *Organometallics* **1986**, *5*, 1389.

131. Kai, Y.; Kasai, N.; Kakudo, M. *Bull. Chem. Soc. Jpn.* **1972**, *45*, 3403.

132. Sharma, V.; Simard, M.; Wuest, J. D. *Inorg Chem.* **1991**, *30*, 579.

133. Müller, E.; Bürgi, H, -B. *Helv. Chim. Acta* **1987**, *70*, 520.

134. Robinson, G. H.; Lee, B.; Pennington, W. T.; Sangokoya, S. A. *J. Am. Chem. Soc.* **1988**, *110*, 6260.

135. Kushi, Y.; Fernando, Q. *J. Am. Chem. Soc.* **1979**, *92*, 91.

136. Benn, R.; Rufinska, A.; Lehmkuhl, H.; Janssen, E.; Krüger, C. *Angew. Chem.* **1983**, *95*, 808.

137. Sierra, M. L.; Srini, V.; de Mel, J.; Oliver, J. P. *Organometallics* **1989**, *8*, 2486.

138. Kai, Y.; Yasuoka, N.; Kasai, N.; Kakudo, M. *Bull. Chem. Soc. Jpn.* **1972**, *45*, 3397.

139. van der Steen, F. Thesis, Rijksuniversiteit te Utrect, **1991**.

140. Hansen, K. W.; Bartell, L. S. *Inorg. Chem.* **1965**, *4*, 1775.

141. Yanovski, A. I.; Kozunov, V. A.; Turova, N. Ya.; Furmanova, N. G.; Struchkov, Yu. T. *Dokl. Akad. Nauk. SSSR* **1979**, *244*, 119.

142. Dzugan, S. J.; Goedken, V. L. *Inorg. Chem.* **1986**, *25*, 2858.

143. Turova, N. Ya.; Kozunov, V. A.; Yanovskii, A. I.; Bokii, N. G.; Struschkov, Yu. T.; Tarnopolskii, B. L. *J. Inorg. Nucl. Chem.* **1979**, *41*, 5.

144. Hon, P. K.; Pfluger, C. E. *J. Coord. Chem.* **1973**, *3*, 67.

145. McCleland, B. W. *Acta Crystallogr. Sect. B* **1975**, *31*, 2496.

146. Muetterties, E. L.; Guggenberger, L. J. *J. Am. Chem. Soc.* **1972**, *94*, 8046.

147. Healey, M. D.; Barron, A. R. *J. Am. Chem. Soc.* **1989**, *111*, 398.

148. Ruhland-Senge, K.; Power, P. P. *Inorg. Chem.* **1991**, *30*, 2633.

149. Robinson, G. H.; Zhang, H.; Atwood, J. L. *Organometallics* **1987**, *6,* 887.

150. Weiss, A,; Plass, R.; Weiss, A. *Z. Anorg. Allg. Chem.* **1956**, *283*, 390.

151. Burford, N.; Royan, B. W.; Spence, R. E. v. H.; Rogers, R. D. *J. Chem. Soc., Dalton Trans.* **1990**, 2111.

152. Burford, N.; Spence, R. E. v. H.; Rogers, R. D. *J. Chem. Soc., Dalton Trans.* **1990**, 3611.

153. Robinson, G. H.; Sangokoya, S. A. *J. Am. Chem. Soc.* **1988**, *110*, 1494.

154. Haaland, A.; Stokkeland, O.; Weidlein, J. *J. Organomet. Chem.* **1975**, *94*, 353.

155. Brauer, D. J.; Stucky, G. D. *J. Am. Chem. Soc.* **1969**, *91*, 5462.

156. Bailey, M.; Hendershot, D. G.; Kumar, R.; Oliver, J. P., Abstracts of Papers, XIVth International Conference on Organometallic Chemistry, Detroit, **1990**, p170.

157. Range, K.J.; Huebner, H. *Z. Naturforsch., B* **1973**, *28*, 353.

158. Jumar, R.; de Mel, V. S. J.; Oliver, J. P. *Organometallics* **1989**, *8*, 2488.

159. Robinson, G. H; Self, M. F.; Pennington, W. T.; Sangokoya, S. A. *Organometallics* **1988**, *7*, 2424.

160. Schmidbauer, H.; Kline, H. F. *Angew. Chem.* **1966**, *78*, 750.

161. Smith, M. B. *J. Organomet. Chem.* **1970**, *22*, 273.

162. Rempfer, B.; Oberhammer, H.; Auner, N. *J. Am. Chem. Soc.* **1984**, *108*, 3893.

163. Schmidbauer, H.; Lauteschläger, S.; Müller, G. *J. Organomet. Chem.* **1985**, *33*, 281.

164. Brown, H. C.; Wallace, W. J. *J. Am. Chem. Soc.* 1953, **75**, 6279.

165. Gundersen, G.; Haugen, T.; Haaland, A. *J. Organomet. Chem.* **1973**, *54*, 77.

166. Laubengayer, A. W.; Lengnick, G. F. *Inorg. Chem.* **1966**, *5*, 503.

167. Krieg, V. Weidlein, J. *J. Organomet. Chem.* **1970**, *21*, 281.

168. Lehmkuhl, H.; Olbrysch, O.; Nehl, H. *Liebigs Ann. Chem.* **1973**, 708.

169. Brendhaugen, K.; Haaland, A.; Novak, D. P. *Acta Chem. Scand.* **1974**, *A28*, 45.

170. Dohmeier, C.; Mattes, R.; Schnökel, H. *J. Chem. Soc., Chem. Commun.* **1990**, 358.

171. Shen, Q., Ph. D. Thesis, Oregon State University, Corvallis, **1974**.

172. Templeton, D. H.; Carter, G. F. *J. Chem. Phys.* **1954**, *58*, 940.

CHAPTER

2

Organoaminoalanes: Unusual Al–N Systems

Gregory H. Robinson

Howard L. Hunter Chemistry Laboratory,
Department of Chemistry, Clemson University
Clemson, South Carolina 29634-1905

2.1. Introduction

The preparation of borazine by Stock and Pohland in 1926[1] is arguably the singular most significant development in main group chemistry of this century. Not only was the π-bonding capability of group 13 elements brought to the fore with the preparation of this boron–nitrogen analog of benzene, but the ability of main group elements to form interesting ring compounds was also realized. In addition, mercury-flow valves and inefficient vacuum pumps notwithstanding, the techniques, principles, and methods developed by Stock and coworkers[2] in their quarter century of seminal work with boron compounds provided a mechanism by which other highly volatile or reactive inorganic substances could be prepared, manipulated, and isolated. The organometallic chemistry of aluminum in general, and that of the highly reactive and pyrophoric aluminum alkyls in particular, is perhaps the greatest beneficiary of the *Stock investigations*. Indeed, aluminum alkyls underscore a unique enigma in organoaluminum chemistry: although organoaluminum moieties are critical in a number of organic and catalytic processes, many substantial points of structure and bonding relating to the metal–ligand interface remain largely ambiguous.

The organometallic coordination chemistry of aluminum, traditionally held to be generally unremarkable, progressed considerably in the last decade. As reports of unusual high coordination number (i.e., five and six), and to a much lesser extent low coordination number (i.e., three) aluminum complexes have become nearly routine in recent years, the coordination of aluminum has been shown to span a wide spectrum thereby suggesting that four-coordinate tetrahedral may no longer be considered the exclusive coordination mode of this metal in organometallic complexes.

The interaction of organoaluminum moieties with amines has provided a plethora of novel Al–N compounds. These compounds have become known, albeit technically inaccurate, as aminoalanes. Importantly, the organo derivatives of these compounds, organoaminoalanes, have afforded valuable knowledge, particularly from a solid state perspective, as to the structural organometallic coordination chemistry of aluminum. This chapter will examine the coordination chemistry of organoaminoalanes relative to four classes of amines: (1) simple monodentate amines; (2) sterically demanding aryl amines; (3) multidentate open-chain amines; and (4) macrocyclic amines. As the interaction of aluminum moieties with amines is one of the most extensively investigated areas of organoaluminum chemistry, this contribution is not intended to represent an exhaustive survey but rather to offer a synthetic and structural representation of general trends in a rather large and diverse field.

A word on molecular structure is in order. While this chapter will rely heavily on the solid-state structures of Al–N compounds as determined by single crystal X-ray diffraction techniques, structural characterization by modern ^{27}Al NMR is not to be dismissed as this method has progressed significantly in recent years[3–5] and is now a powerful tool for studies of organoaluminum systems in solution. Indeed, ^{27}Al NMR is often the technique of choice in aqueous systems. Unambiguous structural and quantitative bonding information, however, is often best obtained from single crystal X-ray diffraction studies.

2.2. Interaction of Aluminum Alkyls with Simple Amines: Al–N Cages

Certainly the most fundamental reaction in this regard is that of trimethylaluminum with ammonia. The initial studies on this reaction were performed by Wiberg.[6] The monomeric addition product, $Me_3Al{:}NH_3$, was reported to undergo condensation with an increase in temperature ultimately resulting in crude aluminum nitride with concomitant exhaustive alkane elimination.

Studies of the conversion of $Me_3Al{:}NH_3$ to aluminum nitride have continued in recent years.[7–10]

Davidson and Brown,[11] in one of the earliest studies of the interaction of organoaluminum moieties with monodentate amines, prepared the $Me_3Al{:}N(H)Me_2$ addition product from reaction of dimethylamine with trimethylaluminum. This product was proposed to undergo condensation to give the dimeric product $[Me_2Al\text{-}NMe_2]_2$ (Equation 2.1).

$$2Me_2Al{:}N(H)Me_2 \longrightarrow [Me_2Al\text{–}NMe_2]_2 + 2Me\text{–}H \tag{2.1}$$

As noted subsequently by Laubengayer *et al.,*[12] the preparation of $Me_3Al{:}N(H)Me_2$ underscored two important points: (1) Reaction of amines with aluminum alkyls at low temperatures affords Lewis acid–Lewis base, Al:N addition products; (2) Accompanied by an increase in temperature, Al:N addition products readily undergo intermolecular condensation, with concomitant alkane elimination, affording products possessing extensive Al–N interaction.

Cohen *et al.*[13] examined a number of addition products resulting from reaction of ethylaluminum dichloride and diethylaluminum chloride with stoichiometric quantities of ammonia. While the Al–C bond in $EtAlCl_2$ was not cleaved by excess NH_3 below room temperature, between 70 and 150°C the addition products yielded a white substance, insoluble in most organic solvents and liquid NH_3, of empirical formula Cl_2AlNH_2. Distillation yielded an off-white, highly cross-linked, polymeric solid. At temperatures above 400°C, NH_4AlCl_4 and HCl were detected as volatile decomposition products. Reaction of an excess of NH_3 with Et_2AlCl reportedly produced a white substance of proposed composition $Et_2AlCl{\cdot}(NH_3)_2$, which was insoluble in hydrocarbon solvents and ether, but soluble in liquid NH_3. Based on X-ray powder photographs, IR spectra, molecular weight determinations, and solubility data, the authors suggested that the product was not $Et_2AlCl{\cdot}(NH_3)_2$, but rather $[Et_2Al(NH_3)_2]Cl$. The composition of $[Et_2Al(NH_3)_2]Cl$ was subsequently supported by conductivity studies in liquid ammonia.[14]

McCaughlin *et al.*[15] examined the reaction of trimethylaluminum with methylamine. Heating of the $Me_3Al{:}N(H)_2Me$ addition product to 50°C for several hours followed by fractional sublimation affords structural isomers of cyclotri-μ-methylamidotris(dimethylaluminum), $[Me_2Al\text{-}N(H)Me]_3$ (Equation 2.2).

$$Me_3Al{:}N(H)_2Me \longrightarrow \tfrac{1}{3}[Me_2Al\text{–}N(H)Me]_3 + Me\text{–}H \tag{2.2}$$

The two isomers were observed to slowly interconvert on the NMR time scale.[16] Although a number of crystallographic difficulties were encountered, elucidation of the solid-state structure of these products proceeded unambiguously to reveal *cis-* and *trans-* stereoisomers of $[Me_2Al\text{-}N(H)Me]_3$. Figure 2.1 shows that both $[Me_2Al\text{-}N(H)Me]_3$ stereoisomers contain non-planar Al_3N_3 cores. As demanded by the C_3 symmetry present in the molecule, the Al_3N_3

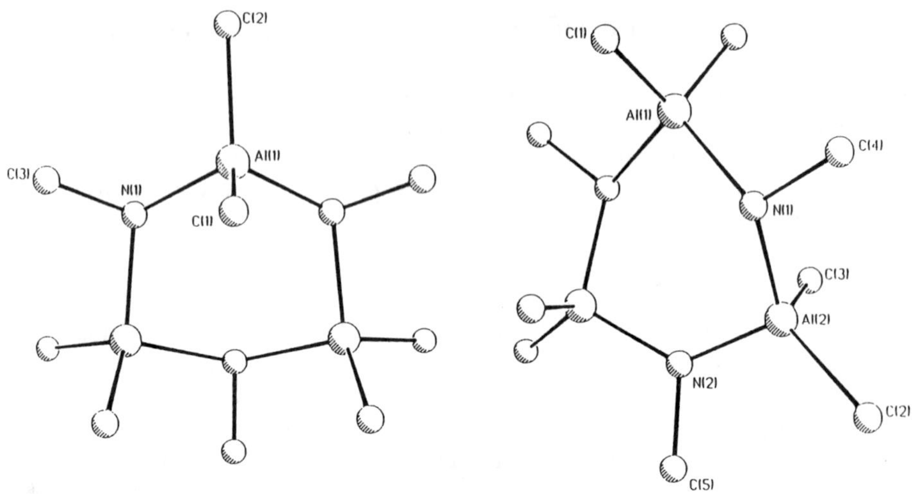

Figure 2.1 *Cis*- and *trans*- isomers of $[Me_2Al\text{-}N(H)Me]_3$.

core of *cis*-$[Me_2Al\text{-}N(H)Me]_3$ was found to reside in a chair conformation whereas that of *trans*-$[Me_2Al\text{-}N(H)Me]_3$ was found to reside in a distorted boat conformation. The major difference between the *cis*- and *trans*-isomers is found at one of the nitrogen atoms in the Al_3N_3 six-membered ring: whereas all three azamethyl groups in the *cis*-isomer resided on the same side of the N_3 plane, in the *trans*-isomer, one of the azamethyl groups resided on the opposite side of the N_3-plane. The mean Al–N bond distance for both isomers was determined to be 1.990 (5) Å and the mean Al–C interaction was 1.973 (5) Å. The energy differences between the two isomers were taken to be negligible as the observed conformations were ascribed to be predominantly a packing effect.

Cyclodi-μ-dimethylamidobis(dimethylaluminum), $[Me_2Al\text{–}NMe_2]_2$, was also reported by McCaughlin *et al.*[15] in the $[Me_2Al\text{–}N(H)Me]_3$ study. This dimeric compound, **1**, isolated from reaction of dimethylamine with trimethyl-

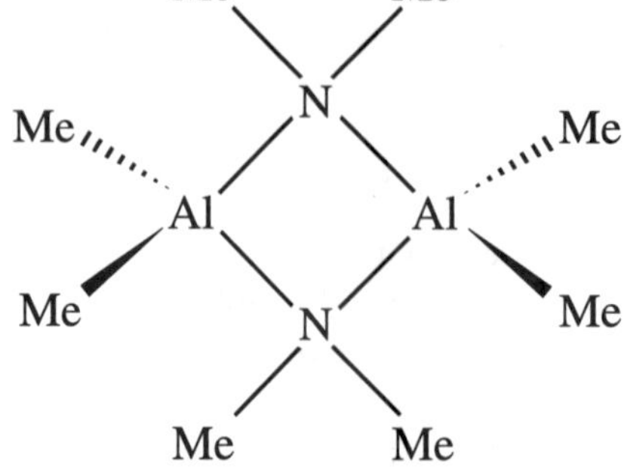

1

aluminum, was shown to possess approximate D_{2h} point symmetry. The X-ray structure of $[Me_2Al{-}NMe_2]_2$ was also independently confirmed by Hess *et al.*[17]

Although at odds with the report by Wiberg,[18] which described products as non-volatile glassy materials, thermal decomposition studies of $[Me_2Al{-}N(H)Me]_3$ in organic solvents were reported by Alford *et al.*[19] to produce two crystalline products. Reported to be extremely susceptible towards oxidation, these materials were soluble in aromatic solvents and could be sublimed under vacuum. However, as chemical analyses and cryoscopic molecular weight measurements proved problematic, it was difficult to accurately determine the formula of these compounds. Based principally on solubility and NMR data, one compound was proposed to be the Al–N cage $[Me_2Al{-}N(H)Me]_3[MeAl{-}NMe]_4$ whereas IR data suggested the $[MeAl{-}NMe]_8$ octamer as the second compound.

Thermal decomposition of the $Me_3Al{:}N(H)_2Me$ addition compound at 215°C in toluene affords a white crystalline substance of the general formula $[MeAl{-}NMe]_n$.[20] Although molecular weight measurements initially suggested a mixture of heptameric and octameric products, NMR data and a single crystal X-ray diffraction study confirmed that thermal decomposition of $Me_3Al{:}N(H)_2Me$ yields the $[MeAl{-}NMe]_7$ heptamer (Equation 2.3).[21]

$$21\ N(H)_2Me + 21\ AlMe_3 \xrightarrow[-21Me-H]{70^\circ C}$$

$$7[Me_2Al{-}N(H)Me]_3 \xrightarrow[-21Me-H]{215^\circ C} 3[MeAl{-}NMe]_7 \qquad (2.3)$$

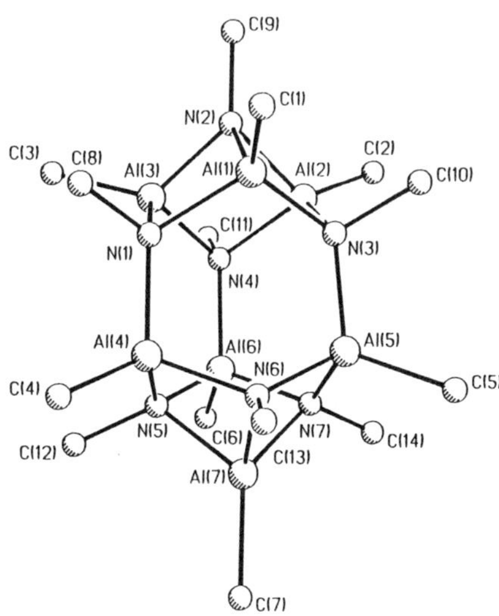

Figure 2.2. X-ray structure of $[MeAl{-}NMe]_7$.

The X-ray structure of [MeAl-NMe]$_7$, as shown in Figure 2.2, reveals an unusual cage compound with each aluminum atom bonding to one methyl group and three nitrogen atoms and each nitrogen atom bonding to one methyl group and three aluminum atoms.

The heptameric product approaches C_{3v} symmetry with the rotation axis containing Al(7) and N(2). All methyl groups are directed along the perimeter of the compound effectively surrounding a highly aggregated Al_7N_7 core. Indeed, the compound contains three Al_3N_3 six-membered rings and six Al_2N_2 four-membered rings. The mean bond distances and angles for Al–N, Al–N–Al, and N–Al–N are 1.91 (4) Å, 89 (1), and 90 (2)°, respectively.

It was subsequently confirmed that decomposition of Me_3Al:N(H)$_2$Me affords a mixture of products.[22] Thermal decomposition of Me_3Al:N(H)$_2$Me was shown to proceed necessarily in a stepwise fashion from the trimeric [Me_2Al:N(H)Me]$_3$ amido product, through the [Me_2Al–N(H)Me]$_2$[MeAl–NMe]$_6$ intermediate, to the major product, the heptameric [MeAl–NMe]$_7$ imido compound. Evidence of a minute quantity of a substance, possibly [MeAl–NMe]$_8$, was also noted. The structure of the [Me_2Al–N(H)Me]$_2$[MeAl–NMe]$_6$ intermediate product was determined by single crystal X-ray diffraction (Figure 2.3).

The mean Al–N bond distance in [Me_2Al–N(H)Me]$_2$[MeAl–NMe]$_6$ was determined to be 1.910 (6) Å. The Al–N bonds adjacent to Al_2N_2 four-membered rings are noticeably short at distances of 1.893 (5) and 1.864 (6) Å. The authors ascribed this to the presumably greater *s* character of the exocyclic bond relative to the greater *p* character of bonds in the ring. The mean Al–C bond distance is 1.97 (2) Å. The six-membered rings reside in boat conformations. The authors concluded that the [Me_2Al–N(H)Me]$_2$[MeAl–NMe]$_6$

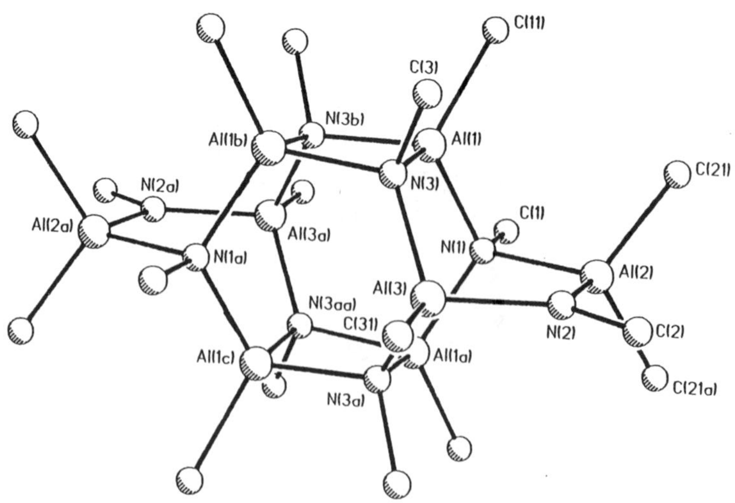

Figure 2.3. X-ray structure of [Me_2Al–N(H)Me]$_2$[MeAl–NMe]$_6$.

compound was, indeed the same as that isolated earlier, but then incorrectly identified as $[Me_2Al{-}N(H)Me]_3[MeAl{-}NMe]_4$. The gallium analog of this compound was also reported in the same study. The heptamer converts into the octamer only slowly at 200°C. The authors suggested that the relative amounts of heptamer and octamer formed from the thermal decomposition of $[Me_2Al{-}N(H)Me]_3$ in any given preparation was largely a function of kinetic factors in the initial stages of pyrolysis.

2.3. Dimers, Trimers, and Tetramers: Organoaluminum Chemistry of Sterically Demanding Aryl Amines

The capability of bulky ligands to stabilize organometallic compounds by saturating the coordination sphere of the metal, thereby sterically inhibiting decomposition pathways, has long been recognized as a fundamental cornerstone of organometallic chemistry. Sterically demanding ligands, or more specifically sterically demanding Lewis bases, are now found in the present debate in main group chemistry concerning the extent of π-bonding in organoaluminum compounds.[23–28] The utilization of sterically demanding ligands promotes low coordination number (i.e., three-coordinate) aluminum complexes. The concept of π-bonding is generally held to be enhanced in such compounds. Sterically demanding aryl amines such as 2,6-diisopropylaniline, $H_2N\text{-}Ph(i\text{-}Pr)_2$, **2**, 2,4,6-trimethylaniline (mesitylamine), $H_2N\text{-}Ph(Me)_3$, **3**, and aniline, $H_2N\text{-}Ph$, **4**, have been examined in this regard.

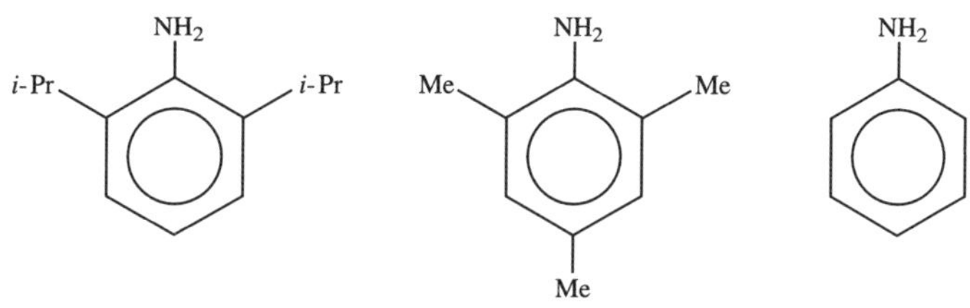

2 **3** **4**

Reaction of Me_3Al with 2,6-diisopropylaniline, as reported by Waggoner *et al.*,[29] readily results in the $[MeAl{-}N(H)Ph(i\text{-}Pr)_2]_2$ dimer. Subsequent heating of the dimer provides the novel trimeric product $[MeAl{-}NPh(i\text{-}pr)_2]_3$ in nearly quantitative yield (Equation 2.4).

The crystal structure of $[MeAl{-}NPh(i\text{-}Pr)_2]_3$ is given in Figure 2.4. The compound resides about a three-fold axis as the Al_3N_3 six-membered ring consitutes a plane. The mean Al–N distance of 1.782 (4) Å is extremely short. The ring angles at the Al and N atoms are 115.3 (5) and 124.7 (5)°, respectively. The carbon atoms of the methylaluminum units, as well as the *ipso*-carbon atoms of the amines, are coplanar with the Al_3N_3 plane. The phenyl rings of the amine units are approximately orthogonal to the Al_3N_3 plane.

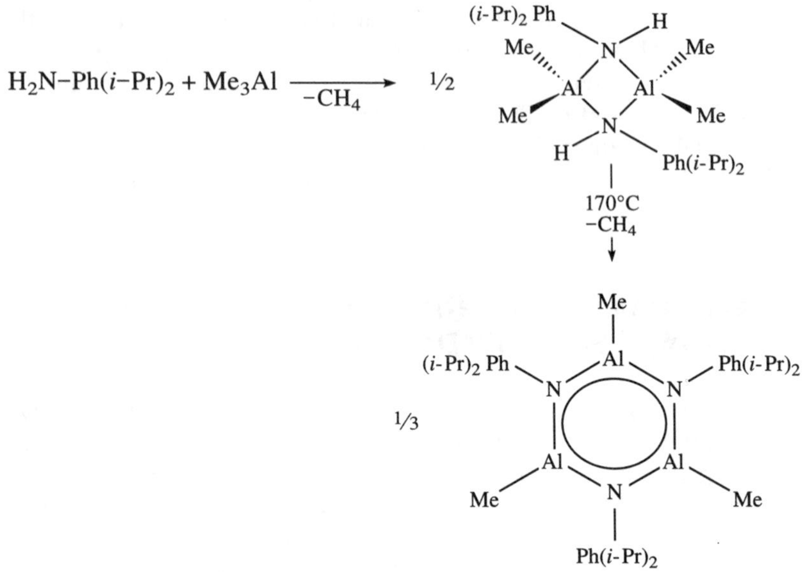

(2.4)

The extremely short Al–N distances coupled with the three-coordinate trigonal planar geometry of the aluminum and nitrogen atoms, legitimately raised the specter of delocalization, π-bonding, and comparisons of [MeAl–NPh(*i*-Pr)$_2$]$_3$ with borazine. The term "alumazene" was used by the authors in describing this compound. The ^{27}Al NMR spectrum consisted of a broad

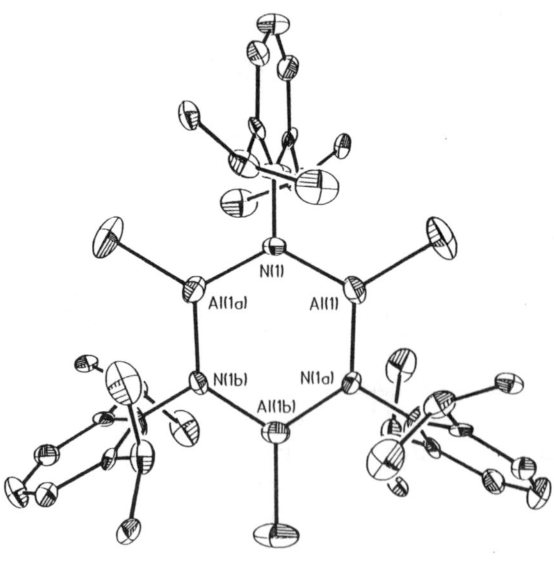

Figure 2.4 X-ray structure of [MeAl–NPh(*i*-Pr)$_2$]$_3$.

singlet at $\delta = 168$ ($\nu_{1/2} = 8800$ Hz). No unusual shifts in the ^{1}H NMR of $[MeAl{-}NPh(i\text{-}Pr)_2]_3$ were observed which could be attributed to the presence of a ring current (thus indicating aromatic character). However, preliminary calculations indicated substantial delocalization within the Al_3N_3 ring.

With the preparation of $[MeAl{-}NPh(i\text{-}Pr)_2]_3$, it is interesting to note that reaction of ethylenimine, $(CH_2CH_2)NH$, with Me_3Al yielded the first structurally characterized trimeric aluminum-nitrogen compound, the ethyleniminodimethylaluminum trimer $[Me_2Al{-}N(CH_2CH_2)]_3$, **5**.[30] The compound resides about a decidedly puckered Al_3N_3 six-membered ring. The mean Al–N bond distance is 1.91 Å and the mean Al–C bond distance is 1.96 Å. The mean Al–N–Al and N–Al–N bond angles are 119.9 and 102°, respectively.

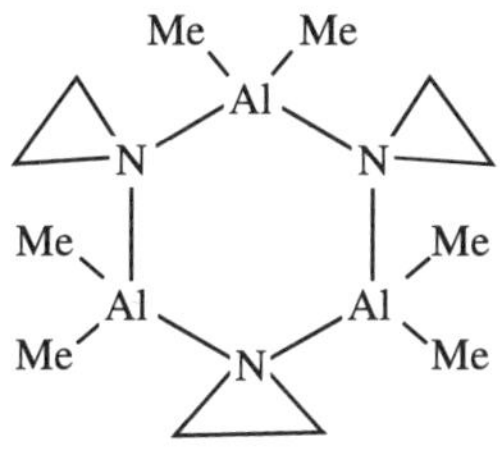

5

The organolauminum-nitrogen tetramer $[MeAl{-}NPh(Me)_3]_4$ was isolated from reaction of $H_2N{-}Ph(Me)_3$ with Me_3Al, after necessarily proceeding through an Al_2N_2 dimer (Equation 2.5).[31]

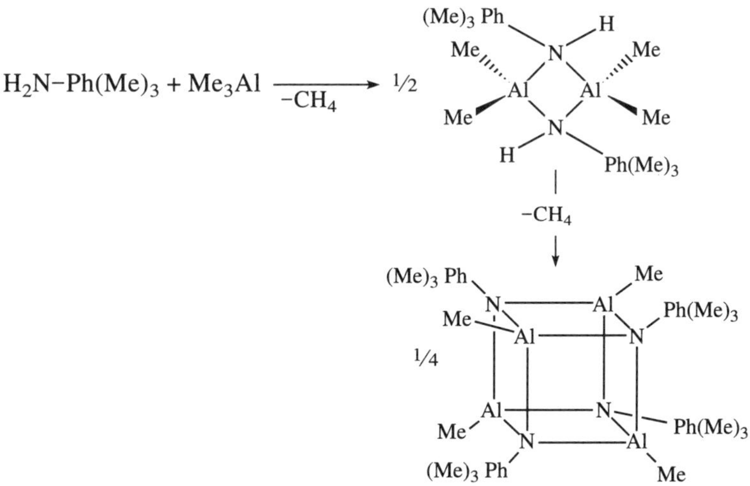

(2.5)

The X-ray crystal structure of the tetramer is given in Figure 2.5. It is inter-sting to view $[MeAl{-}NPh(Me)_3]_4$ as a fusing of two Al_2N_2 dimers resulting in a virtually perfect Al_4N_4 cube: each aluminum atom bonds to three nitrogen atoms (in addition to one carbon atom) while each nitrogen atom bonds to three

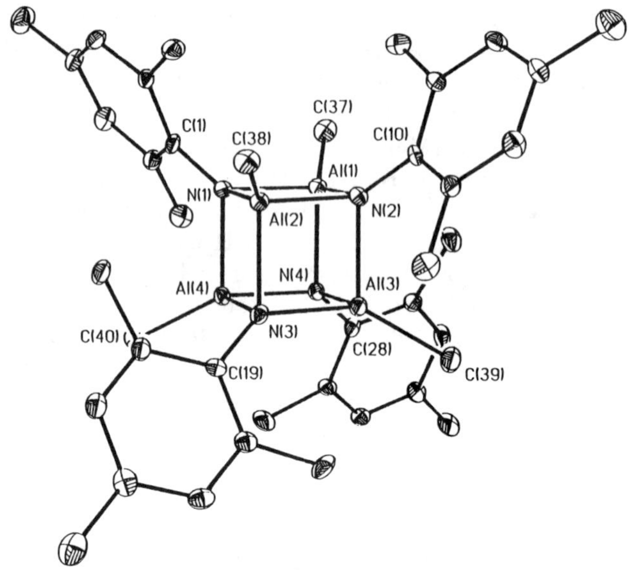

Figure 2.5. X-ray structure of $[MeAl{-}NPh(Me)_3]_4$.

aluminum atoms (in addition to a $Ph(Me)_3$ group). The geometry about each aluminum and nitrogen atom is roughly tetrahedral; the mean Al–N distance in $[MeAl{-}NPh(Me)_3]_4$ (1.948 (7) Å) is unremarkable. The mean Al–C bond distance is 1.949 (3) Å.

The literature reveals only a few Al_4N_4 cubanes which have been characterized by single crystal X-ray diffraction techniques. The compounds $[HAl{-}N(i\text{-}Pr)]_4$ and $[MeAl{-}N(i\text{-}Pr)]_4$, prepared from reaction of $LiAlH_4$ or Me_3Al with isopropylamine, respectively, are two such examples.[32] The mean Al–N bond distances for the hydridoaluminum and methylaluminum derivatives were found to be 1.913 (2) and 1.923 (1) Å, respectively. The basic Al_4N_4 structure of these two compounds is, however, quite similar to that observed for $[MeAl{-}NPh(Me)_3]_4$.

A particularly intriguing reaction concerns that of trimethylaluminum with aniline.[33] A white crystalline material, formed on reaction of Me_3Al with $H_2N{-}Ph$, was not unambiguously characterized but approximated to be $[Me_2Al{-}N(H)Ph]_n$. Importantly, subsequent heating of this material for 24 h at 180°C followed by slow cooling yields the novel Al_6N_6 hexamer hexakis(methyl-μ_3-phenylimidoaluminum), $[MeAl{-}NPh]_6$, (Equation 2.6).

Although $[HAl{-}N(i\text{-}Pr)]_6$[34] and $[HAl{-}N(n\text{-}Pr)]_6$[35] are among a number of hexameric Al_6N_6 compounds which have been reported, $[MeAl{-}NPh]_6$ is significant as it was prepared by direct reaction of an aluminum alkyl (Me_3Al) with an amine ($H_2N{-}Ph$). Furthermore, $[MeAl{-}NPh]_6$ was the first organoaluminum Al_6N_6 hexameric imide to be characterized by X-ray diffraction. Although inclusion of solvent molecules in the crystal lattice contributed to

$$H_2N{-}Ph + Me_3Al \xrightarrow[-CH_4]{} {}^{1}\!/_{n}[Me_2Al{-}N(H)Ph]_n \qquad (2.6)$$

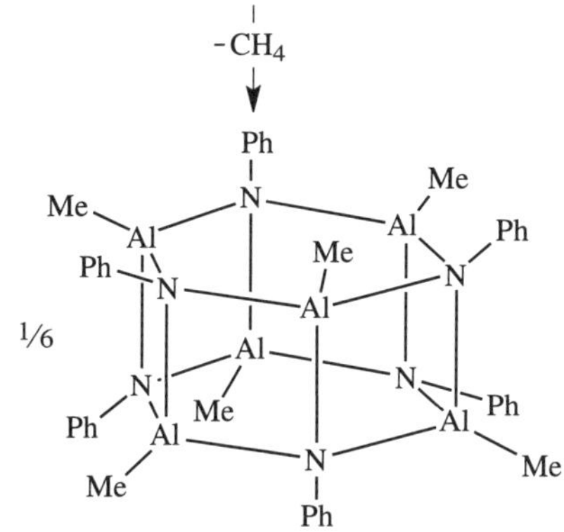

less than ideal X-ray diffraction data, the solid-state structure of $[MeAl{-}NPh]_6$ was determined unambiguously. Indeed, the hexameric nature of $[MeAl{-}NPh]_6$ was confirmed in the gas phase by mass spectrometry. The Al_6N_6 cage, containing planar four- and six-membered rings, resides about a crystallographically imposed S_6 axis. The phenyl rings of the nitrogen atoms are almost orthogonal to the respective Al_3N_3 six-membered rings. As expected, the Al–N bonds in the six-membered rings (1.912 (6) Å) were found to be significantly shorter that the transverse bonds (1.951 (6) Å) joining the two Al_3N_3 rings. Bond distance values for this heaxamer were similar to the values observed for other Al–N hexamers.

Based on the fact that reactions of R_3Al with primary amines initially afford Al_2N_2 dimers, which in turn serve as precursors to higher Al–N oligomers, it is reasonable that $[Me_2Al{-}N(H)Ph]_n$ was dimeric (i.e. $n = 2$). Perhaps an enhanced solubility of this product precluded its complete characterization. Given the relatively harsh reaction conditions of this system, exhaustive condensation of this dimeric species could produce the hexamer in low yield.

A similar organoaluminum Al_6N_6 hexamer, $[MeAl{-}N(i{-}Pr)]_6$, has been prepared from reaction of $[HAl{-}N(i{-}Pr)]_6$ with Me_3Al in refluxing benzene.[36] In addition, a number of other products were isolated from this system including $[\{MeAl{-}N(i{-}Pr)\}_5\{HAl{-}N(i{-}Pr\}]$ and a mixture of $[\{MeAl{-}N(i\text{-}Pr)\}_4\{HAl{-}N(i{-}Pr)\}_2]$ isomers. The degree of alkylation by the aluminum alkyls was dependent on reaction conditions and on the nature of the aluminum alkyl.

Although the subtle differences in reactivity between aluminum alkyls and gallium alkyls have been well documented, parallel reactions of trimethylgallium with these sterically demanding amines are noteworthy. Analogous reactions of Me_3Ga with $H_2N{-}Ph(Me)_3$,[31] afforded signficantly different products that those isolated with Me_3Al. Whereas the $[MeAl{-}NPh(Me)_3]_4$ tetramer was isolated from reaction of Me_3Al with $H_2N{-}Ph(Me)_3$ (as discussed earlier), the unusual *ortho*-metallated isomeric products $[MeGa{-}N(H)Ph(Me)_2\text{-}\mu\text{-}(CH_2)]_2$, **6**, and $[Me_2Ga\{N(H)Ph(Me)_2CH_2\}_2\text{-}\mu\text{-}\{Ga\}]$, **7**, were prepared from the corresponding reaction with Me_3Ga. Furthermore, the previously discussed

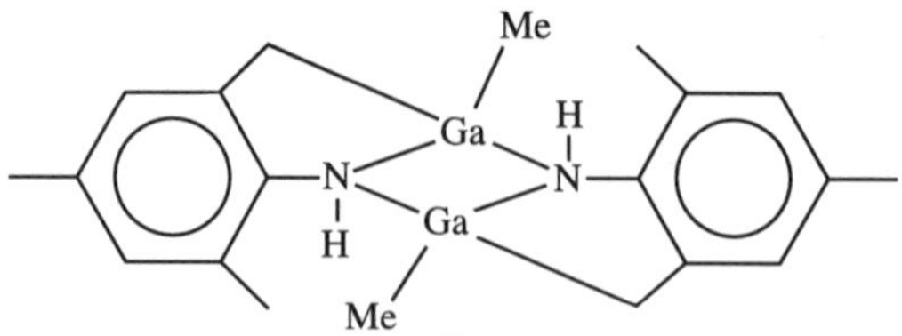

6

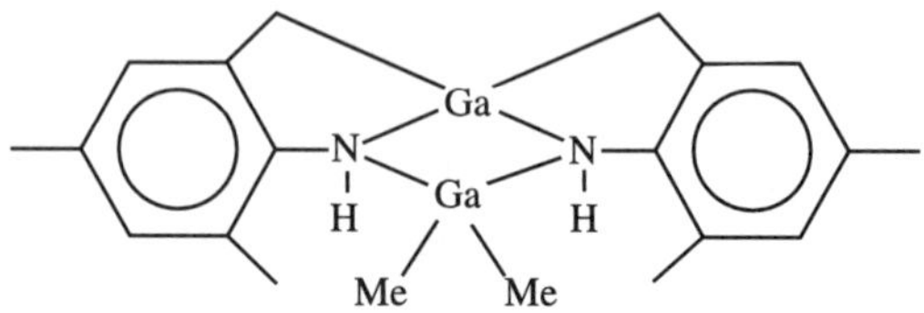

7

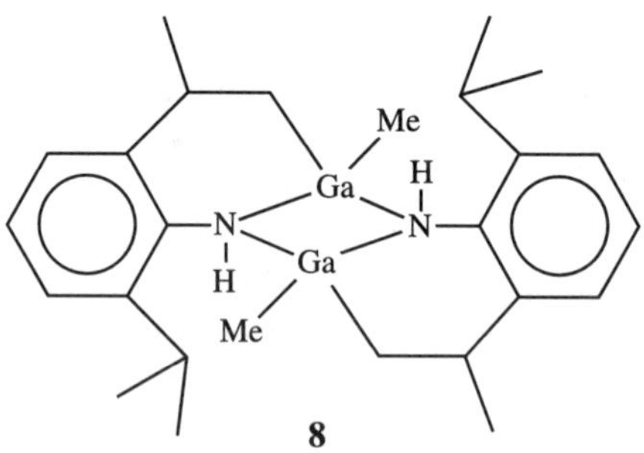

8

trimeric compound $[MeAl{-}NPh(i{-}Pr)_2]_3$ differs considerably from the metalled organogallium product, $[MeGa{-}N(H)Ph(i{-}Pr)\text{-}\mu\text{-}\{C(Me)CH_2\}]_2$, **8**, isolated from the parallel reaction with Me_3Ga. The formation of these organogallium products may be viewed as a consequence of the apparent hesitance of gallium to form higher aggregates coupled with the spacial orientation of the ligand aliphatic phenyl substituents and the gallium centers.

Aliphatic metallations of sterically bulky amines have also been observed with aluminum. Hitchcock *et al.*[37] reported that reaction of Me_3Al with tri-*t*-butylaniline, $H_2N{-}Ph(t\text{-}Bu)_3$, initially afforded the unremarkable dimeric species $[Me_2Al{-}N(H)Ph(t\text{-}Bu)_3]_2$ (Equation 2.7).

$$H_2N{-}Ph(t{-}Bu)_3 + Me_3Al \longrightarrow \tfrac{1}{2}[Me_2Al{-}N(H)Ph(t{-}Bu)_3]_2 + Me{-}H \quad (2.7)$$

Subsequent heating of $[Me_2Al{-}N(H)Ph(t\text{-}Bu)_3]_2$ resulted in elimination of methane and concomitant metallation of one of the methyl groups of the *t*-Bu substituents to give the unusual cyclometallated dimeric product $[MeAl{-}N(H)Ph(t{-}Bu_2)\text{-}\mu\text{-}(CMe_2CH_2)]_2$, **9**.

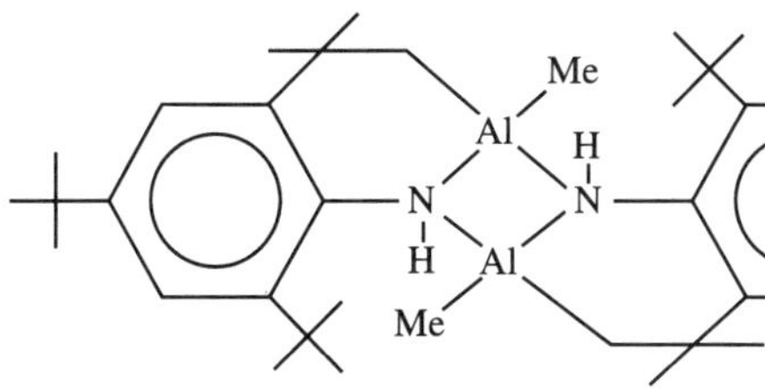

9

Metrical values for this compound were found to be within the range expected for Al_2N_2 compounds. However, the environment about the four-coordinate aluminum atoms may be described as severely distorted tetrahedral: N–Al–N: 88.1 (3)°; N–Al–C: 99.8 (3)°; N–Al–C': 120.9 (3)°; C–Al–C': 115.5 (4)° (the conformation about the nitrogen atoms in the Al_2N_2 ring may be similarly described). The Al_2N_2 four-membered ring is planar. The two six-membered N–Al–C_4 rings reside in twisted boat conformations (one on either side of the Al_2N_2 fragment). The authors cited steric constraints and the geometric positioning of the *t*-Bu groups, over electronic effects, as significant factors in the formation of the product. A compound similar to [MeAl–N(H)Ph(*t*–Bu_2)-μ-(CMe_2CH_2)$]_2$ has been isolated from reaction of (*i*–Bu$)_2$AlH with H_2N–Ph(*t*–Bu$)_3$ by Robinson *et al.* (Dillingham *et al.* unpublished).

Metallations of aromatic systems have also been observed. Reaction of Me_3Al with HN–Ph_2 was shown to afford the interesting *ortho*-metallated product [{Me(μ–NPh_2)Al}$_2$NPh(μ–C_6H_4)][38] (Figure 2.6). This compound may be considered a simple [MeAl–$NPh_2$$]_2$ dimer to which a bridging diphenyl-

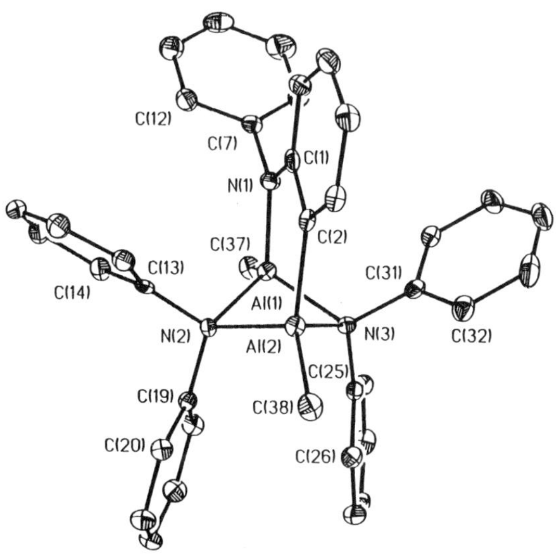

Figure 2.6. X-ray structure of [{Me(μ-NPh_2)Al}$_2$NPh(μ-C_6H_4)].

amine unit has been added across the two aluminum atoms via the nitrogen atom and an *ortho*-carbon atom of one of the phenyl rings. Similar *ortho*-metallations have been observed with iminodibenzyl and diisobutylaluminum hydride.[39] Although the mechanism of such aromatic metallations by aluminum alkyls remains unclear, such behavior is often observed under sterically demanding conditions along the dimer–trimer reaction pathway.

Before proceeding it is appropriate to assess the factors and rationalize the formation of the products isolated from reaction of primary amines with trialkylaluminum species. Certainly, the formation of Al_2N_2 dimers would appear to be a necessary juncture along the reaction coordinate as all higher Al–N oligomers may be prepared directly from the Al_2N_2 dimer. The particular Al–N oligomer formed, however, appears to be largely dependent on three factors: (1) Steric bulk of the alkyl groups on the amine; (2) The spacial positioning of the aryl substitutents relative to the aluminum alkyl moieties; and (3) Reaction conditions. The progression along the reaction coordinate from dimer to trimer, tetramer, *ortho*-metallated product, and hexamer loosely parallels these three factors. Beginning with the dimer (while keeping the alkyl

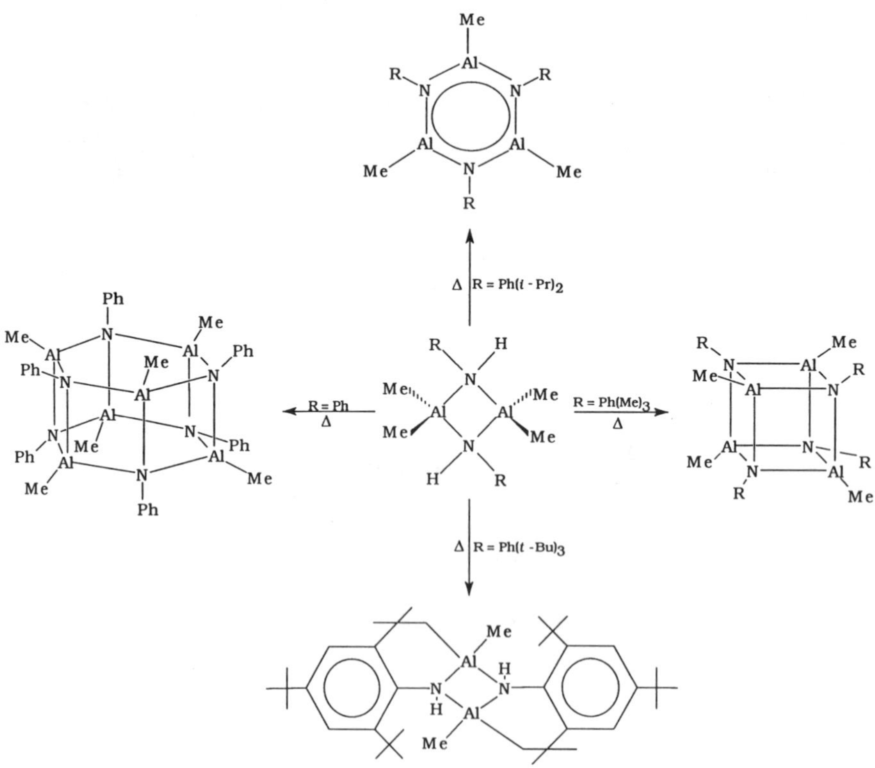

Figure 2.7. Reaction scheme for the $[Me_2Al–N(H)R]_2$ dimer.

group on aluminum constant as Me), if the amine is a bit bulky (and under mild reaction conditions), a trimeric product may be isolated. Under similarly mild reaction conditions, a more sterically demanding amine would favor a tetramer. Should the amine have substitutents which sufficiently approach the aluminum alkyl moiety, a metallated dimer may be formed. Lastly, under sufficiently harsh reaction conditions with an amine of appropriate bulk, the system may favor the low yield preparation of the Al_6N_6 hexamer. This reaction pathway for the Al_2N_2 dimer is illustrated in Figure 2.7.

2.4. Reaction of Organoaluminum Moieties with Multidentate Open-chain Amines: High Coordination Number Aluminum Complexes

Whereas the interaction of organoaluminum species with simple amines has been extensively investigated, the corresponding chemistry of multidentate open-chain amines, a seemingly logical extension, has not been developed to a comparable extent. The fact that the nitrogen atoms of multidentate open-chain amines are constrained to reside in close proximity to each other, often separated only by two or three $-CH_2-$ fragments, is generally an attractive feature as this could affect saturation of the coordination sphere of the aluminum atom(s). Moreover, as multidentate open-chain amines often possess labile azahydrogen atoms, organoaluminum condensation products are possible. The organoaluminum chemistry of the tridentate open-chain amines diethylenetriamine, **10**, and 3,3'-iminobispropylamine, **11**, have been examined in this regard.

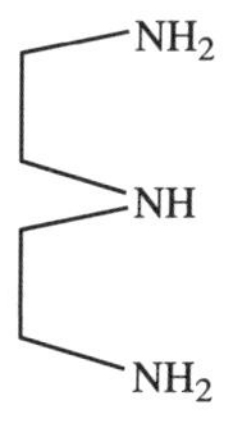

10

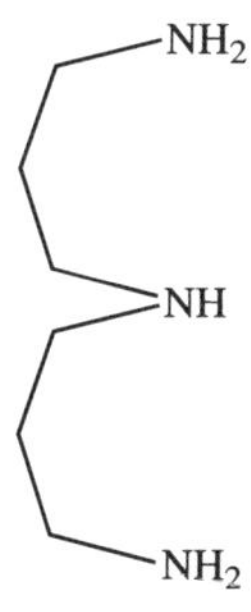

11

Reaction of Me_3Al with diethylenetriamine results in $[(MeAl)_2C_8H_{20}N_6][AlMe_2]_2$.[40] The structure of $[(MeAl)_2C_8H_{20}N_6][AlMe_2]_2$, confirmed by X-ray diffraction techniques, is as shown in **12**. The dimeric product consists of two diethylenetriamine units bridged by a series of four methylaluminum units: two Me_2Al units and two MeAl units. The molecule resides about a center of symmetry located at the center of the central Al_2N_2 four-membered ring. The major point of note concerning $[(MeAl)_2C_8H_{20}N_6][AlMe_2]_2$, however, is the coordination of the two central aluminum atoms. Both central aluminum atoms are five-coordinate, bonding to four nitrogen atoms in addition to a remaining methyl unit. As each respective set of four nitrogen atoms constitutes a basal plane, the methyl group resides in an axial position. The coordination of both central aluminum atoms may be described as square pyramidal. The Al–N distance in the square pyramid at 2.02 (3) Å is quite long. The Al–C distance of 1.962 (4) Å in the methylaluminum fragment is comparable to the distance of 1.947 (4) and 1.956 (4) Å observed for the dimethylaluminum fragment. The Al···Al contacts of 2.907 (1) and 3.20 Å do not suggest significant Al···Al interaction.

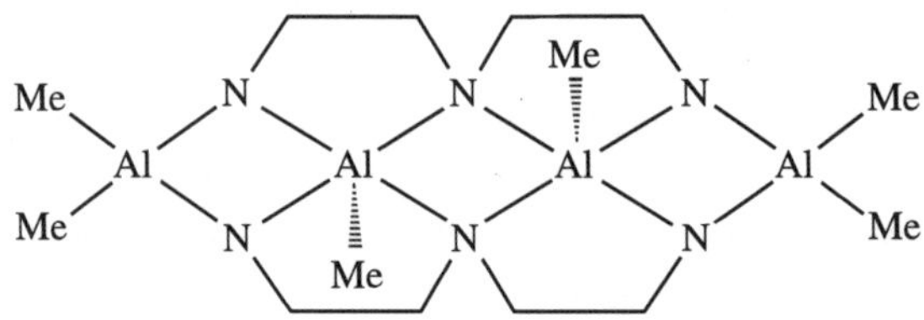

12

The analogous reaction of the propyl congener of diethylenetriamine, 3,3'-iminobispropylamine, with Me_3Al, affords the dimeric product $[(MeAl)_2C_{12}H_{28}N_6][AlMe_2]_2$ (Figure 2.8).[41] The two central most Al atoms are

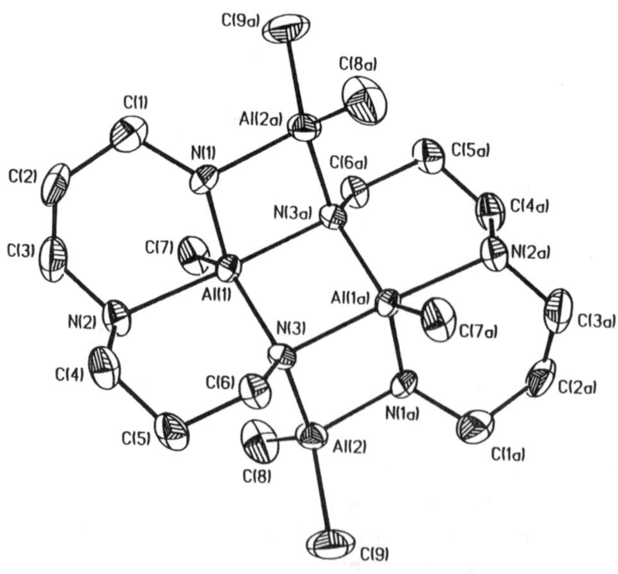

Figure 2.8. X-ray structure of $[(MeAl)_2C_{12}H_{28}N_6][AlMe_2]_2$.

five-coordinate, bonding to four nitrogen atoms and one methyl group. The atoms Al(1), N(3), C(7), and N(1) constitute a basal plane with N(3a) and N(2) residing in axial positions. The coordination of the two five-coordinate Al atoms is trigonal bipyramidal. The N(2)–Al(1)–N(3a) bond angle is 169.0 (2)° and the N(2)–Al(1)–C(7), N(2)–Al(1)–N(3), and N(1)–Al(1)–N(3) bond angles are 89.8 (2), 91.6 (2), and 86.2 (2)°, respectively. The axial bond distances Al(1)–N(2) and Al(1)–N(3a) are exceedingly long at 2.193 (4) and 2.065 (4) Å. The equatorial Al–N bond distances are considerably shorter at distances of 1.967 (4) and 1.878 (4) Å for Al(1)–N(1) and Al(1)–N(3), respectively. The Al–C bond distance involving the five-coordinate Al atom (1.988 (6) Å) was comparable to the Al–C bond distances in the independent dimethylaluminum units (1.987 (7) and 1.974 (7) Å).

The fact that the central Al atoms in the 3,3'-iminobispropylamine product assumes a trigonal bipyramidal geometry whereas the corresponding aluminum atoms in the diethylenetriamine product favors a square pyramidal geometry is noteworthy. This point may be a manifestation of the greater inherent flexibility of 3.3'-iminobispropylamine relative to diethylenetriamine (owing to the longer hydrocarbon chain). The geometry of five-coordinate aluminum atoms in organoaluminum complexes seems to depend markedly on the ligands involved. It appears that a trigonal bipyramidal geometry is often favored in situations where more flexible ligands are employed. Square pyramidal geometries are frequently observed with more inflexible ligands.

The propensity of the tridentate amines to form dimers with Me_3Al piqued interest relative to the corresponding chemistry of tetradentate amines. The tetradentate open-chain amine *N,N'*-bis(3-aminopropyl)ethylenediamine, $C_8H_{22}N_4$, reacts smoothly at room temperature with two equivalents of Me_3Al to form $[MeAl{\cdot}C_8H_{19}N_4][AlMe_2]$.[42] The structure of $[MeAl{\cdot}C_8H_{19}N_4][AlMe_2]$, **13**, also provides evidence for the preference of trigonal bipyramidal over square pyramidal in organoaluminum complexes.

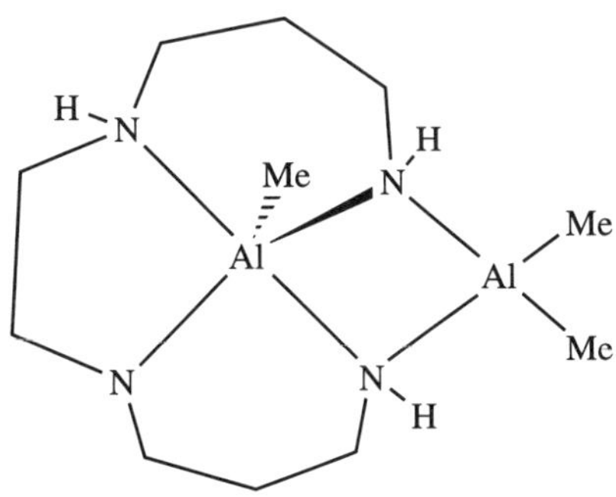

13

The central aluminum atom, two nitrogen atoms, and the methyl group comprise a basal plane. The coordination sphere of the central aluminum atom is completed by two nitrogen atoms residing in axial positions. The bond angle between the axial nitrogen atoms and the central aluminum atom was found to be 169.2 (1)°. The bond angles involving the five-coordinate Al atom and those atoms in the plane are 117.2 (1), 93.2 (1) and 112.4 (1)° for N–Al–N, N–Al–C, and C–Al–N, respectively. The coordination about the central aluminum atom was described as distorted trigonal bipyramidal. As expected, the Al–N bond distanced from the central Al atom to the axial atoms are quite long at 2.135 (2) and 2.055 (2) Å. The $Al–N_{equatorial}$ bonds were noticeably shorter at 1.826 (2) and 1.987 (3) Å. Expectedly, the shortest Al–N interaction (1.826 (2) Å) involved the three-coordinate N atom. The Al–C bond was 1.987 (3) Å. Valence Shell Electron Pair Repulsion (VSEPR) theory would support these experimental results.

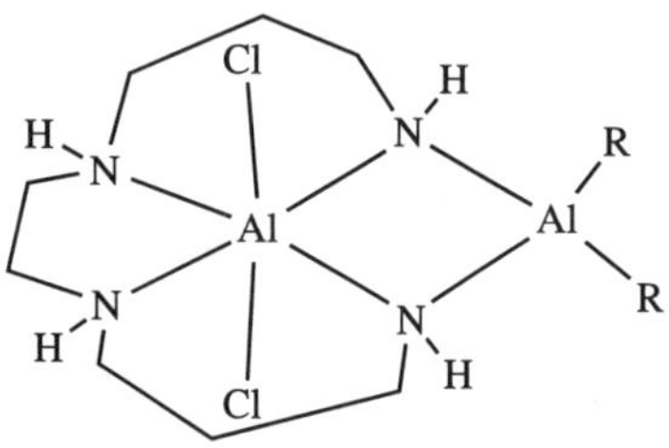

14

N,N'-bis(3-aminopropyl)ethylenediamine has proven quite versatile as reactions of alkylaluminum halides with this tetradentate open-chain amine have afforded interesting octahedral complexes containing six-coordinate aluminum atoms. Reaction of R_2AlCl (R = Et,[43] *i*-Bu[44]) with this amine results in the high yield preparation of $[Cl_2Al{\cdot}C_8H_{20}N_4][AlR_2]$, **14**. The formation of these compounds, particularly in the absence of dialkylaluminum chloride fragments, is significant and principally ascribed to disproportionation of the $[R_2AlCl]_2$ dimer. A shift in the equilibrium of dialkylaluminum chloride dimers, prompted by the presence of sufficiently strong Lewis bases, to provide the stronger Lewis acid species (i.e., $RAlCl_2$) and a trialkylaluminum species (i.e., R_3Al) was initially suggested by Ziegler.[45] Such a process would necessarily proceed through a transient mixed alkyl-halo bridged species. Predicated on the availability of the stronger Lewis acid, cleavage of such a dimeric species to provide $RAlCl_2$ and R_3Al, as opposed to R_2AlCl, would be favored, **15**. Thus, $RAlCl_2$ was taken to have reacted along the interior of the amine, while the R_3Al species reacted with the primary amine units (both operations eliminating alkane). The central aluminum atom is six-coordinate bonding to the four nitrogen atoms (which constituted a plane) and two chlorine atoms

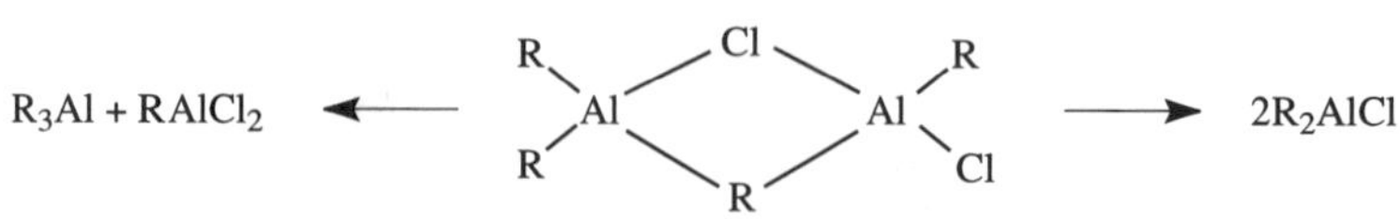

15

residing in axial positions. Metrical values for the Et and *i*-Bu derivatives were: Cl–Al–Cl: 175.1 (1)°; 175.4 (1)°; Al–Cl: 2.368 (2) A; 2.380 (3) Å; Al–N_{mean}: 1.97 (3) Å and 1.973 (4) Å, respectively.

One of the most interesting compounds in this area concerns the reaction of Me_3Al with triethanolamine, $N(CH_2CH_2OH)_3$, to produce the compound $[\{N(CH_2CH_2O)_3]Al_2Me_3\}_2$.[46] The $\{[N(CH_2CH_2O)_3]Al_2Me_3\}_2$ compound, **16**, was unique in that it represented the first structurally characterized example of a six-coordinate aluminum alkyl, and may be regarded as a dimer in which two amine units are bridged by a series of four organoaluminum fragments. The compound, residing about a center of symmetry, contains an Al_4O_6 core. The two six-coordinate aluminum atoms are in distorted octahedral environments having a mean O–Al–O axial bond angle of 165.8 (8)°. The Al(μ-O)$_2$Al bridges, largely due to steric constraints, are asymmetric with distances of 1.982(9) and 1.837 (8) Å. The longer distance, expectedly, involves the central aluminum atoms. Of particular importance is the Al–C bond distance involving the six-coordinate aluminum atom. Interestingly, this distance of 1.99 (1) Å is comparable to the analogous Al–C distances found in five-coordinate Al complexes.

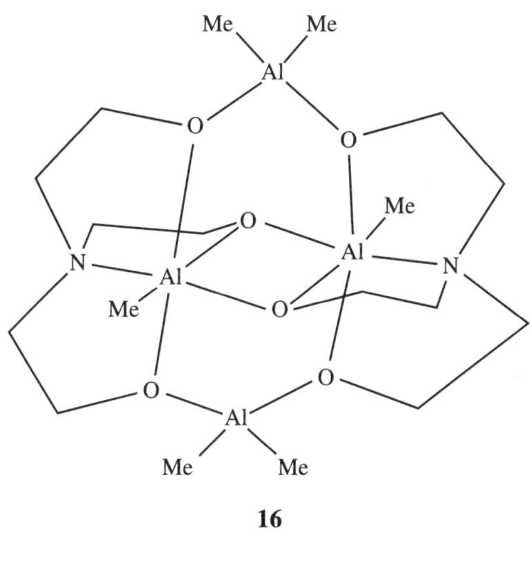

16

A product similar to $\{[N(CH_2CH_2O)_3]Al_2Me_3\}_2$, the major difference being that the methyl groups bonded to the two six-coordinate Al atoms were replaced by chlorine atoms, has been prepared from reaction of diisobutylaluminum hydride with triethanolamine in the presence of methylene chloride.[47] The presence of chlorine atoms in this molecule is viewed as a consequence of solvent participation by methylene chloride.

At this point it is noteworthy that the interaction of organoaluminum moieties with primary amines, while often affording higher oligomers, rarely results in an increase in the coordination number of the associated aluminum atoms beyond four. This is contrasted with the results presented earlier in which multidentate open-chain amines appear to be much more conducive to facilitating higher coordination number aluminum complexes. Oligomers greater than dimers are seldom encountered.

2.5. Organoaluminum Chemistry of Macrocyclic Amines

The first structural report of an organoaluminum-macrocyclic amine product concerned [$EtAl \cdot C_{22}H_{22}N_4$],[48] prepared from the stepwise insertion of Et_3Al into the central cavity of the macrocyclic tetradentate amine 5,14-dihydro-6,8,15,17-tetramethyldibenzo [b,*i*]-[1,4,8,11] tetraazacyclotetradecinato in hexane (Equation 2.8).

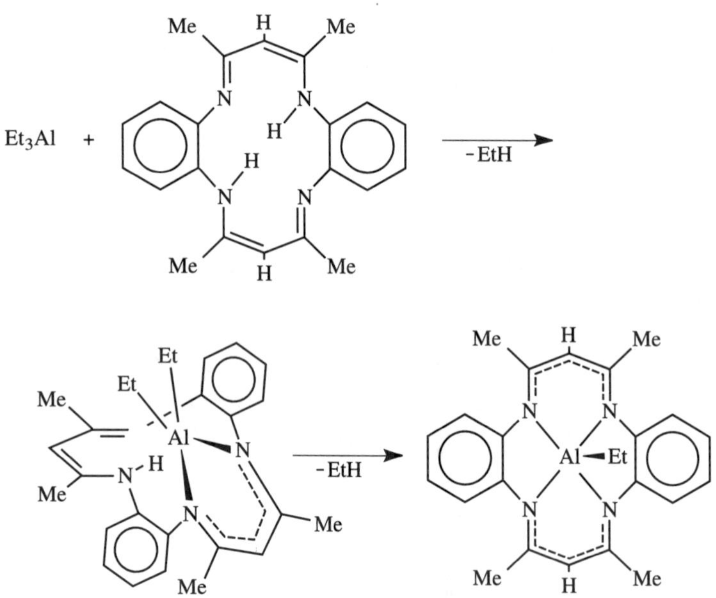

(2.8)

This organoaluminum-macrocyclic amine complex is unusually stable to Al–Et dissociation and demetallation as it remains unchanged with heating up to 300°C (under an inert atmosphere). This unusual stability was attributed to a number of synergic factors including: (a) Simple alkyl dissociation leaving a square planar four-coordinate aluminum species would be energetically prohibited;[49] (b) An unusually good fit of the aluminum ion in the macrocyclic cavity; and (c) A high ionic contribution to the stability afforded by the metal ion and the dianionic ligand. However, the Al–C bond was shown to react readily with Ph–OH, H_2O, and O_2 affording [$PhOAl \cdot C_{22}H_{22}N_4$], [$HOAl \cdot C_{22}H_{22}N_4$], and $[OAl \cdot C_{22}H_{22}N_4]_2$, respectively. Perhaps most intriguing, the Al–Et bond was found to be photosensitive undergoing homolytic bond cleavage on irradiation with visible light. Irradiation of [EtAl·$C_{22}H_{22}N_4$] in $CHCl_3$ affords the chloro complex [ClAl·$C_{22}H_{22}N_4$ and ethane. Time-resolved and conventional ESR studies of [EtAl·$C_{22}H_{22}N_4$][50] were also examined, and photochemical cleavage of the Al–C bond in the organoaluminum-porphyrin complex EtAl·TPP[51] (TPP = tetraphenylporphyrinato) was observed.

The coordination of the aluminum atom in [EtAl·$C_{22}H_{22}N_4$] is noteworthy as it is five-coordinate residing in a square pyramidal geometry. The mean Al–N bond distance was 1.967 Å with the aluminum displaced 0.57 Å from the N_4-plane. The Al–C bond distance of 1.976 (3) Å is comparable to distances observed in four-coordinate aluminum centers.

Although the macrocyclic tetradentate secondary amine, 1,4,8,11-tetraazacyclotetradecane or [14]aneN_4, **17**, (cyclam)[52] was reported thirty years before the preparation of oxygen-based crown ethers,[53,54] its ability to form stable complexes with a variety of transition metal ions was not recognised until much later.[55-58] Indeed, the X-ray structure of [14]aneN_4, confirming a hydrogen-bonding-stabilized endodentate conformation of the macrocyclic ligand, was only reported in 1988.[59]

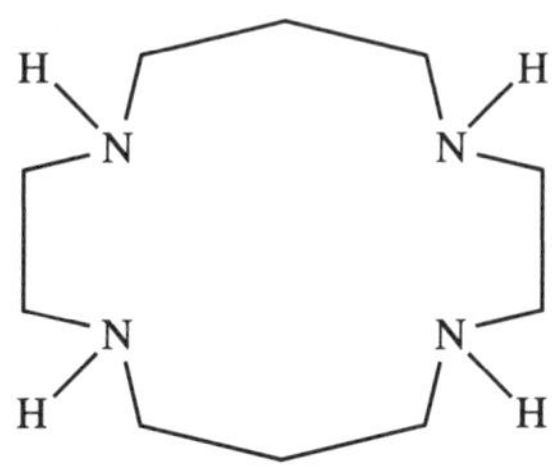

17

The examination of the organoaluminum chemistry of oxygen-based crown ethers, prompted by studies concerning organoaluminum-based liquid inclusion compounds, began in the early 1980s.[60,61] Investigations concerning the corresponding chemistry with macrocyclic azacrown ethers followed a few years later. Macrocyclic azacrown ethers were particularly interesting in that they offered Lewis base sites in addition to the possibility of condensation products (as was observed in simple and multidentate open-chain amine systems). Moreover, the extent to which the macrocyclic effect would affect the coordination chemistry of aluminum could also be ascertained.

Trimethylaluminum reacts smoothly with [14]aneN_4 in toluene producing $[(MeAl)_2$·[14]ane$N_4[AlMe_3]_2$.[62] It is noteworthy that this product is fundamentally different from the aminoalanes reported by Smith as $[(MeAl)_2$·[14]aneN_4 $[AlMe_3]_2$ contains a single amine unit. The previously reported products contained a minimum of two simple amine units. As shown by the X-ray crystal structure of $[(MeAl)_2$·[14]ane$N_4[AlMe_3]_2$ (Figure 2.9), the azacrown experienced significant distortion. The molecule resides about a crystallographic center of symmetry. As required by symmetry, the Al_2N_2 fragment is planar although slightly asymmetric with Al–N distances of 1.959 (4) Å and 1.913 (4)

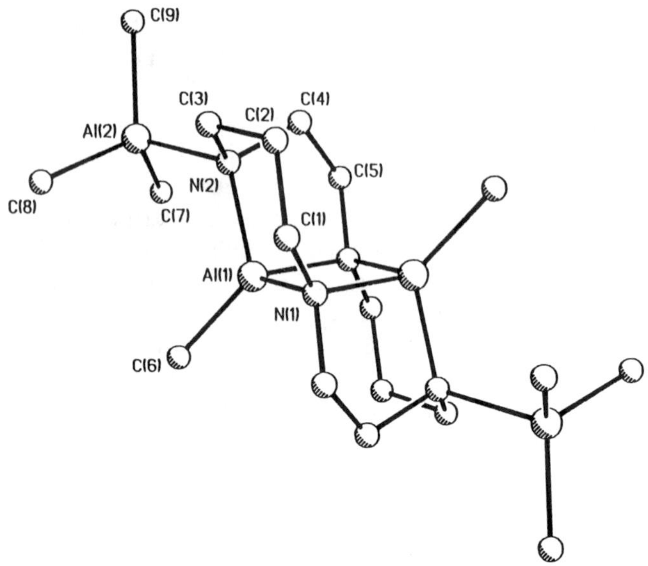

Figure 2.9. X-ray structure of $[(MeAl)_2[14]aneN_4][AlMe_3]_2$.

Å. The N(1)–Al(1)–N(2) bond angle is 101.3 (2)° while Al(1)–N(1)–Al(1a) is 89.9 (2)°. Perhaps a consequence of steric demands, the shorter Al–N distance was associated with Al(1) (Al(1)–N(2), 1.886 (4) Å), while the longer interaction involves the Me_3Al–N adducts (Al(2)–N(2), 2.042 (4) Å). The Al···Al separation across the Al_2N_2 four membered ring is 2.735 (2) Å.

The synthesis and structure of the heterobimetallic complex $[(MeAl)_2\cdot[14]aneN_4][GaMe_3]_2$,[63] observed to be isostructural with $[(MeAl)_2\cdot[14]aneN_4][AlMe_3]_2$, is significant as it is the first mixed-group 13 metal azacrown complex. The fact that Me_3Ga formed adducts whereas the two trimethylaluminum units interacted along the macrocyclic cavity and underwent condensation in $[(MeAl)_2\cdot[14]aneN_4][GaMe_3]_2$ was due to the greater Lewis acidity of Me_3Al relative to Me_3Ga. In addition, the slightly larger macrocyclic amine $[15]aneN_4$ has been examined. Reaction of Me_3Al with this macrocyclic amine yielded the $[15]aneN_4$derivative $[(MeAl)_2[15]aneN_4][AlMe_3]_2$.[64]

The organoaluminum chemistry of $[14]aneN_4$ proved fruitful as a variety of novel products have been prepared. Although interesting disproportionations of alkylaluminum halides[65] and hydrides[66] were observed in reactions with $[14]aneN_4$, perhaps the most interesting organoaluminum product is that isolated from reaction of $[14]aneN_4$ with Me_3Al in the presence of $ZrCl_4$ (Equation 2.9) to produce $[Al\cdot[14]aneN_4\cdot AlMe_2][Me_2AlCl]_2$.[67]

$$4Me_3Al + [14]aneN_4 \xrightarrow{ZrCl_4} [Al\cdot[14]aneN_4\cdot AlMe_2]\text{–}[Me_2AlCl]_2 + 4Me\text{-}H + Me_2ZrCl_2 \quad (2.9)$$

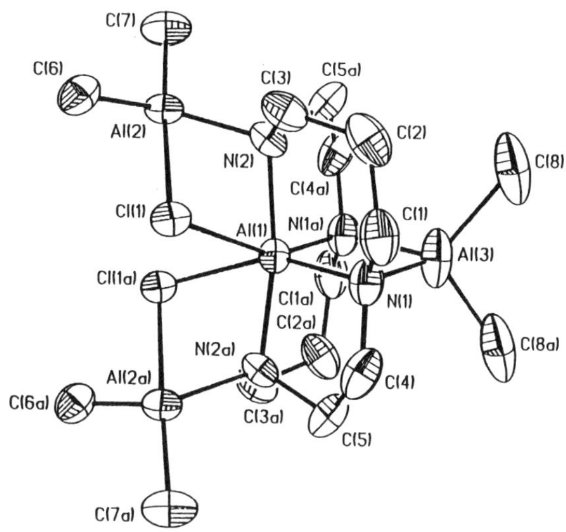

Figure 2.10. X-ray structure of $[Al{\cdot}[14]aneN_4AlMe_2][Me_2AlCl]_2$.

While it is evident that $ZrCl_4$ played a pivotal role in the formation of this compound, zirconium by-products were not isolated. The $ZrCl_4$ exchanged some of its chlorine atoms with a Me_3Al moiety to afford the Me_2AlCl units. The considerable amount of blackening residue in the flask was taken to be some measure of $[MeZrCl]_n$. Mass balance would suggest the equation shown in Equation 2.9. The X-ray crystal structure of $[Al{\cdot}[14]aneN_4AlMe_2][Me_2AlCl]_2$ is given in Figure 2.10.

A number of points are worth noting regarding structure and bonding in $[Al\text{-}[14]aneN_4{\cdot}AlMe_2][Me_2AlCl]_2$. The molecule resides about a two-fold axis (coincident with Al(1) and Al(3). Particularly significant is the coordination about Al(1). This atom is six-coordinate residing in an octahedral environment bonding to the four nitrogen atoms of the amine in addition to two chlorine atoms that are bridging two dimethylaluminum units. Al(1) is coplanar with N(1), N(1a), Cl(1), and Cl(1a), while N(2) and N(2a) reside in axial positions. The N(2)–Al(1)–N(2a) bond angle is nearly linear at 170.0 (2)°. The Cl(1)–Al(1)–N(1) bond angle is 177.0 (1)°. The Al–C bond distances are slightly within expected ranges (1.942 (4), 1.965 (4), and 1.986 (6) Å, for Al(2)–C(6), Al(2)–C(7), and Al(3)–C(8), respectively). The Al(1)–Cl(1) and Al(2)–Cl(1) bond distances in two other Cl_2Al-crown ether complexes containing octahedral aluminum atoms: $[Cl_2Al{\cdot}12\text{-crown-}4]^+$ (2.202 (5) Å) and $[Cl_2Al{\cdot}18\text{-crown-}6]^+$ (2.148 (3) Å).[68] Indeed, the Al–Cl distances in this compound are longer than the Al–Cl distances reported for the seven-coordinate aluminum atom in $[Cl_2Al{\cdot}benzo\text{-}15\text{-crown-}5]^+$ (2.202 (5) and 2.197 (7) Å).[69]

The organoaluminum chemistry of the tetramethyl derivative of $[14]aneN_4$, $[Me_4[14]aneN_4]$ has also been examined. As condensation was not possible,

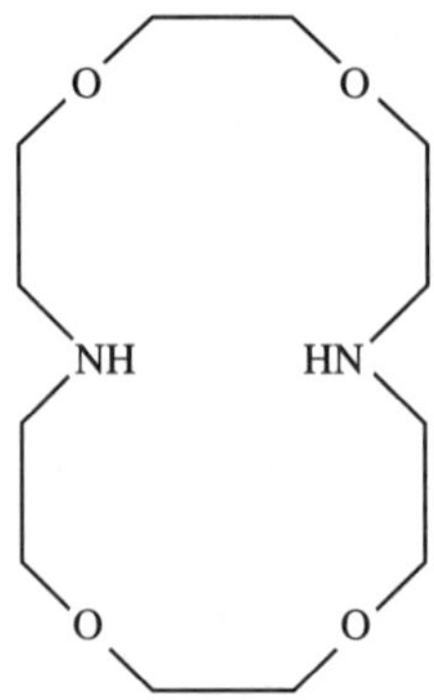

18

reaction of Me_3Al with $[Me_4[14]aneN_4]$ afforded the addition product $[Me_3Al]_4[Me_4[14]aneN_4]$.[70]

The azacrown 1,4,10,13-tetraoxa-7,16-diazacyclooctadecane, also referred to as diaza-18-crown-6, is particularly noteworthy as it possesses both nitrogen and oxygen donor species, **18**. Reaction of Et_2AlCl with diaza-18-crown-6 results in $[(EtAl)_2$-diaza-18-crown-6$][EtAlCl_3]_2$.[71] The molecular structure of the $[EtAl_2$·diaza-18-crown-6$]^{2+}$ cation (Figure 2.11) is noteworthy. The $[(EtAl)_2$·diaza-18-crown-6$]^{2+}$ cation resides about a center of symmetry. Although the mean Al–N and Al–O bond distances are generally unremarkable

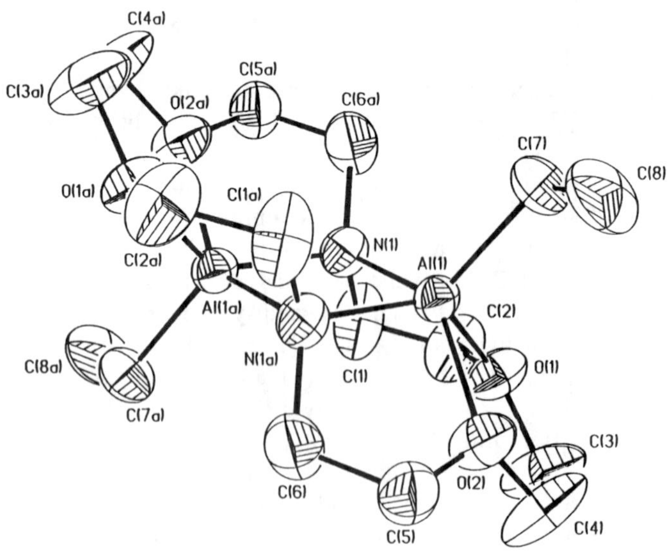

Figure 2.11. X-ray structure of $[(EtAl)_2$·diaza-18-crown-6$]^{2+}$ cation.

(1.968 (6) and 1.955 (6) Å, respectively), the coordination of the two Al atoms was quite striking. The coordination sphere of each aluminum atom consists of two nitrogen atoms and two oxygen atoms of the azacrown, in addition to an ethyl group. As the nitrogen and oxygen atoms constitute a basal plane, the coordination of the aluminum atoms in the $[(EtAl)_2{\cdot}$diaza-18-crown-$6]^{2+}$ cation is described as five-coordinate square pyramidal. Indeed, the coordination of the aluminum atoms in this cation was very similar to the coordination of the aluminum atom in the organoaluminum–Schiff base compound [EtAl·salen].[72] At a distance of 1.944 (5) Å, the Al(1)–C(7) bond distance is comparable to other Al–R bond distances. Dispropotionation of Et_2AlCl, identical to that discussed for R_2AlCl, was an important factor in the formation of $[(EtAl)_2{\cdot}$diaza-18-crown-6]$[EtAlCl_3]_2$.

2.6. Conclusions

Aluminum-nitrogen compounds have become ubiquitous in organoaluminum chemistry. In terms of condensation reactions, the chemistry revolves around the cleavage of two simple fragments: Al–R and N–H. The nature of the reaction products are governed by the following factors: (a) Steric bulk of the amine; (b) Reaction conditions; and (c) The number of nitrogen atoms in the amine. Thus, by controlling these factors Al–N oligomers or high coordination number aluminum complexes may be selectively isolated. Furthermore, the preparation of trimers, tetramers, metallated products, hexamers, or heptamers may readily be approached if one appropriately adjusts the system. Where multidentate open-chain amines and macrocyclic amines are concerned, the preparation of five-coordinate aluminum compounds – both in trigonal bipyramidal and square pyrajmidal geometries are possible—and six-coordinate aluminum complexes are also possible. With regard to five coordinate complexes, trigonal bipyramidal appears to be significantly favored over square pyramidal. If the alkyl units are chosen with care octahderal complexes are readily accessible. In these higher coordination complexes, the longer aluminum bonds always involve the axial substitutents while the shorter bonds involve the equatorial moieties. In addition, it would follow that the equatorial interactions are more covalent in nature whereas the axial interactions are more dative in nature.

The great affinity that exists between aluminum and nitrogen, coupled with the propensity of Al–N compounds to eliminate alkane and form oligomers or high coordination number complexes, have served to bring the coordination chemistry of aluminum, to the fore of main group chemistry. Considering the present emphasis directed towards Group 13-based advanced materials, the organoaluminum chemistry of amines is particularly noteworthy as many workers may have judged this area of chemistry quite mature even twenty years ago.

Acknowledgments

The author gratefully acknowledges the National Science Foundation for financial support of his research on organoaluminum chemistry. In addition, the author is indebted to members of his reseach group for countless readings of this manuscript while it was in the formative stages.

References

1. Stock, A.; Pohland, E. *Ber.* **1926**, *59*, 2215.
2. Stock, A. *Hydrides of Boron and Silicon;* Cornell University Press: Ithaca, New York, **1933**.
3. Benn, R.: Rufinska, A.; Lehmkuhl,; Janssen, E.; Kruger, C. *Angew. Chem. Int. Ed. Engl.* **1983**, *22*, 779.
4. Benn, R.; Rufinska, A. *Angew. Chem. Int. Ed. Engl.* **1986**, *25*, 861.
5. Benn, R.; Rufinska, A. *Angew. Chem. Int. Ed. Engl.* **1986**, *5*, 825.
6. Wiberg, W. *F.I.A.T. Review of German Science* 1939–1945; Inorganic Chemistry Part II, p.159.
7. Interrante, L. V.; Carpenter II, L. E.; Whitmarsh, C.; Lee, W. *Mats. Res. Soc. Symp. Proc.* **1986**, *73*, 359.
8. Bolt, T.D.; Tebbe, F. N. *Mater. Res. Soc. Symp. Proc.* **1988**, *108*, 337.
9. Baker, R. T.; Bolt, T. D.; Reddy, G. S.; Roe, D. C.; Staley, R. H.; Tebbe, F. N.; Vega, A. J. *Mater. Res. Soc. Symp. Proc.* **1988**, *121*, 471.
10. Interrante, L. V.; Sigel, G. A.; Ganbauskas, M.; Hejna, C.; Slack, G. A. *Inorg. Chem.* **1989**, *28*, 252.
11. Davidson, N.; Brown, H. C. *J. Am. Chem. Soc.* **1942**, *64*, 316.
12. Laubengayer, A. W.; Smith, J. D.; Ehrlich, G. G. *J. Am. Chem. Soc.* **1961**, *83*, 542.
13. Cohen, M.; Gilbert, J. K.; Smith, J. D.; *J. Chem. Soc.* **1965**, 1092.
14. Gilbert, J. K.; Smith, J. D. *J. Chem. Soc. (A)* **1968**, 233.
15. McLaughlin, G. M.; Sim, G. A.; Smith, J. D. *J. Chem. Soc., Dalton Trans.* **1972**, 2197.
16. Alford, K. J.; Gosling, A. K.; Smith, J. D. *J. Chem. Soc., Dalton Trans.* **1972**, 2203.
17. Hess, H.; Hinderer, A.; Steinhauser, S. *Z. Anorg. Chem.* **1970**, *377*, 1.
18. Bahr, G., *F.I.A.T. Review of German Science,* 1939-1945. Inorganic Chemistry, Part II, p. 159.
19. Alford, K. J.; Gosling, K.; Smith, J. D. *J. Chem. Soc., Dalton Trans.* **1972**, 2203.
20. Hitchcock, P. B.; McLaughlin, G. M.; Smith, J. D.; Thomas, K. M. *J. Chem. Soc. Chem. Commun.* **1973**, 934.
21. Hitchcock, P. B.; Smith, J. D.; Thomad, K. M. *J. Chem. Soc., Dalton Trans.* **1976**, 1433.
22. Amirkhalili, S.; Hitchcock, P. B.; Smith, J. D. *J. Chem. Soc., Dalton Trans.* **1979**, 1206.
23. Petrie, M. A.; Olmstead, M. M.; Power, P. P. *J. Am. Chem. Soc.* **1991**, *113*, 8704.
24. Healy, M. D.; Ziller, J. W.; Barron, A. R. *Organometallics* **1991**, *10*, 597.

25. Lichtenberger, D. L.; Hogan, R. H.; Healy, M. D.; Barron, A. R. *Organometallics* **1991**, *10*, 609.

26. Power, P. P. *J. Organomet. Chem.* **1990**, *400*, 49.

27. Power, P. P. *Angew. Chem. Int. Engl.* **1990**, *29*, 449.

28. Healy, M. D.; Wierda, D. A.; Barron, A. R. *Organometallics* **1988**, *7*, 2543.

29. Waggoner, K. M.; Hope, H.; Power, P. P. *Angew. Chem. Int. Ed. Engl.* **1988**, *27*, 1699.

30. Atwood, J. L.; Stucky, G. D. *J. Am. Chem. Soc.* **1970**, *92*, 285.

31. Waggoner, K. M.; Power, P. P. *J Am. Chem. Soc.* **1991**, *113,* 3385.

32. Del Piero, G.; Cesari, M.; Dozzi, G.; Mazzei, A. *J. Organomet. Chem.* **1977**, *129*, 281.

33. Al-Wassil, A. -A. I.; Hitchcock, P. B.; Sarisaban, S.; Smith, J. D.; Wilson, C. L. *J. Chem. Soc., Dalton Trans.* **1985**, 1929.

34. Cesari, M.; Perego, G.; Del Piero, G.; Cucinella, S.; Cernia, E. *J. Organomet. Chem.* **1974**, *78*, 203.

35. Del Piero, G.; Cesari, M.; Perego, G.; Cucinella, S.; Cernia, E. *J. Organomet. Chem.* **1977**, *129,* 289.

36. Cucinella, S.; Salvatori, T.; Busetto, C.; Cesari, M. *J. Organomet. Chem.* **1976**, *121*, 137.

37. Hitchcock, P. B.; Jasim, H. A.; Lappert, M. F.; Williams, H. D. *Polyhedron* **1990**, *9*, 245.

38. Waggoner, K. M.; Olmstead, M. M.; Power, P. P. *Polyhedron* **1990**, *9*, 257.

39. Lee, B.; Pennington, W. T.; Robinson, G. H. *Inorg. Chim. Acta,* **1991**, *190*, 173.

40. Robinson, G. H.; Sangokoya, S. A. *J. Am. Chem, Soc.* **1987**, *109*, 6852.

41. Robinson, G. H.; Moise, F.; Pennington, W. T.; Sangokoya, S. A. *Polyhedron* **1989**, *8*, 1279.

42. Robinson, G.H.; Sangokoya, S.A.; Moise, F.; Pennington, W.T. *Organometallics* **1988**, *7*, 1887.

43. Sangokoya, S. A.; Moise, F.; Pennington, W. T.; Self, M. F.; Robinson, G. H.*Organometallics* **1989**, *8,* 2584.

44. Moise, F.; Sangokoya, S. A.; Pennington, W. T.; Robinson, G. H. *J. Coord. Chem.* **1990**, *21*, 129.

45. Ziegler, K. In *Organometallic Chemistry*, Zeiss, H., Ed.; Reinhold: New York, 1960: p 194.

46. Healy, M. D.; Barron, A. R. *J. Am. Chem. Soc.* **1989**, *111*, 398.

47. Lee, B.; Moise, F.; Pennington, W. T.; Robinson, G. H. *J. Coord. Chem.* in press.

48. Goedken, V. L.; Ito, H.; Ito, T. *J. Chem. Soc., Chem. Commun.* **1984**, 1453.

49. Dzugan, S. J.; Goedken, V. L. *Inorg. Chem.* **1986**, *25*, 2858.

50. Tero-KuBota, S.; Oshio, H.; Ito, T.; Goedken, V. L.; Higuchi, J. *Chem. Phys. Lett.* **1986**, *131*, 430.

51. Tero-Kubota, S.; Hoshino, N.; Kato, M.; Goedken, V. L.; Ito, T. *J. Chem. Soc. Chem. Commun.* **1985**, 959.

52. Van Alphen, J. *Recl. Trav. Chim. Pays-Bas* **1937**, *56*, 343.

53. Pedersen, C. J. *J. Am. Chem. Soc.* **1967**, *89*, 2495.

54. Pedersen, C. J. *J. Am. Chem. Soc.* **1967**, *89*, 7017.

55. Bosnich, B.; Mason, R.; Pauling, P. J.; Robertson, G. B.; Tobe, M. L. *Chem. Comm.* **1960**, *6*, 97.

56. Bosnich, B.; Poon, C. K.; Tobe, M. L. *Inorg. Chem.* **1965**, *4*, 1102.

57. Collman, J. P.; Schneider, P. W. *Inorg. Chem.* **1966**, *5*, 1380.

58. Bounsall, E. J.; Koprich, S. R. *Can. J. Chem.* **1970**, *48*, 1481.

59. Robinson, G. H.; Sangokoya, S. A.; Pennington, W. T.; Self, M. F.; Rogers, R. D. *J. Coord. Chem.* **1988**, *19*, 287.

60. Atwood, J. L.; Hrncir, D. C.; Shakir, R.; Dalton, M. S.; Priester, R. D.; Rogers, R. D. *Organometallics* **1982**, *1*, 1021.

61. Atwood, J. L.; Hrncir, D. C.; Priester, R. D.; Rogers, R. D. *Organometallics* **1983**, *2*, 985.

62. Robinson, G. H.; Rae, A. D.; Campana, C. F.; Byram, S. K. *Organometallics* **1987**, *6*, 1227.

63. Robinson, G. H.; Pennington, W. T.; Lee, B.; Self, M. F.; Hrncir, D. C. *Inorg. Chem.* **1991**, *30*, 809.

64. Robinson, G. H.; Appel, E. S.; Sangokoya, S. A.; Zhang, H.; Atwood, J. L. *J. Coord. Chem.* **1988**, *17*, 373.

65. Robinson, G. H.; Sangaokoya, S. A. *Organometallics* **1988**, *7*, 1453.

66. Self, M. F.; Pennington, W. T.; Robinson, G. H. *Inorg. Chim. Acta* **1990**, *175*, 151.

67. Robinson, G. H.; Self, M. F.; Sangokoya, S. A.; Pennington, W. T. *J. Am. Chem. Soc.* **1989**, *111*, 1520.

68. Atwood, J. L.; Elgamal, H.; Robinson, G. H.; Bott, S. G.; Weeks, J. A.; Hunter, W. E. *J. Incl. Phenom.* **1984**, *2*, 367.

69. Bott, S. G.; Elgamal, H.; Atwood, J. L. *J. Am. Chem. Soc.* **1985**, *107,* 1796.

70. Robinson, G. H.; Zhang, H.; Atwood, J. L. *J. Organomet. Chem.* **1987**, *331*, 153.

71. Self, M. F.; Pennington, W. T.; Laske, J. A.; Robinson, G. H. *Organometallics* **1991**, *10*, 36.

72. Dzugan, S. J.; Goedken, V. L. *Inorg. Chem.* **1986**, *25*, 2858.

CHAPTER

3

The Aqueous Coordination Chemistry of Aluminum

Chris Orvig

Department of Chemistry, University of British Columbia
2036 Main Mall, Vancouver, British Columbia V6T 1Z1

3.1. Introduction

Aluminum, the most abundant metallic element on earth, and the third most abundant element overall (after oxygen and silicon), comprises 7 – 8% of the planet's crust. It is ubiquitous in Nature, normally in the form of oxides and aluminosilicate minerals. Principally due to both availability and its pivotal role in a number of catalytic and industrial processes, the non-aqueous inorganic and organometallic chemistry of aluminum has been extensively developed. However, this has not been the case for the aqueous coordination chemistry of aluminum. After many years of relative obscurity, largely due to a perception that hydrolysis rendered it "difficult," this area is now a rapidly advancing field.[1]

With an ionic radius of 0.53 Å, aluminum(III) is a highly-charged, small cation which is easily hydrolyzed in aqueous solutions in the absence of competing ligands. The species in aqueous solutions are pH-dependent although the reduction potential of Al metal ($Al^{3+} + 3e^- \rightarrow Al$, $E° = -1.67$ V) dictates that only the III oxidation state exists in water.[2] At low pH, aluminum(III) remains unhydrolyzed, the main mononuclear species being $Al(H_2O)_6^{3+}$. This

Table 3.1. Log hydrolysis constants for Al^{3+} (25°C, μ= 0.6M-NaCl);[a] water molecules which complete the coordination sphere to 6 are omitted.

$Al^{3+} \rightarrow$	$-\log K_h$
$[Al(OH)]^{2+}$	5.5
$[Al(OH)_2]^+$	11.3
$Al(OH)_3$	17.3
$[Al(OH)_4]^-$	23.5
$[Al_2(OH)_2]^{4+}$	8[b,c]
$[Al_3(OH)_4]^{5+}$	13.6
$[Al_{13}O_4(OH)_{24}]^{7+}$	105.5

[a] Refs 17, 67; [b]μ = 1M-NaClO$_4$[7].
[c] Öhman questions the existence of this species.

species, containing an octahedral array of water molecules,[3–6] readily undergoes proton transfer reactions. As the pH increases, mononuclear species such as $[Al(OH)(H_2O)_5]^{2+}$, $[Al(OH)_2(H_2O)_4]^+$, and $Al(OH)_3$ are formed, in addition to polynuclear species (if the Al concentration is high enough) (Table 3.1).

The available evidence[7–9] indicates that the stable polynuclear species are $[Al_2(OH)_2(H_2O)_8]^{4+}$, $[Al_3(OH)_4(H_2O)_{10}]^{5+}$, and $[Al_{13}O_4(OH)_{24}(H_2O)_{12}]^{7+}$. The dimer and the tridecamer have both been crystallographically characterized.[10,11] The tridecamer contains one Al^{3+} in the center which is tetrahedrally coordinated (AlO_4) by four oxo bridges while the remaining Al atoms are hexacoordinated with bridging hydroxides or water molecules. $Al(OH)_3$ is amphoteric and readily dissolves under alkaline conditions to give the predominant Al-containing product in basic solution, $Al(OH)_4^-$, regardless of concentration. The distributions of hydrolysis products in concentrated (0.1 M) and dilute (10 μM) aluminum solution are shown in Figure 3.1. The solubility of aluminum hydroxide has a minimum of about 10^{-8}M at pH 6. The hydrolysis behavior of Al^{3+} under varying conditions and in the presence of weakly coordinating ligands has been extensively studied [9,12–19] especially by Akitt and co-workers using ^{27}Al NMR in combination with other techniques.[20–30] These methods have recently been reviewed.[31] Despite interest which has spanned over a century, the hydrolysis of Al^{3+} remains a controversial topic of great chemical interest because of the variation in hydrolysis behavior depending on concentration, conditions, other cations, and anions in solution.

3.2. Reaction Kinetics

Aluminum is generally regarded as a *sluggishly labile* metal ion. In general, intermolecular reactions in water (e.g., ligand substitution, H_2O exchange) are relatively slow, and the rates of intramolecular exchange processes (e.g., rearrangements, interconversion of stereoisomers) vary considerably. The

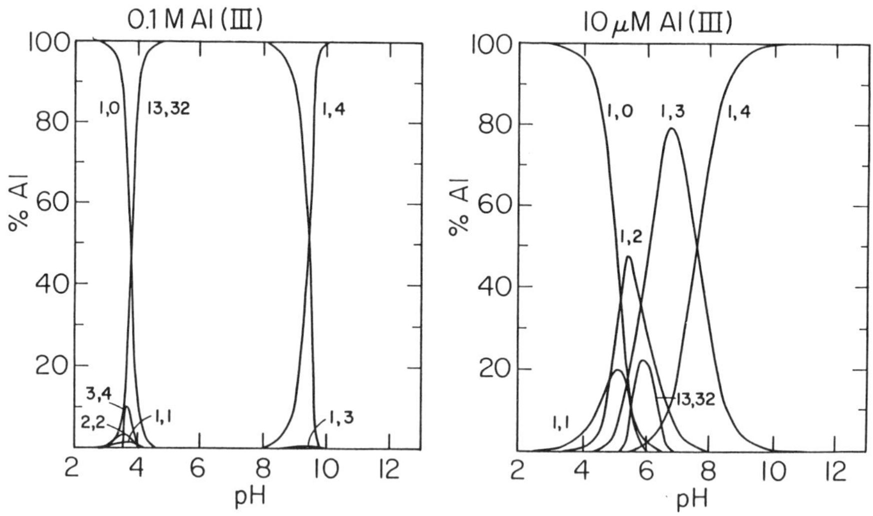

Figure 3.1. Distribution of Al^{3+} hydrolysis products in aqueous solutions at $[Al^{3+}]$ = 0.1 M and 10 μM calculated from the hydrolysis constants in Table 3.1 ($\{x,y\} = \{(Al^{3+})_x (OH^-\ equivalents)_y\}$).

steric bulk of the ligands is largely responsible for the variance of the latter.[32,33] More research has been conducted on the kinetics of rearrangement of octahedral Al complexes in organic solutions than on aqueous solutions of Al complexes. The complicated hydrolysis and polymerization of aluminum, and the lack of convenient ultraviolet or visible spectral absorbance bands to monitor reaction progress, have severely curtailed the study of aqueous reaction rates. Aluminum displays a relatively slow solvent exchange rate with first-order rate constants of up to 10^4 s^{-1}, which is defined by Langford and Gray[34] as class III in the classification of metal ion rates. The rate constant for self-exchange of $Al(H_2O)_6{}^{3+}$ with H_2O at 25°C is 1.29 ± 0.04 s^{-1}. It has been suggested that this reaction follows a dissociative interchange, I_d, pathway.[35] The exchange of bound ligands at metal ion centers roughly follows the aquo exchange rate. Burgess[36] has suggested that the exchange reactions of Al complexes in aqueous solution are I_d in nature, consistent with the small size of aluminum(III) requiring a ligand to dissociate before another may coordinate. The hydrolyzed species $[Al(OH)(H_2O)_5]^{2+}$ undergoes anation reactions faster than the hexaaquo species $[Al(H_2O)_6]^{3+}$ by more than three orders of magnitude (*vide infra*)[37–39] and this has been speculated to result from the mutual ligand interaction due to the anionic hydroxide ligand.[35,40] Studies on the data of the volume of activation also suggest a dissociative mechanism for ligand substitution.[40] Gallium, the chemistry of which closely resembles that of aluminum, was originally suggested to undergo substitution reactions *via* an associative pathway.[6] However, it is now generally accepted that gallium undergoes substitution in a dissociative fashion similar to aluminum.[40]

Table 3.2. Rate constants for the forward reaction of $[Al(H_2O)_6]^{3+}$ with various bidentate ligands (mostly μ=0.1 or 1.0M, 25°C).

Ligand	$k_1(M^{-1}s^{-1})$	Reference
$CH_3C(O)N(OH)H$ (acetohydroxamic acid)	0.17	38
$C_6H_5C(O)N(OH)H$ (benzohydroxamic acid)	0.17	38
$CH_3C(O)N(OH)4\text{-}CH_3C(O)C_6H_4$	0.23	38
$CH_3C(O)N(OH)C_6H_5$	0.15	38
$CH_3C(O)N(OH)4\text{-}CH_3C_6H_4$	0.22	38
Desferriferrioxamine B[c]	0.13	39
Kojic acid	0.83	171
7-Iodo-quinolin-8-ol-5-sulfonic acid (ferron)	0.86	172
Salicylic acid	0.91	173
5-Nitrosalicylic acid	~0.7	174
Fulvic acid	2.3	175
5-Sulfosalicylic acid	1.5	176
Formic acid[a]	9	177
Citric acid	80	148
Acetic acid[b]	2	148
F^- [a]	2.43	178
$(NH_3)_5CoO_2CC_6H_4O^-$ [a]	0.16	179

[a] monodentate; [b] bis(monodentate) complex.
[c] Acting as a bidentate ligand.

Conventional spectrophotometry, stopped-flow, pressure-jump and ^{27}Al NMR techniques have all been utilized in the study of ligand substitution reactions of Al^{3+} in aqueous solution. The ^{27}Al NMR method has been employed by Crumbliss and co-workers[37,38] for the study of Al complexes of bidentate hydroxamic acids. The method is based on the equation k_{obsd} $(s^{-1}) = \pi(h_{ex}-h_0)$ where h_{ex} and h_0 are the half-heights of the ^{27}Al NMR resonances of $[Al(H_2O)_6]^{3+}$ in the presence and absence of exchanging ligand, respectively. This work has been reviewed by Crumbliss.[39]

The rates shown in Table 3.2 are for a variety of (mostly bidentate) ligands determined by different techniques and are therefore, only roughly comparable. These rates are those for the simple forward reaction (k_1 according to the Eigen–Wilkins scheme — chosen to allow for a rough comparison between ligands):

$$Al^{3+} + HL \xrightarrow{k_1} AlL^{2+} + H^+ \tag{3.1}$$

Except for formate and citrate (Eigen–Tamm scheme[41]), most of the rate constants are consistent with the water exchange rate. Thus, complex formation is dominated by the water exchange. The Eigen–Wilkins mechanism,[42] or dissociative ion pair mechanism, is now generally accepted for ligand substitution reactions of aluminum (and other hydrated metal ions) in aqueous solution (k_1 in this scheme is as in reaction 3.1):

$$Al(H_2O)_6{}^{3+} + HL \underset{(1)}{\overset{K_{os}}{\rightleftharpoons}} Al(H_2O)_6(HL)^{3+} \underset{(2)}{\overset{k_1}{\longrightarrow}} Al(H_2O)_5(HL)^{3+} \underset{(3)}{\longrightarrow} Al(H_2O)_5L^{2+}$$

The first step is the rapidly established formation of an outer-sphere ion pair complex (K_{os}). The ligand then penetrates into the coordination shell of the aluminum, this is the rate-determining step (k_1). Finally, H_3O^+ is eliminated in the third and final step. If K_{os} [HL] <<1, then $k_{forward}$ in the first step is equal to $K_{os}k_1$. The value of K_{os} may then be obtained from the Fuoss equation.[43] Thus, k_1 can be compared with the water exchange rate. A possible mechanism may then be proposed. Saito and Nagasawa[40] studied the volume of activation ($\Delta V^{\neq}$) for the exchange of solvent molecules and for the exchange of acetylacetonate ligands at Al^{3+}. They found $\Delta V^{\neq}$ to be positive for several monodentate solvents suggesting an I_d mechanism. Other reactions with monodentate ligands also show positive $\Delta V^{\neq}$.

From a consideration of the hydrolysis of Al^{3+}, the $AlOH^{2+}$ species is normally taken to be the (faster) parallel path for Al^{3+} substitution:

$$Al(H_2O)_6{}^{3+} \overset{K_h}{\rightleftharpoons} Al(H_2O)_5(OH)^{2+} + H^+$$

$$Al(H_2O)_5(OH)^{2+} + HL \rightleftharpoons Al(H_2O)_5L^{2+} + H_2O$$

There is no exchange rate of $AlOH^{2+}$ available in the literature thus far (to our knowledge) and it is difficult to draw conclusions without this comparison. Although dissociative pathways dominate most substitution reactions of Al^{3+}, the study of the substitution kinetics of aluminum is still a very fertile area indeed.

3.3. Complexation in Water

The hydrolysis of aluminum can be partially or completely prevented by complexation with ligands. Aluminum(III) is a hard acid and, as such, prefers hard bases such as oxygen or fluorine. It forms strong hydroxide complexes as part of its hydrolysis behavior, but it also complexes strongly with fluoride.[44] Other inorganic ligands such as sulfate, thiocyanate and hydrogen carbonate also form complexes with aluminum, but not nearly as strongly as does fluoride. The very high affinity of F^- for Al^{3+} cannot be incorporated into multidentate ligands so it is oxygen (the next most electronegative element) which is incorporated into chelating ligands with high formation constants for Al^{3+}.

Complexes of aluminum with organic oxygen-containing ligands possess high stability constants.[45–47] Stability constants for aluminum(III) with various ligands are summarized in Table 3.3; these constants are mostly for the first complexation reaction of Al^{3+} with the completely deprotonated ligand (Al^{3+} + $L^{n-} \rightarrow [AlL]^{(3-n)+}$) although many ligands form *bis*- or *tris*-ligand species.

Table 3.3. Log stability constants[a] for Al^{3+} with various ligands (T=25°C, μ=1M unless otherwise noted). The values are neither comprehensive nor directly comparable in a quantitative sense but allow for rough comparisons of ligand affinity. *K* values are for the reaction of Al^{3+} with the completely deprotonated ligand (unless otherwise noted):

$$Al^{3+} + L^{n-} \xrightarrow{K} [AlL]^{(3-n)+}$$

Ligand	Structural formula	log *K*
Formic acid	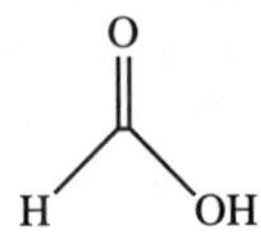	1.36
Acetic acid	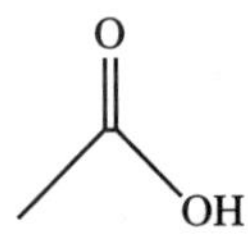	1.51
Propanoic acid	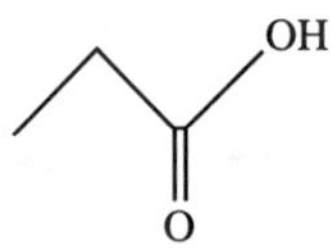	1.69
Butanoic acid	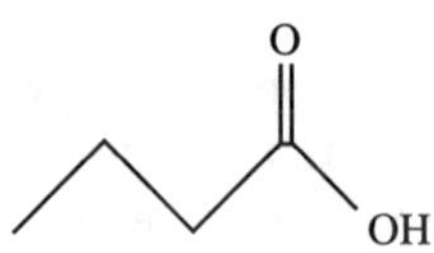	1.58
Benzoic acid	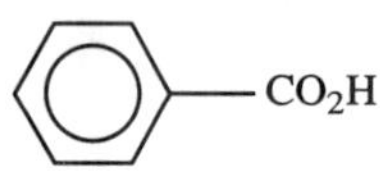	12.09[a]
Hydroxide ion	OH^-	8.31
Hydrogen thiocyanate	HNCS	0.42[b]
Hydrogen fluoride	HF	6.09
Hydrogen sulfate ion	HSO_4^-	1.48

(Continued)

Table 3.3. *(Continued)*

Ligand	Structural formula	log *K*
Hydrogen carbonate	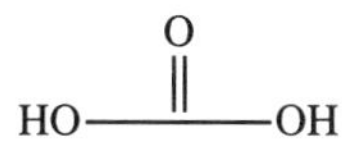	−7.87[c]
Iminodiacetic acid	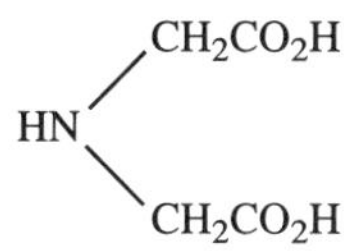	8.10[d]
N-(2-hydroxyethyl)imino diacetic acid	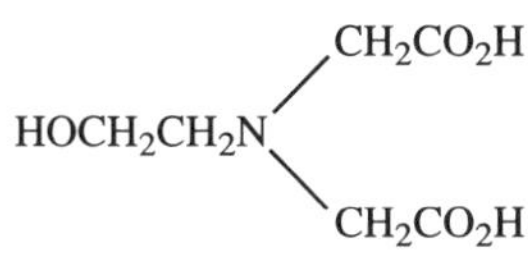	7.74[e]
Pyridine-2,5-dicarboxylic acid	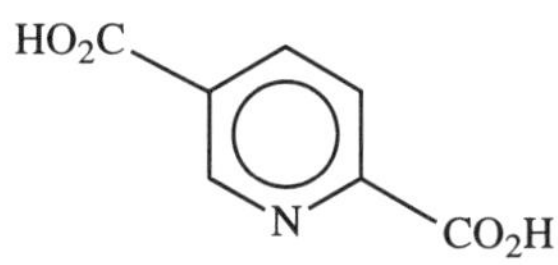	3.95[d]
Pyridine-2,6-dicarboxylic acid	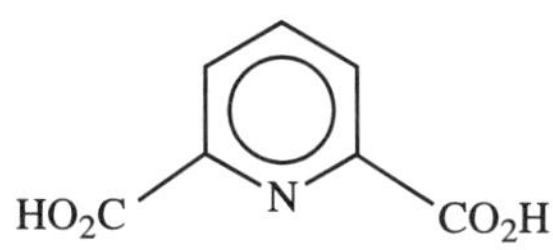	4.87[d]
7-Iodo-8-hydroxyquinoline-5-sulfonic acid	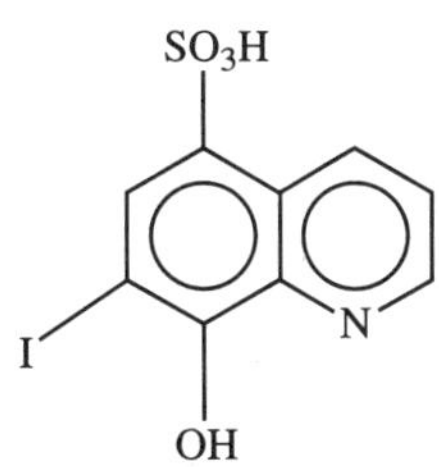	7.60[e]
2-Amino-2-propylphosphonic acid	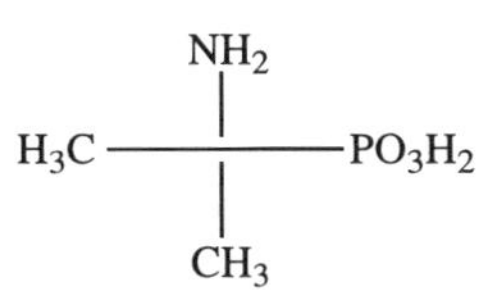	11.42[e]

(Continued)

Table 3.3. *(Continued)*

Ligand	Structural formula	log *K*
1,2-Dihydroxybenzene-4-sulfonic acid	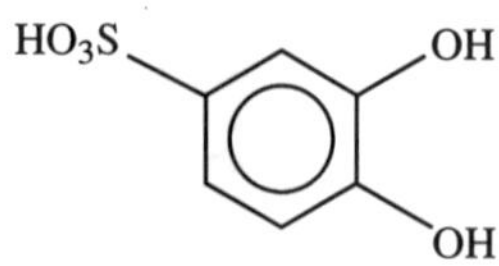	16.6[f]
Tiron	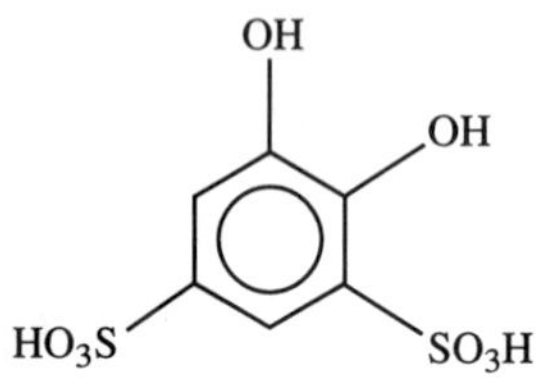	14.9
Solochrome Violet R	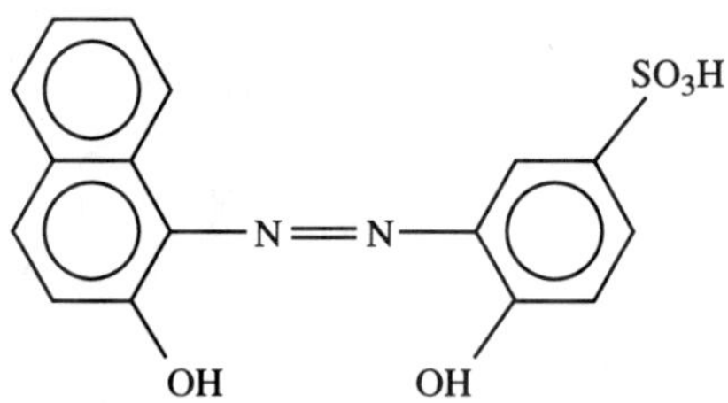	18.4[g]
3-Hydroxy-2-naphthoic acid	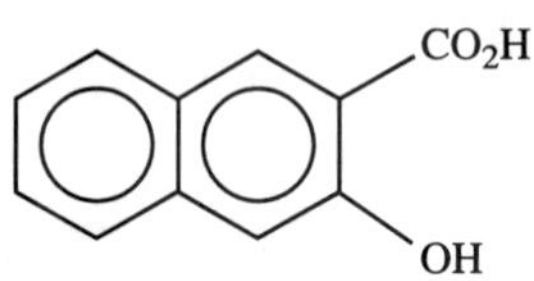	12.94[d]
Chromotropic acid	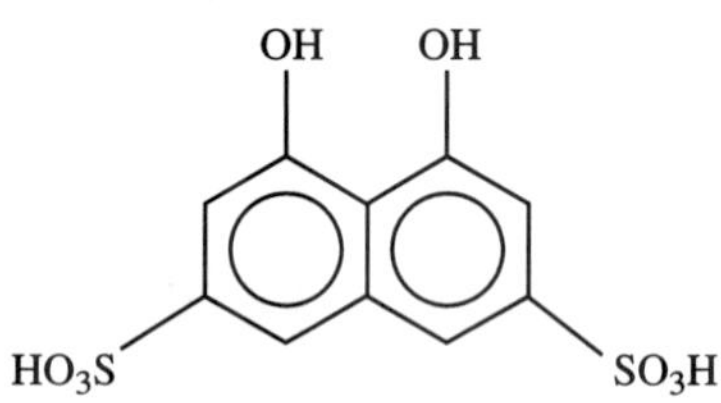	1.5[h]
Nitroacetic acid	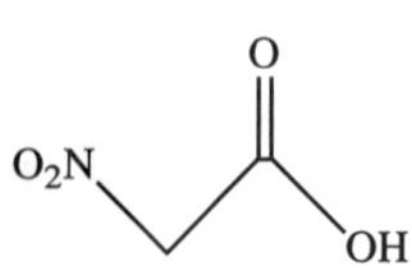	0.48[i]

(Continued)

Table 3.3. *(Continued)*

Ligand	Structural formula	log *K*
Oxalic acid	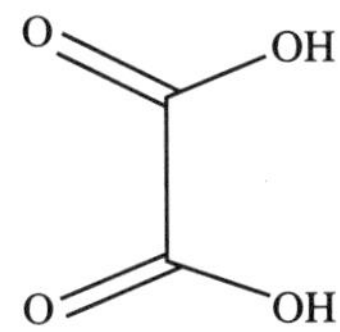	6.1
Phthalic acid	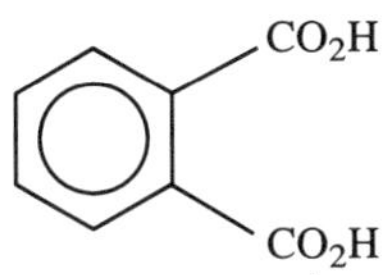	2.94[s]
Oxydiacetic acid	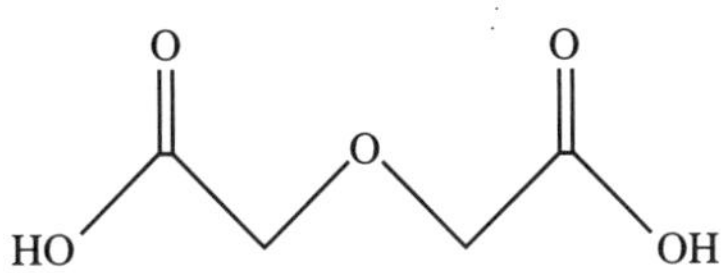	3.16[d]
Thiodiacetic acid	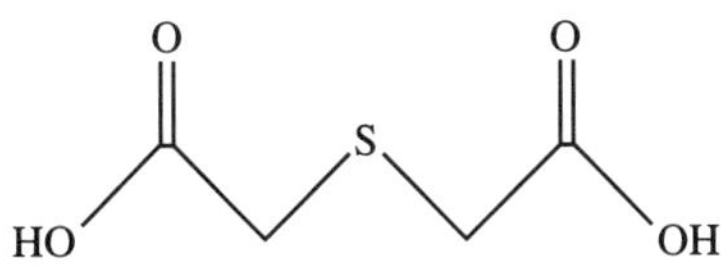	1.93[d]
Salicyclic acid	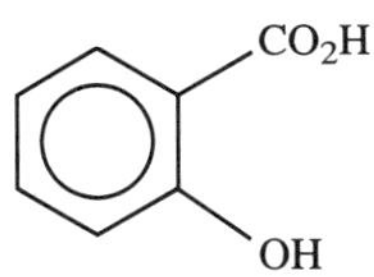	12.9[f]
5-Sulfosalicylic acid	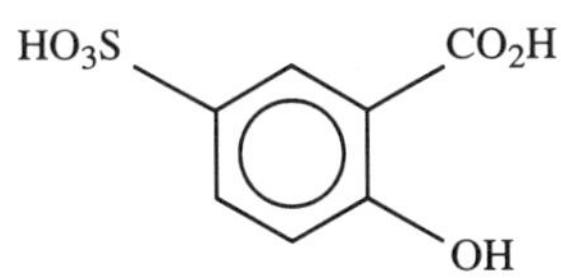	12.3[f]
8-Hydroxy-7-(arylazo)-quinoline-5-sulfonic acid	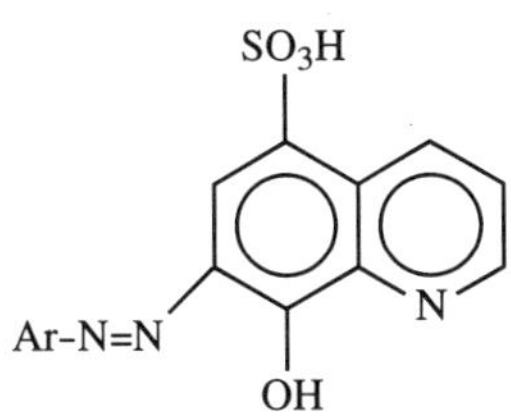	9.86[e]

(Continued)

Table 3.3. *(Continued)*

Ligand	Structural formula	log *K*
2-Oxobutanoic acid	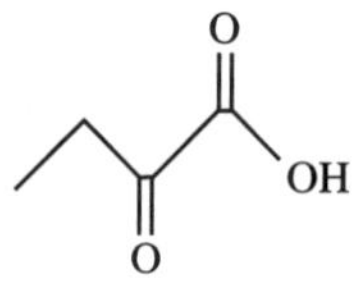	2.09[e]
Acetoacetic acid	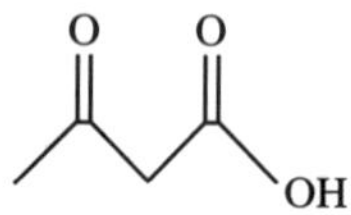	1.21[e]
Phenylserine	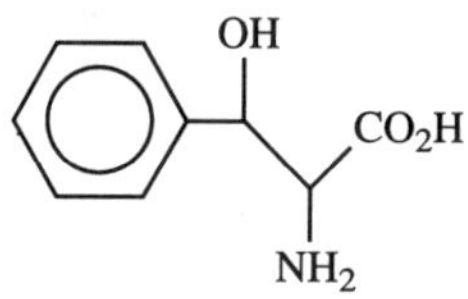	3.98[e]
L-Ascorbic acid	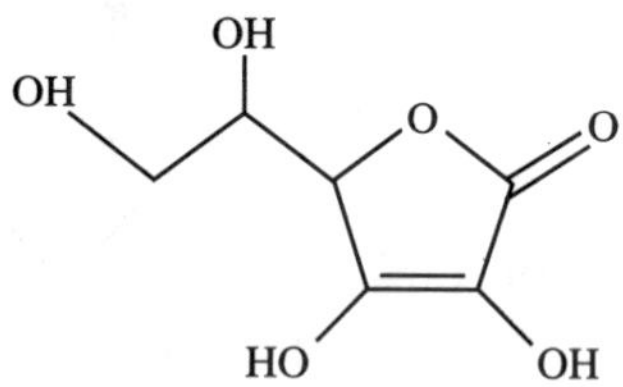	1.89[j]
Squaric acid	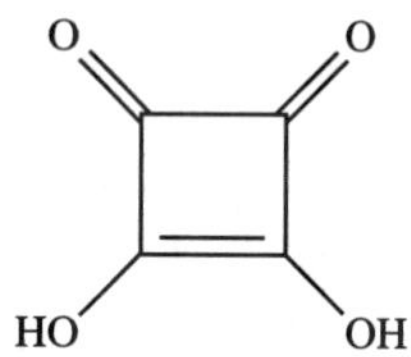	2.83[k]
Acetohydroxamic acid	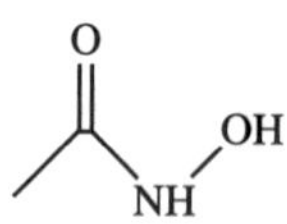	7.95[f]
N-Methyliminodiacetic acid	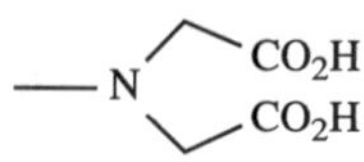	7.55[d]
3-Hydroxy-5,7-disulfo-2-naphthoic acid	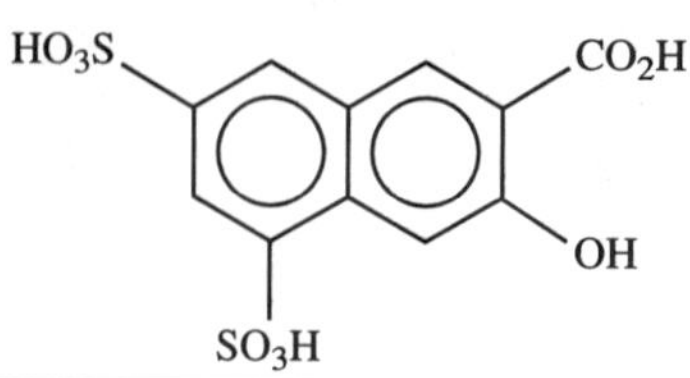	10.81[d]

(Continued)

Table 3.3. *(Continued)*

Ligand	Structural formula	log *K*
2,3-Dihydroxynaphthalene-6-sulfonic acid	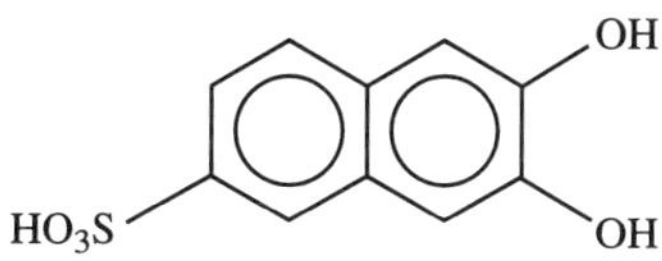	15.11[d]
Sulfoxine	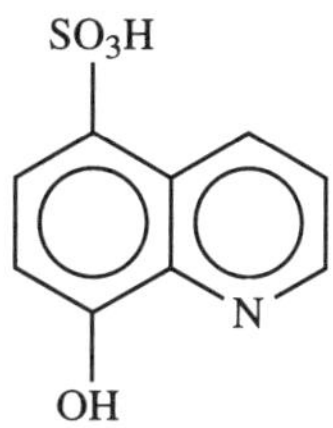	8.7[e]
Oxine	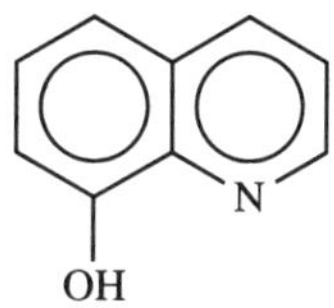	30.6[l]
2-(Phenylazo)chromotropic acid	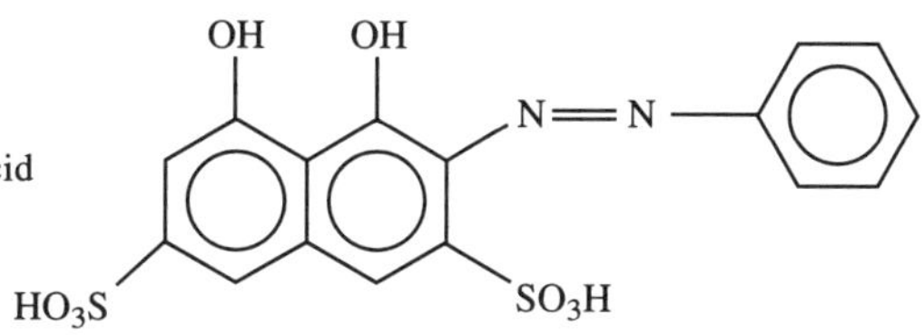	3.77[h]
Acetylacetone	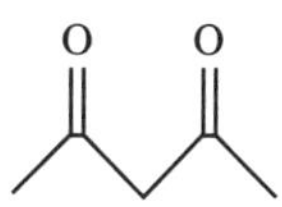	8.1[m]
Malonic acid	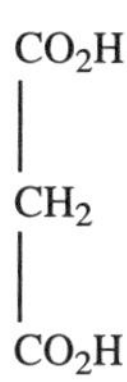	6.26[n]
Succinic acid	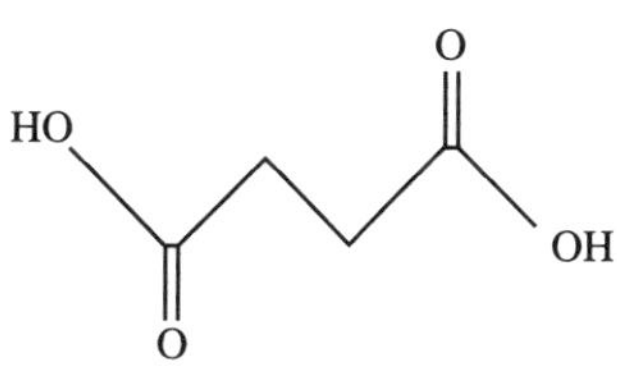	3.20[d]

(Continued)

Table 3.3. *(Continued)*

Ligand	Structural formula	log *K*
3,4-Dihydroxybenzoic acid	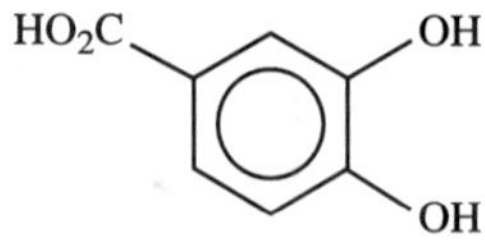	5.77[p]
Gallic acid	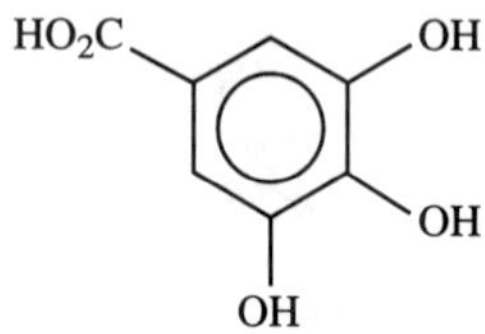	14.24[q]
2-(3,4-Dihydroxyphenyl)- 2-(1,1-benzopryran)-3,5,7-triol	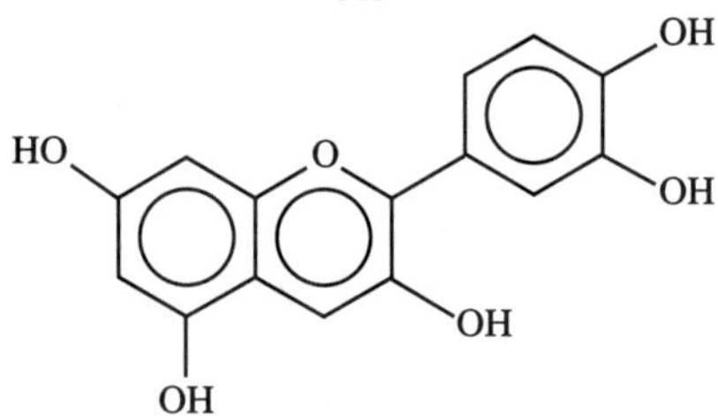	17.10[r]
3-Hydroxy-7-sulfo-2-naphthoic acid	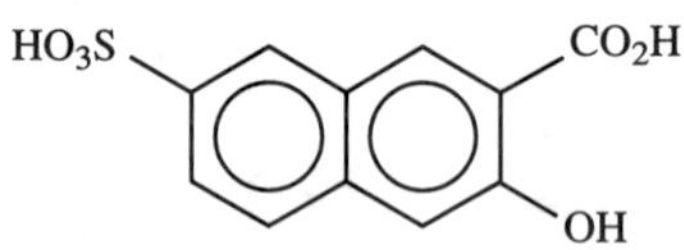	11.89[e]
1,2-Dihydroxynaphthalene-4-sulfonic acid	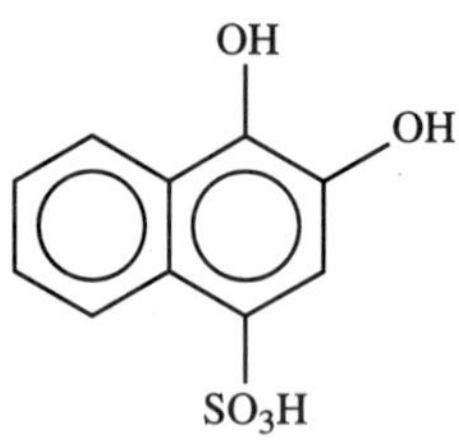	5.9[s]
N,N-Bis(2-hydroxyethyl)glycine	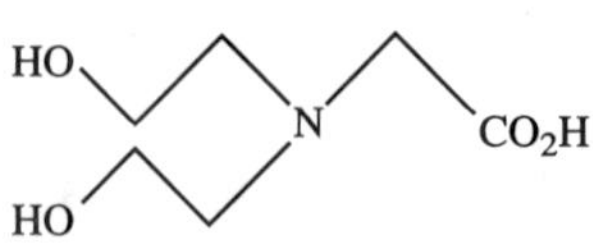	3.38[h]

(Continued)

Table 3.3. *(Continued)*

Ligand	Structural formula	log *K*
N-(Phosphonomethyl)-iminodiacetic acid	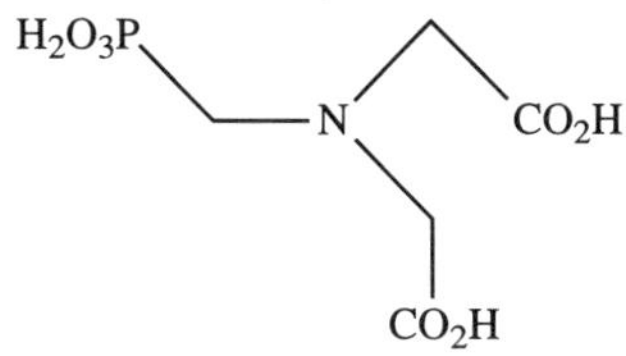	14.7[e]
N-(2-Hydroxyethyl)-iminodiacetic acid	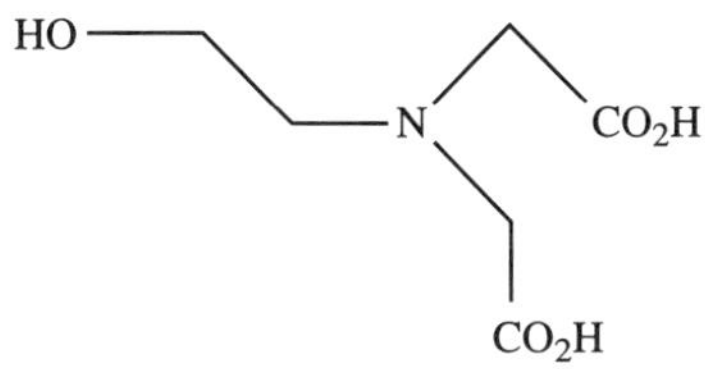	7.74[e]
Iminobis(methylene-phosphonic acid)	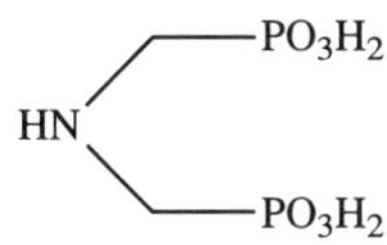	6.92[p]
D-Gluconic acid	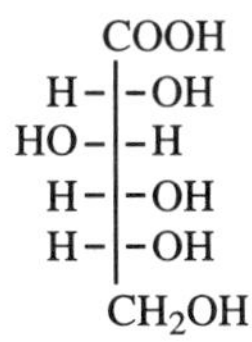	1.98[e]
Saccharic acid	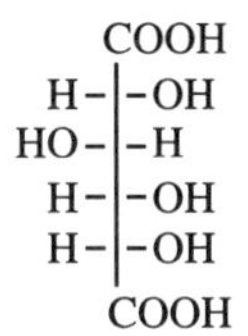	1.57[e]
Tartaric acid	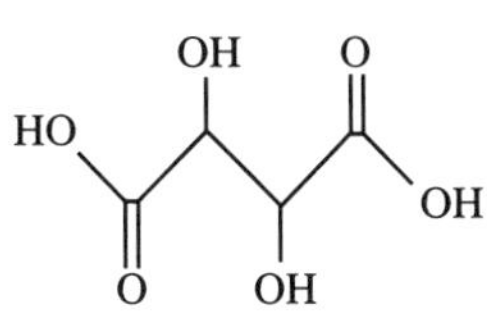	5.66[s] 1.18[w]
1-Oxopropane-1,2-dicarboxylic acid	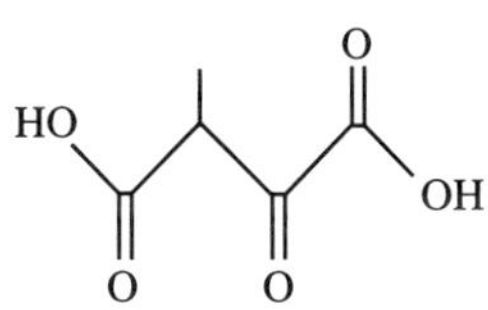	6.16[e]

(Continued)

Table 3.3. *(Continued)*

Ligand	Structural formula	log *K*
Propane-1,2,3-tricarboxylic acid	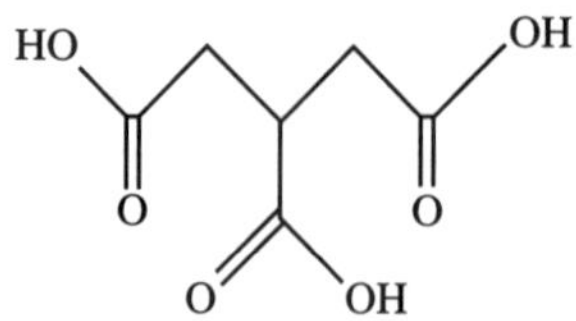	5.44[n]
Citric acid	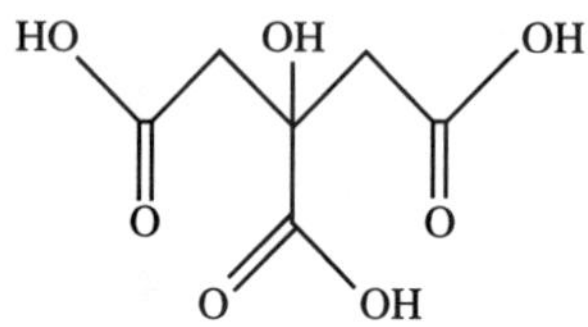	7.87
BAMTPH	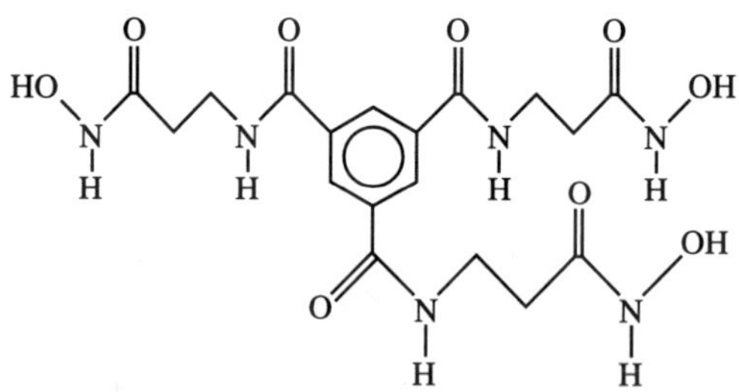	21.85[e]
Desferriferrioxamine-B = DFB	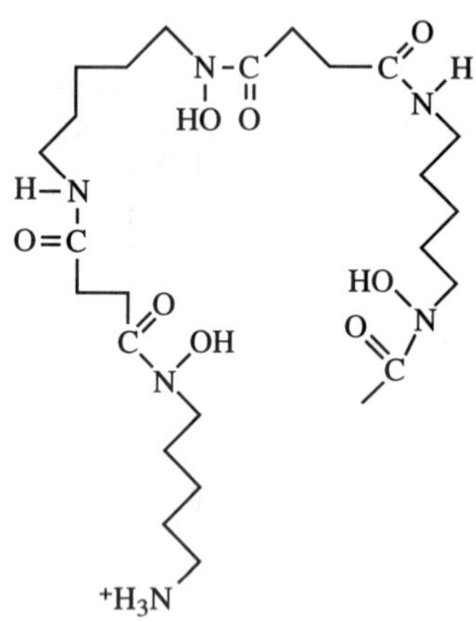	24.14[e]

(Continued)

Table 3.3. *(Continued)*

Ligand	Structural formula	log *K*
1,7-dihydroxy-4-sulfo-2-naphthanoic acid	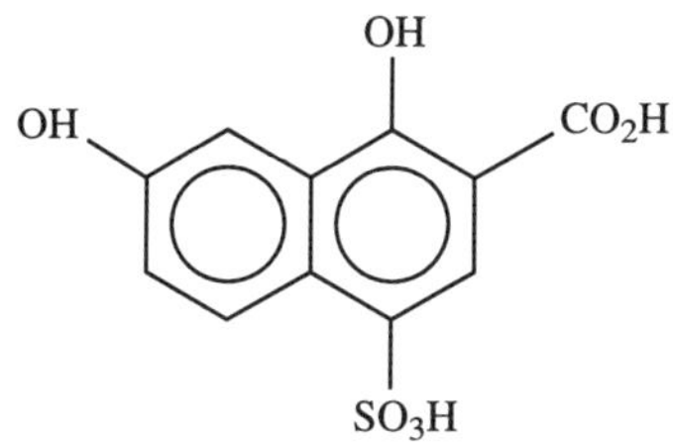	14.95[d]
Aspartic acid	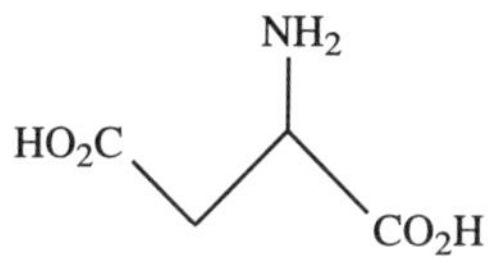	2.17[t]
Glutamic acid	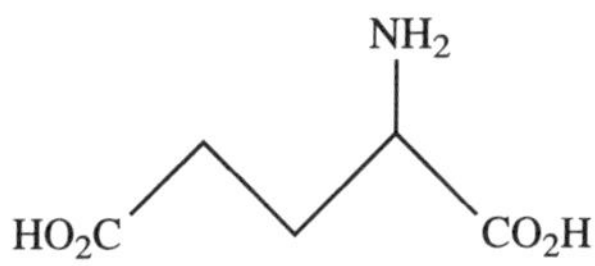	2.30[t]
Pyridoxal-5-(dihydrogenphosphate)	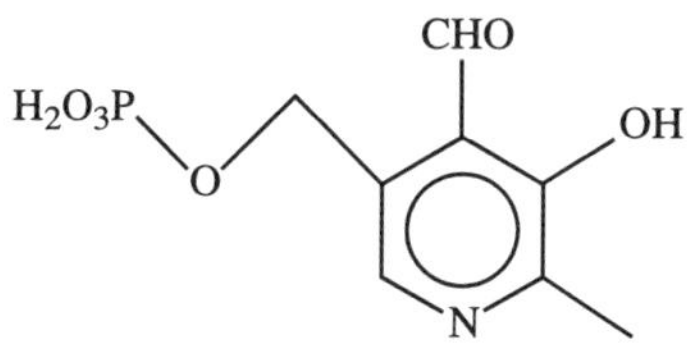	10.48[u]
Pyridoxal	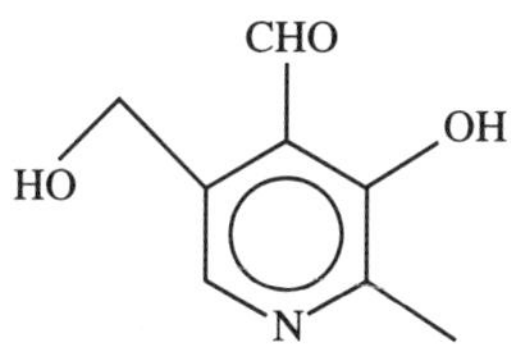	2.45[e]
Nitrilotriacetic acid	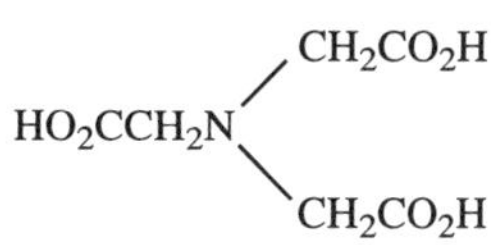	11.1[s]
Amino(phenyl)methylene-diphosphoric acid	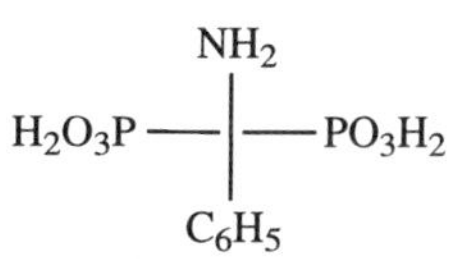	18.79[e]

(Continued)

Table 3.3. *(Continued)*

Ligand	Structural formula	log *K*
Ethylenebis[imino-(2-hydroxyphenyl)methylene (methyl)-phosphonic acid]	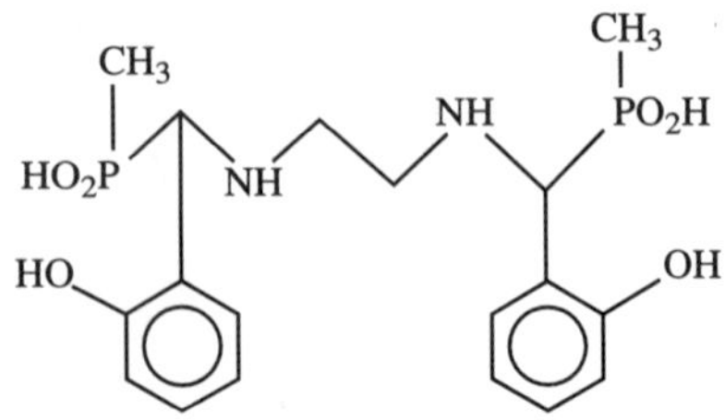	20.0[e]
N-(2-hydroxyethyl)-ethylenedinitrilo-*N,N',N'*-triacetic acid	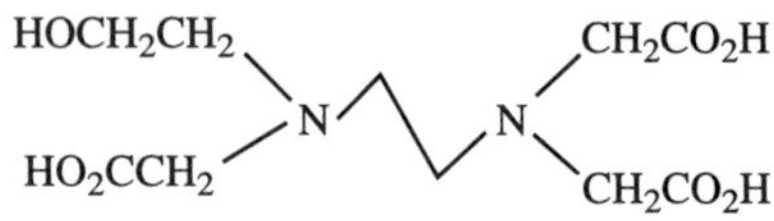	14.4[e]
EDTA	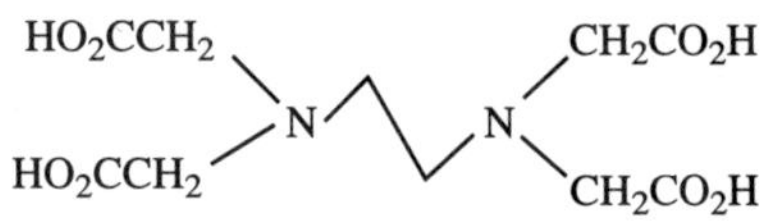	16.5[e]
Trimethylenedinitrilo tetraacetic acid	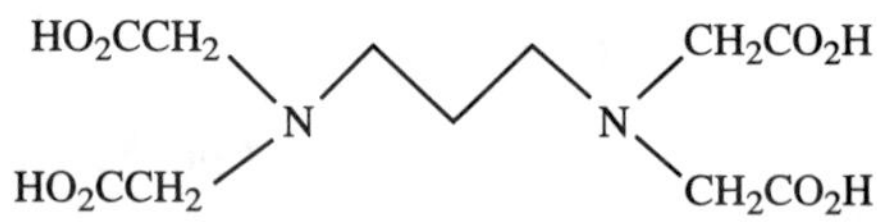	16.32[f]
CDTA	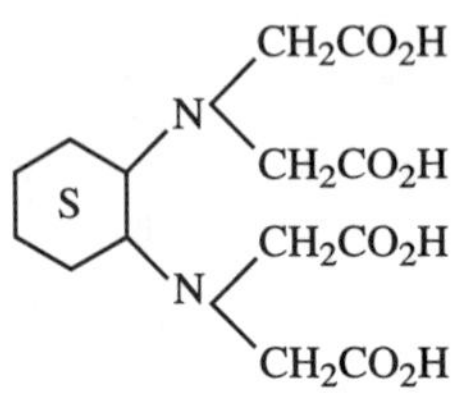	19.6[e]
(2-Hydroxytrimethylene)-dinitrilotetraacetic acid	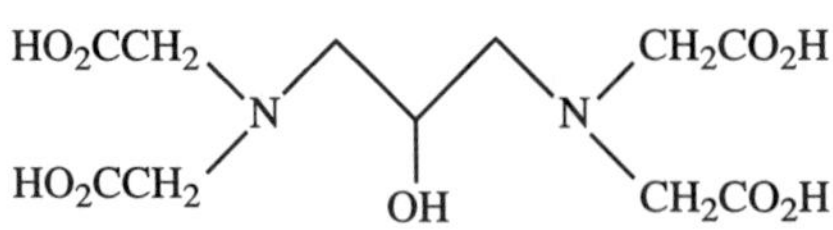	15.2[e]
EGTA	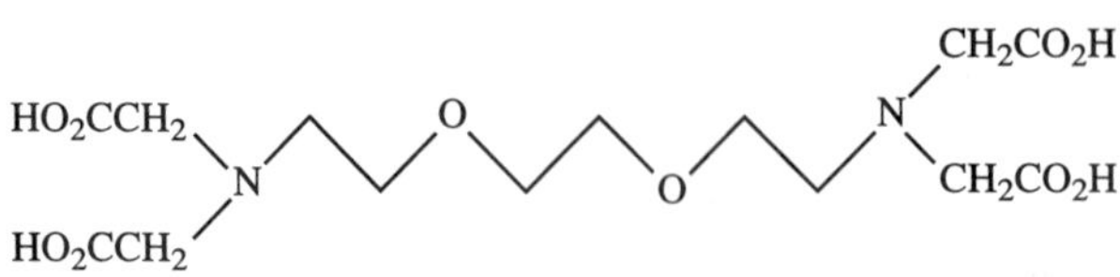	13.9[v]

(Continued)

Table 3.3. *(Continued)*

Ligand	Structural formula	log *K*
DTPA	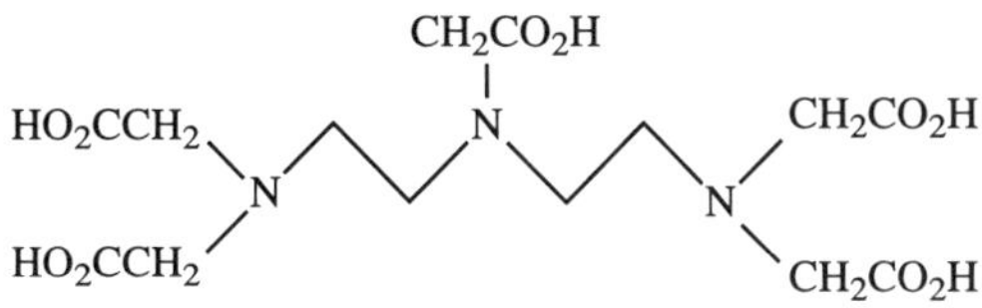	18.7[e]
TTHA	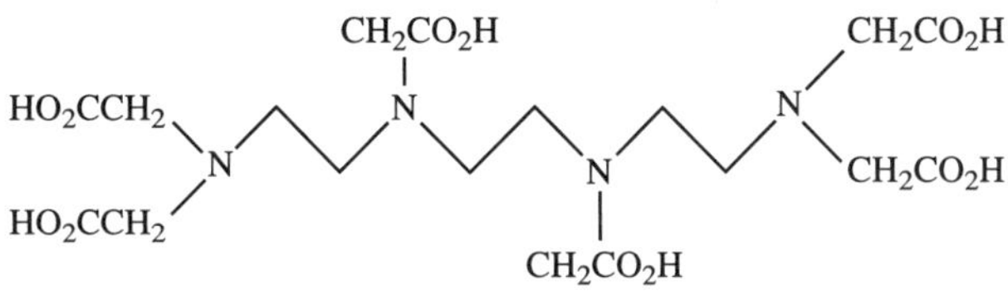	21.0[e]
Semi-Xylenol Orange	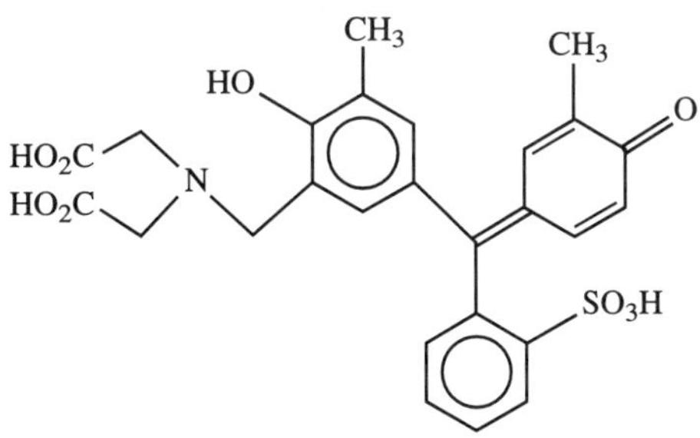	16.7[e]
Semi Methylthymol Blue	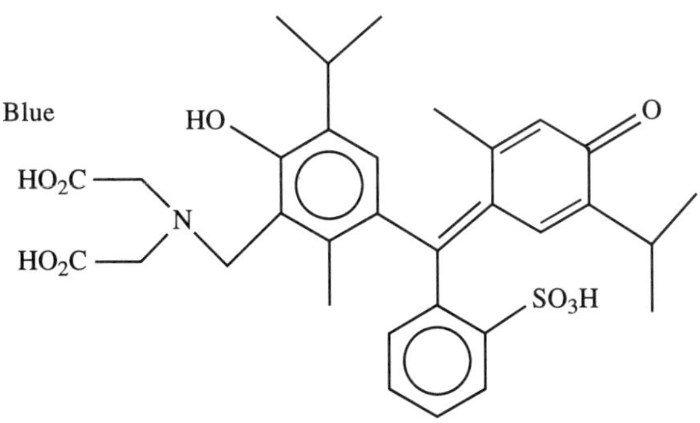	17.9[e]

(Continued)

Table 3.3. *(Continued)*

Ligand	Structural formula	log *K*
Xylenol orange	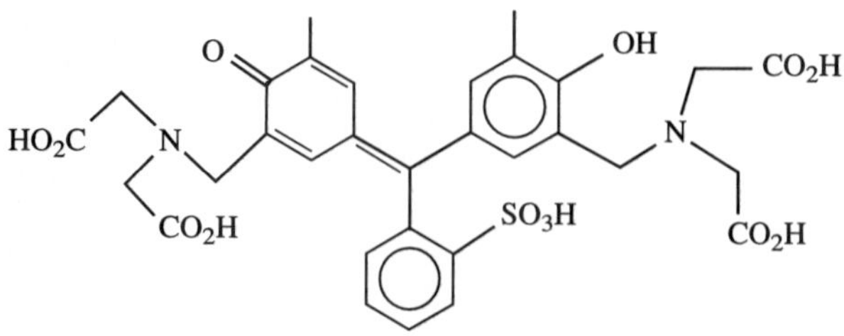	16.3[e]
Methylthymol blue	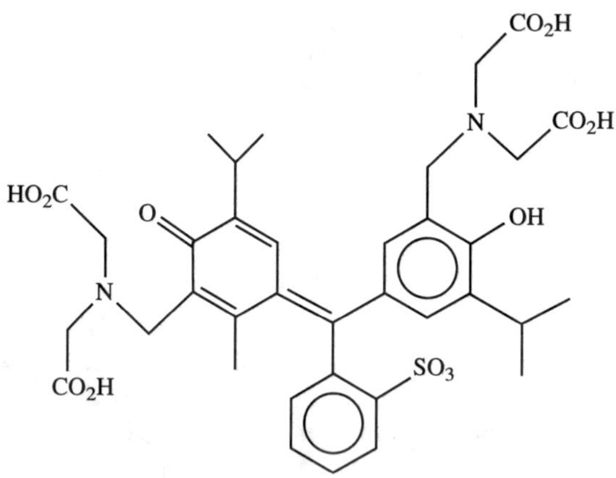	16.2[e]
3-Hydroxyglutamic acid	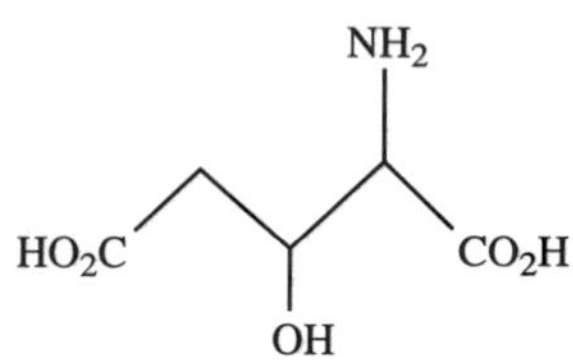	4.87[e]
L-Phosphoserine	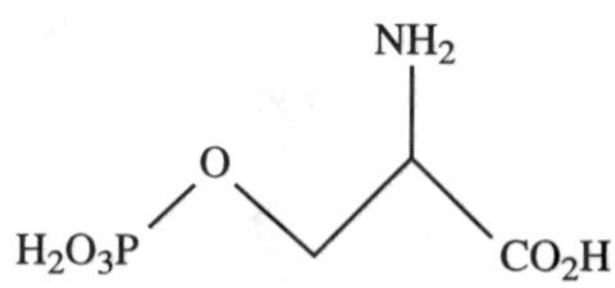	4.79[e]
DL-Amino-3-phosphopropanoic acid	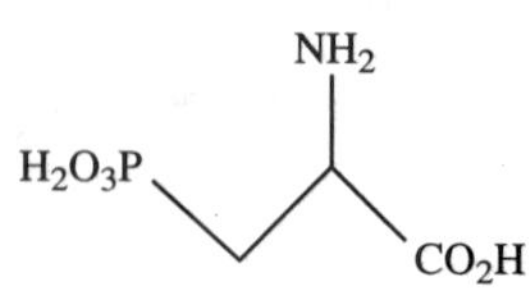	13.12

(Continued)

Table 3.3. *(Continued)*

Ligand	Structural formula	log *K*
Glyphosphate	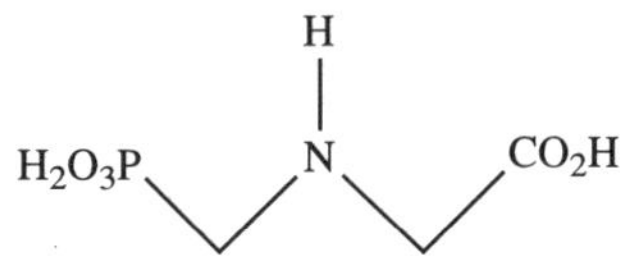	13.70[e]

[a]25°C, 0.5M; [MOHL]/[MOH][L]. [b]23°C, 0M. [c]25°C, 0.5M; $[M_2(OH)_2L][H]^3/[M]^2[HL]$. [d]25°C, 0.5M. [e]25°C, 0.1M. [f]20°C, 0.1M. [g]25°C, 0M. [h]20°C, 0.1M; [ML][H]/[M][HL]. [i]18°C, 0.4M. [j]25°C, 0.1M; [MHL]/[M][HL]. [k]25°C, 0.3M. [l]20°C, 0.1M; $[ML_3]/[M][L]^3$. [m]30°C, 0.1M. [n]37°C, 0.15M. [p]25°C, 0.1M; [ML]/[MOHL][H]. [q]25°C, 0.6M; [MHL]/[M][HL]. [r]25°C, 0.1M; $[MH_2L]/[M][H_2L]$. [s]25°C, 0.6M. [t]25°C, 0.5M; [MHL]/[M][HL]. [u]22°C, 0.1M. [v]25°C, 0.2M. [w]25°C, 0.1M; $[M(H_{-1}L)][H]/[M][L]$.

This table is not intended to be comprehensive since the compilation by Martell and Smith[47] accomplishes this for *all* metal ions. Indeed, the constants in Table 3.3 are a selection of those available from this six-volume series, from the published work of Martell and co-workers,[48–53] and from the excellent studies by the Swedish group on (mostly) weakly complexing ligands (these ligands are associated with natural waters).[17,54–73] The Swedish group has discovered many monoligand dialuminum species Al_2L (and occasionally trinuclear species) with weakly binding ligands such as monocarboxylates and carbonate. These species presumably have bridging CO_2^- groups.

Aluminum cations bind firmly with negatively charged organic functional groups in the acidic to neutral pH range, although few ligands can compete with hydroxide in the very high pH regime (pH > 11.5). This is a consequence of the high stability of the $Al(OH)_4^-$ complex and of the great concentration of hydroxide in a very basic solution. It has been pointed out that the stability of negative O-donor complexes closely correlates with the affinity of the metal ions for hydroxide.[52] As most ligands are weak acids there is competition between protons and Al^{3+} for coordination at low pH, therefore the neutral to weakly acidic regime is that of highest ligand binding.

Table 3.3 shows that Al^{3+} forms complexes of the greatest stability with ligands containing strongly basic O atoms (phenolate O^-, carboxylates, phosphonate O^-) and N atoms (amines) which are appropriately spaced to form five-membered or (less preferably) six-membered metallacycles (the polyaminocarboxylates EDTA, EGTA, DTPA etc.). The catecholates (*vide infra*) form the strongest complexes with trivalent metal ions[52] although this binding strength is considerably modified by proton competition in neutral aqueous solution. Phosphate is a ligand of considerable importance in biological systems; delineation of the complexes formed with Al^{3+} have long been hampered by the insolubility of $AlPO_4$ and much effort has been expended on the study of this system without a definitive result having been achieved.[2,74–80]

Because of the chelate effect, complexes of higher stabilities are demonstrated by bidentate and multidentate ligands, and the oxygen donor groups are usually carboxylate, hydroxyl, or carbonyl (in conjunction with hydroxyl); amines also play an important role in metal chelation when alternated with carboxylate in aminocarboxylates such as EDTA or DTPA. Carboxylates are poor ligands for Al^{3+} unless they are arranged to allow multidentate chelation.

Although the hydrolysis of aluminum makes determination of formation constants quite problematic, the existence of good linear relationships between the formation constants of complexes of different metals containing the same hard oxygen donor groups (catecholates, phenolates, hydroxamates, phosphonates, carboxylates, salicylates, β-diketonates, hydroxide) allows the estimations of possible values for aluminum complexes and the design of new chelating ligands.[52]

The 100% natural abundance, a small quadrupole moment, and a high nuclear spin serve to make ^{27}Al NMR a powerful tool.[8,81] The linewidths of ^{27}Al NMR spectra depend greatly on the geometry around the ^{27}Al nucleus. Since aluminum is quadrupolar, asymmetry in the ligand field produces an increase in the NMR linewidth; the resonances of tri- or pentasubstituted species are expected to be broader than those of tetra- or hexacoordinated species. When the symmetry is quite high (e.g., $[Al(H_2O)_6]^{3+}$),[12,18,21–23] the ^{27}Al spectral peak is sharp with a linewidth of ca. 3Hz (at 0 ppm, the reference for ^{27}Al). High symmetry octahedral complexes never exceed 100 Hz.[81] The linewidth variation for aluminum complexes is from 3 to 6000 Hz with a chemical shift range of 450 ppm. The combination of linewidth and chemical shift serves as a reasonably diagnostic probe for studying the chemical and physical environment about the aluminum atom in a coordination complex.

^{27}Al NMR has been widely applied to the study of complexes under a variety of pH conditions. The basic parameters are chemical shift, δ, and line width at half-height, $W_{1/2}$. These quantities depend on ligand type (complexes with different donor atoms appear at different chemical shifts), and coordination number (chemical shift is determined by the number of the ligands — octahedral complexes appear near 0 ppm, whereas tetrahedral complexes appear further downfield). The effect of substituting one or two water molecules by a bidentate organic ligand is almost additive[82] though such a relationship is not always observed. Line widths of complexes are mostly dependent on the symmetry of ligands around the ^{27}Al nucleus; the more symmetric the complex, the more narrow is the line width. Many aqueous aluminum complexes show slow enough ligand exchange rates (on the NMR time scale) that different species can be observed. ^{27}Al NMR has been used, for example, to examine the aluminosilicate species formed in solution of Al^{3+} and silicate anion.[83] Table 3.4 lists the chemical shifts and line widths of the partially hydrolyzed aluminum species.

Polycarboxylate, hydroxycarboxylate, and polyaminocarboxylate ligands are of special interest in aqeous aluminum chemistry for the following reasons: (a) many occur in natural waters and may be important in the transport of

Table 3.4. ^{27}Al NMR chemical shifts and line widths of Al-containing hydrolyzed species.

Species		δ (ppm)	$W_{1/2}$(Hz)
$Al(H_2O)_6^{3+}$		0	3–20
$[Al(OH)]_4^-$		80	20–150
$[Al_{13}O_4(OH)_{24}]^{7+}$	(AlO_4)	62.5	50
	(AlO_6)	~ 5	~ 3500
$[Al_2(OH)_2]^{4+}$		4	500, 700

aluminum; (b) most of their aluminum complexes are water soluble; (c) most show strong affinity towards aluminum; and (d) some exist in the human body. For example, citric acid is the main low-molecular-weight carrier of aluminum in plasma.[84–87] ^{27}Al NMR spectroscopy can be informative in the study of the hydrolysis of these types of complexes, although not all peaks can be assigned to individual species. Table 3.5 summarizes some of the complexes that have been investigated. The ^{27}Al NMR spectral changes are explained by the gradual displacement of water by ligand, and of the latter by hydroxide, as pH increases from the very acidic realm to the very basic.

The data for lactate suggest that the ligand is coordinated to aluminum and the hydrolytically stable range is pH 5 – 7.[88] The crystal structure of tris(lactato)aluminum(III)[89] verified that the aluminum is surrounded by a distorted octahedral coordination sphere where the oxygen atoms of the carboxylate and the hydroxyl coordinate to Al; both oxygen atoms are deprotonated. ^{17}O NMR spectra showed the inequivalence of the two oxygen atoms in the carboxylate group while only one of the oxygen atoms in that group is coordinated. Such phenomena can also be observed in the complexes of simple mono- or bidentate carboxylate ligands.[90] Potentiometric studies have demonstrated the existence of *mono*-, *bis*-, and *tris*(lactato)aluminum(III) species in water (as a function of pH) in addition to dinuclear species at higher concentrations.[56,59] The binding of both lactate and acetate to Al^{3+} in water is weak, however, as is the binding of all carboxylates. Less information is available for multidentate ligand complexes because of the complicated coordination around the aluminum (lower symmetry, broader lines). However, these ligands form quite strong complexes. Citrate is, arguably, the most extensively investigated, and controversial ligand for aluminum in aqueous solution.[2,19,49,65,67,84,88,91–93] ^{27}Al NMR spectra showed that the aluminum–citrate system is pH-dependent in the same way as the aluminum–lactate system, but its structure cannot be unequivocally defined in solution since numerous coordination possibili-ties, as well as protonation possibilities, exist for three carboxyl groups and one hydroxyl group. Potentiometric studies suggest that neutral AlL is a significant species at $2 \leq pH \leq 5$.[2,84] The trimeric complex $[NH_4]_4[Al_3(H_{-1}Cit)_3(OH)(H_2O)]$ was crystallized from aqueous solution between pH 7 and 9.[94] The citrate hydroxy group is deprotonated, as was

Table 3.5. ^{27}Al NMR data on carboxylate complexes in water.

Ligand	pH	δ (ppm) ($W_{1/2}$ (Hz), species)
Oxalate[a]	0.5	11.9 (?), 6.8(AlL^+), 0
	1	12.5 (AlL_2^-)
	1–7	16.3 (AlL_3^{3-})
	7–9	16.3, 13.3 (?)
	10	80.1
Citrate[a]	<2	6.9 (500), 0
	2	14 (500, $AlH_{-1}L^-$), 21.1(300, AlL), 28 (250), 9.3 (sh, $AlHL^+$)
	3–6	14, 21.1, 28, 6.5
	3–11	14, 21.1, 28
	10	80.1
Tartrate[a]	<0.5	11(600), 0
	2	23(>1KHz) fast exchange
	2–8	0
	8–12	33.4 $Al(H_{-6}L_3)^{9-}$
Malonate[a]	<1	1 (250), 0
	1–9	2.2 (150, AlL_3^{3-})
	7–12	2.2, ppt.
dl-Malate[a]	<2	7.6 (300), 0
	2–8	−0.7 (200), 12.0 (400), 18.9 (400) (contains AlL, $Al(H_{-1}L)^-$, $Al(H_{-2}L)^{2-}$)
dl-Mandelate[a]	0–1	7.5 (400), 0
	2–8	26.6 (700, AlL_3^{3-})
Glycolate[b]	<1	9 (400, AlL^{2+}), 0
	2–4	17.5 (1400, AlL_2^+), 9, 0
	4–7	27.0 (2200, AlL_3), 22.0
	8–12	60.5, 55.5, 34.0
	>12.5	80.7
EDTA[c]	2–4	32 (AlL), 0
	4–7	32
Lactate[c]	3	14 (1200, comb. of AlL^{2+} and AlL_2^+), 0
	3–5	24 (AlL_3), 14
	5–7	24
	8	~ 53
	8–10	80(120), 60
	11.5	80, 53

[a] Refs. 81, 93; [b] Ref. 90; [c] Ref. 88.

suggested by Motekaitis and Martell[49] concerning the displacement of aliphatic hydroxyl protons from ligands in which they are in positions favorable for metal ion coordination, and the deprotonated hydroxyl bridges two aluminum atoms. The crystal structure appears to be conserved in solution ($3 < pH < 9$) and its hydrolysis behavior is the same as that obtained directly from a mixture of aluminum(III) and citrate.

As EDTA is potentially a hexadentate ligand, there is considerably less ambiguity in this coordination behavior than in that of citrate. EDTA forms an octahedral coordination complex with Al and produces a remarkably stable complex in the pH range 4–7. Only one peak, 32 ppm, is observed in the ^{27}Al

NMR spectrum. The nitilotriacetate complex of Al (NTA/Al) was once regarded as a chemical species of primary environmental relevance and NTA was once used as a detergent building block. Its strong coordinating properties towards metal ions mobilize them from the sediments of lakes. An Al-NTA neutral molecule shows more toxicity than simple aluminum salts. Although there are no ^{27}Al NMR data available for this system, a crystal structure of $[Al(NTA)(H_2O)_2]_2$ as an acetone–water solvate showed that the complex is a simple neutral distorted octahedral species, and NTA behaves as a tetradentate ligand.[95] It has been accurately shown to form very strong Al complexes.[69]

Maltol (Hma) is used as a flavoring additive in a number of breads and cakes. The complex $Al(ma)_3$ is found to be quite soluble in aqueous solution at physiological pH.[82] The chelate complex is of neutral charge and has been found to be biologically mobile and extremely neurotoxic.[96] This combination of properties (water solubility and neutral charge) and the varied biological roles that may be played by aluminum has prompted a more thorough study of the complexes formed by aluminum with several 3-hydroxy-4-pyrones and 3-hydroxy-4-pyridinones.[82,97–100] 3-Hydroxy-4-pyridinone complexes were also examined[97,99,100] because 3-hydroxy-4-pyrone ligands lacked an easily functionalizable site on the ring and the pyridinones can vary in their lipophilicity by changes in the substituents on the ring nitrogen atom (without altering the high formation constants for Al).[98]

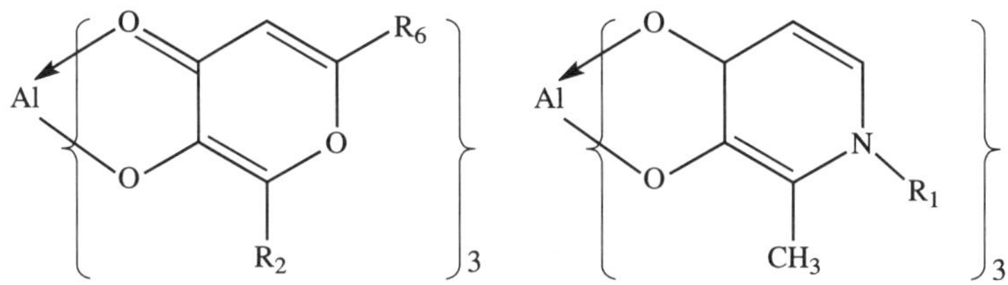

$Al(pa)_3$ $R_2 = R_6 = H$
$Al(ma)_3$ $R_2 = CH_3, R_6 = H$
$Al(ka)_3$ $R_2 = H, R_6 = CH_2OH$

$Al(mpp)_3$ $R_1 = H$
$Al(dpp)_3$ $R_1 = CH_3$
$Al(mepp)_3$ $R_1 = C_2H_5$
$Al(mhpp)_3$ $R_1 = C_6H_{13}$
$Al(ppp)_3$ $R_1 = C_6H_5$

Variable pH ^{27}Al NMR spectra present a very clear explanation of the formation of aluminum maltol complexes in water over the pH range 1.7–11 (Figure 3.2). At low pH, the three peaks at 26, 13 and 0 ppm were assigned to $[Al(ma)_2(H_2O)_2]^+$, $[Al(ma)(H_2O)_4]^{2+}$, and $[Al(H_2O)_6]^{3+}$, respectively.[82] When pH is between 4 and 9, only one peak (39 ppm for $Al(ma)_3$) is seen. At higher pH, $Al(ma)_3$ undergoes basic hydrolysis to form the aluminate ion $[Al(OH)_4]^-$. Acetohydroxamate,[93,101,102] oxalate,[103] lactate,[88] isomaltol,[104] and a variety of hydroxycarboxylate ligands[88,93] undergo similar hydrolysis behavior; however, the pyrone complexes (and pyridinone complexes) show the broadest range of

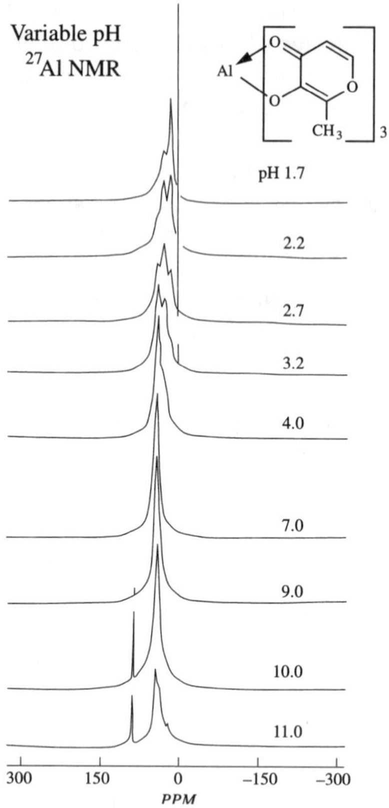

Figure 3.2. Variable pH ^{27}Al NMR of *tris*(maltolato)aluminum(III). Reprinted with permission from *Inorg. Chem.* **1987**, *26*, 2171. Copyright 1987 American Chemical Society.

stability to hydrolysis of the bidentate ligands. Variable pH ^{27}Al NMR spectra for $Al(mpp)_3$ and $Al(dpp)_3$[97] showed similar pH-dependent behavior and demonstrated that these complexes were more stable to hydrolysis that the aluminum maltol compound at similar concentrations (stable pH range 3.5 – 9 for $Al(mpp)_3$, 3–9 for $Al(dpp)_3$. The chemical shifts are all about 38 ppm, beyond the common range of AlO_6 (−40 to +20 ppm). The complexes are not subject to a rigorously octahedral field and the chemical shift reflects this deviation. Other exceptions are the octahedral tris(hydroxamato) complexes[102] and the alumichrome trihydroxamate peptides,[101] which are observed at 36–42 ppm. Values of $W_{1/2}$ for $Al(mpp)_3$ and $Al(dpp)_3$ are 660 and 680 Hz, respectively (narrower than 900 Hz for $Al(ma)_3$). This was explained by the extensive hydrogen-bonding characteristic of the *fac* configuration.[97] The higher stabilities of the *tris*(pyridinonato)aluminum(III) complexes compared with those of the pyronates in Table 3.6 are explained by the greater ability of the N-containing ring to delocalize positive charge in the formation of the

Table 3.6. Stability constants of pyrone,[73] and pyridinone[98,100] complexes of Al^{3+}

$$Al^{3+} + L^- \rightleftharpoons AlL^{2+} \quad \beta_1$$
$$Al^{3+} + 2L^- \rightleftharpoons AlL_2^{+} \quad \beta_2$$
$$Al^{3+} + 3L^- \rightleftharpoons AlL_3 \quad \beta_3$$
$$\beta_n = \frac{[ML_n^{(3-n)+}]}{[M^{3+}][L^-]^n}$$

Ligand	$\log\beta_1 = \log K_1$	$\log\beta_2 = \log K_1K_2$	$\log\beta_3 = \log K_1K_2K_3$
Hka	7.23	13.72	19.27
Hma	8.25	15.80	22.47
Hmpp	11.87	22.54	32.05
Hdpp	11.91	22.83	32.25
Hmepp	11.75	22.52	32.17
Hmhpp	11.51	22.49	31.71
Hppp	11.36	21.78	30.7

complex. The hydroxy oxygen atoms become harder. It has been found that Hdpp, under physiological conditions (pH 7.4, 100 μM citrate, 50 μM transferrin vacant sites), is more effective at complexing aluminum than EDTA, maltol or catechol.[105]

3.4. Schiff Base Complexes of Aluminum

Group 13 (IIIA) Schiff base complexes are surprisingly rare when compared to the large number of complexes that are known for transition metals. Most of the known aqueous aluminum Schiff base chemistry has resulted from mechanistic studies of the decarboxylation, transamination, and dephosphonylation of α-amino acids or α-keto acids with pyridoxal (and analogs) pertinent to the chemistry of vitamin B_6.[53,106–114] It is thought that Al^{3+} can serve as a template for the synthesis of Schiff bases by enhancing the reactivities of aldimine and ketimine forms and that Al–Schiff base complexes may be intermediate in vitamin B_6 catalysis.[108,114] The formed Schiff bases are easily hydrolyzed in water without the aid of aluminum(III). Some pertinent stability constants are given in Table 3.7.

The optimum pH value for the formation of a pyridoxal-amino acid Schiff base is 7–8. ^1H and ^{31}P NMR spectral studies have shown that there can be quite a variety of equilibrium species present depending on the pyridoxal and amino acid substituents, and the pH. Kinetic measurements[112] have shown that the most active species in transamination and dephosphonylation (when R = $CH_2PO_3^{2-}$) are those in which the carboxylate group of the amino acid moiety of the Schiff base is coordinated to the metal ion and when the phosphonate (R = $CH_2PO_3^{2-}$) is not.

Table 3.7. Stability constants of Al–Schiff base complexes $K = [AlL]/[Al][L]$ (25°C, 1*M*).

Ligand	Structure	log *K*
APP-PLP	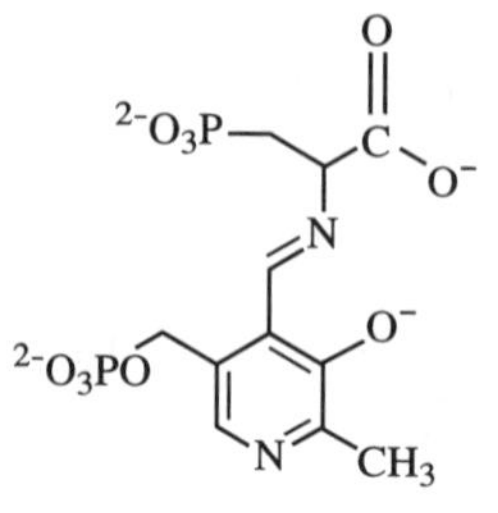	17.12[a]
DPL-3APP	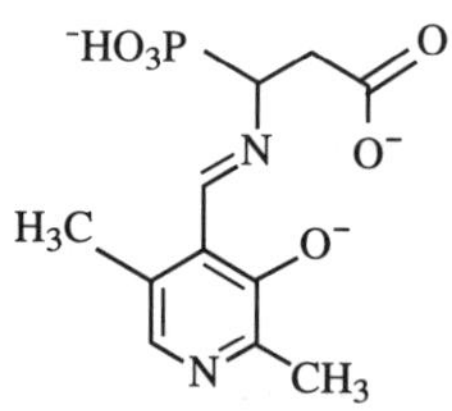	18.68[b]

[a] Ref. 111; [b] Ref. 53.

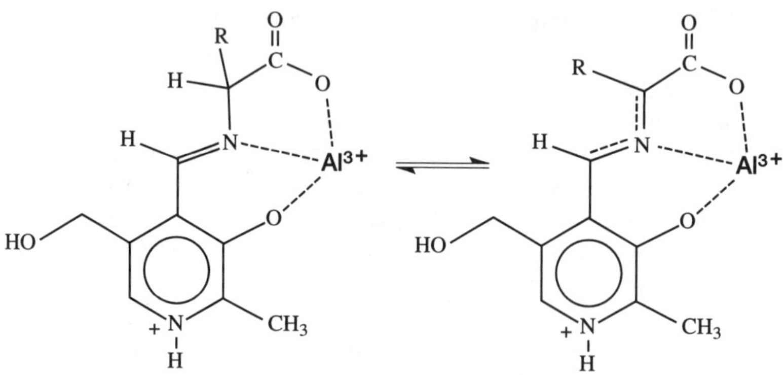

1:1 pyridoxylidene(aminoacido)aluminum(III) complex — the possible Al(III) chelate intermediate in transamination

It is generally held that Schiff base formation occurs in two distinct steps: (1) amine addition to a carbonyl (yielding an intermediate carbinolamine) and (2) dehydration to an imime.[115] In metal–vitamin B_6–amino acid systems, NMR has failed, with one exception,[111] to confirm the presence of a carbinoalamine intermediate. Martin[113] has postulated that Al^{3+} replacement of the azomethine proton deactivates the carbanion intermediate in the transamination.

Evans and Jakubovic[116,117] have studied less complex Schiff bases by ^1H NMR and found that harder Schiff bases form more stable Al(III) complexes. The pAl value (Table 3.8) (pAl = $-\log_{10}$ $[Al^{3+}]_{free}$) at pH 7.4 for type N_2O_4, N_3O_3, N_2O_4 decreased across the series despite the fact that the hexadentate ligands N_3O_3 and N_2O_4 benefit from the chelate effect at physiological pH. Clearly, the most important effect is the thermodynamic effect of coordinating more hard oxygen atoms to aluminum in N_2O_4 salsaps. It was also found that the pAl value is similar to those of related hydroxamic acid siderophore complexes and considerably greater than EDTA (pAl = 15.9)[118] and DTPA (pAl = 15.6),[119] so it may have application as a sequestering agent. Here pAl values are of more practical significance than normal stability constants since they express the effective binding strength and directly compare the efficiency of different ligands (with differing denticities) at a particular pH (most often 7.4).

We have thus far only considered the main two types of geometry around aqueous Al^{3+}, tetrahedral and octahedral. Complexes containing five-coordinate aluminum atoms in square pyramidal geometry are rare in aqueous systems and are more commonly found in non-aqueous systems. A most

Table 3.8. pAl Values for Schiff base complexes of Al in D_2O ($-\log_{10}[Al^{3+}]_{free}$ at pH 7.4).

Ligand	Structure	pAl
Salsaps(N_2O_4)	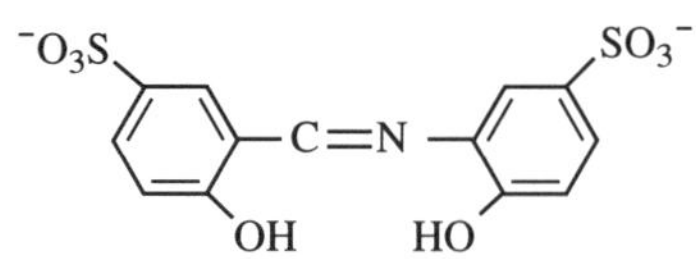	17.5[a]
Saltames(N_3O_3)	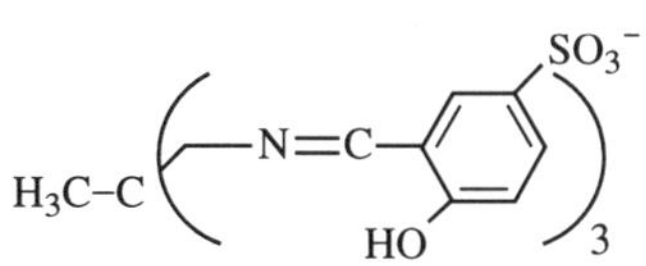	16.9[a]
Saltriens(N_4O_2)	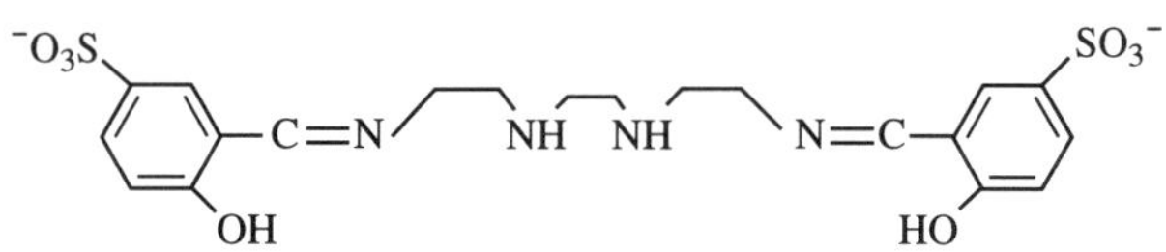	15.9[b]

[a] Ref. 117; [b] Ref. 116.

notable series of complexes has been isolated with Schiff bases[120,121] and the dibenzo [14]N_4 macrocycle $C_{22}H_{24}N_4$[122]. These complexes comprise a tetradentate ligand which occupies the base of the square pyramid and an ethyl group at the apex. Furthermore, these complexes are notable for the remarkable stability of the ethyl ligand as a function of the tetradentate ligand. Comparison of complexes of the tetradentate Schiff base ligands acen, salen, salpn, and salophen showed that the reactivity depends on the flexibility of the backbone connecting the salicylaldehyde chelates.[120] The EtAl(salophen) complex is the most rigid because it is totally conjugated and its hydrolysis is extremely slow. The EtAl complexes of the more flexible $salpn^{2-}$, $salen^{2-}$, and $acen^{2-}$ ligands are hydrolyzed quickly. The Schiff base–ethyl complexes are stable in the solid state but react with moisture or acid to hydrolyze the ethylaluminum linkage. The dibenzo[14]–ethyl complex is stable to alkyl hydrolysis even under recrystallization from water-containing solvents. These complexes may point the way to aluminum alkyls (which are usually highly pyrophoric) that are stable in aqueous solution.

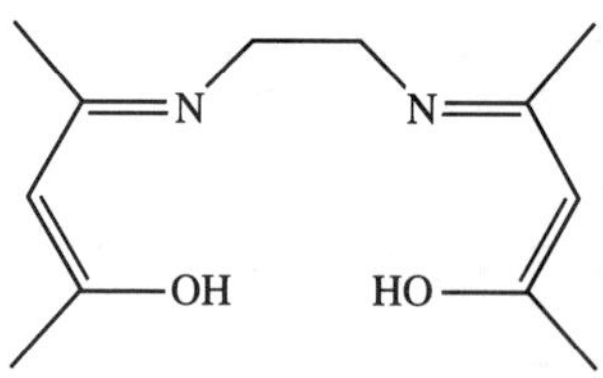

H_2acen

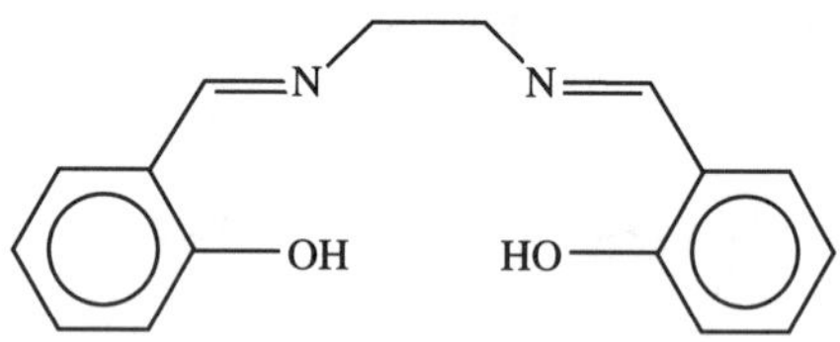

H_2salen

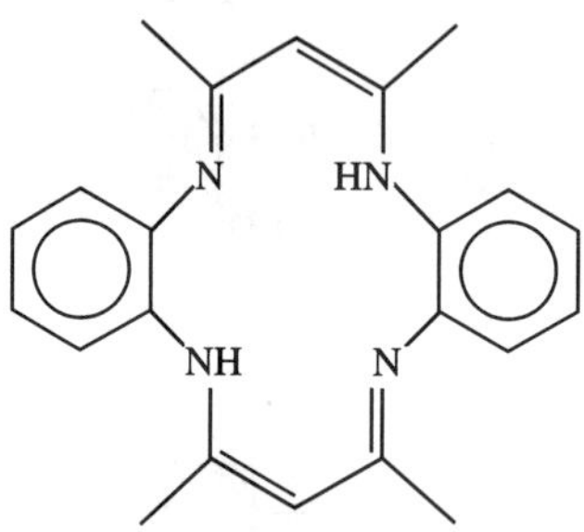

$C_{22}H_{24}N_4$

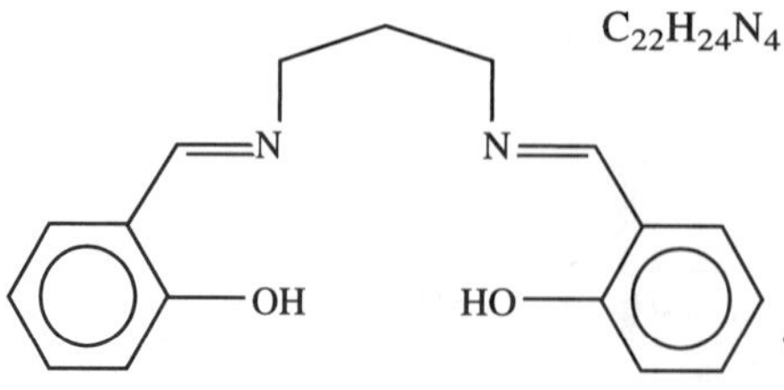

H_2salpn

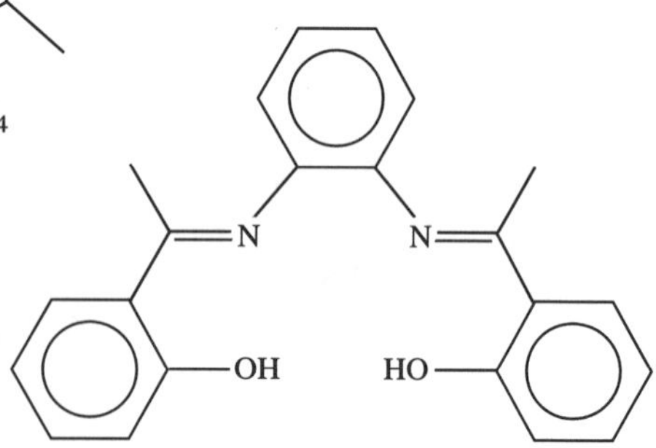

H_2salophen

3.5 Applications and Implications

The most important applications of aqueous aluminum c
in the 1990s arise from the ubiquity of the metal in Nature an
spread use in the food processing and metals industries.[123] With the eve
increasing acidity of our environment and the discovery of various diseases which involve aluminum (or cause its accumulation), a better knowledge of its aqueous chemistry is imperative. The great similarity between aluminum(III) and ferric ion (an element essential for O_2 transport and electron transport) has been noted and discussed in detail by Crumbliss and Garrison.[39] There are a number of books on the bioinorganic and environmental aspects of Al chemistry.[124–128]

The chemistry of aluminum under hydrolysis conditions (moderate acidity) and in the presence of naturally chelating ligands (such as phosphate[74–80]) has become of increasing importance in recent years.[129] The concentration of aluminum in fresh waters is generally below 0.02 mg/l,[130] but it can be greatly elevated if the pH drops to below 5.[131] A marked increase is caused by acid rain[131–134] which releases aluminum from mineral stores and introduces it into the aquatic environment where it can move up the food chain.[135] The identified aluminum-containing species in acidified surface waters are complexes of fluoride, citrate, oxalate, and sulfate.[9] It is well-known that Al-binding ligands, such as aliphatic organic acids, phenols, and hydroxamate siderophores, are a concern in microbial environments (such as soils), in addition to those already mentioned. Much interest is currently focused on the poorly characterized, but extremely relevant, fulvic and humic acids as soil-borne Al-chelating agents,[136] these compounds are believed to contain phenol and aliphatic organic acid groups.

Al/sulfonated-phthalocyanine complexes are currently of great interest in photochemotherapy, combining the water-solubility of sulfonate groups with the photoactivity of the normally insoluble Al-phthalocyanines.[137] Work on the elimination of Al toxicity to fish in acidified waters by co-ligands such as silicate has shown that ternary complexes can play an important role in the ecosystem and should generate some coordination chemistry of ternary complexes of Al.[138,139]

North Americans ingest approximately 36 mg/day of aluminum, but this number has large standard deviations.[130] The total body burden of aluminum in a normal 70 kg individual is 20–50 mg and the aluminum concentrations in serum and in tissue, with the exception of lung, are quite low, about 0.005 and 1.5 mg/l, respectively. Generally, aluminum plays no biological role; it has never been shown to be essential for any biological process. Elevated levels of aluminum, however, are associated with several neurological and osteological disorders. The most common disease associated with aluminum is Alzheimer's disease.[96,140] Elevated levels of the metal are often found in the brain of patients afflicted with this disease. A high serum aluminum concentration (0.55 mg/l) is found in kidney dialysis patients unless great care is taken to clude the metal from the dialysis water. These patients tend to accumulate Al and, in order to control urinary phosphate and oxalate, they are often given Al-containing antacids as part of their treatment. In recent years concern about aluminum has extended to its concentrations in acidified waters (and toxicity to fish), tea, infant formula, antacids, and to its leaching from cooking utenisls and juice containers.[138,141,142] Indeed, a correlation has been discovered between aluminum concentrations in drinking water and the incidence of Alzheimer's disease.[143] Although very little aluminum is absorbed in healthy humans and the main route of its elimination is via the feces, there is considerable evidence that aluminum can enter the circulation when appropriately complexed in the stomach. The discovery that aluminum uptake in rats and humans is mediated by citrate[144–146] has meant that the aqueous coordination chemistry of Al has taken on a new importance and relevance. Particularly, a correct view of aqueous Al/citrate coordination chemistry in dilute solution has become of tantamount importance. In the 1980s, the aqueous coordination chemistry of Al/citrate complexes was extensively investigated;[5,65,67,84-87,91,92,94,147,148] it appears that the neutral AlL species may be responsible. From aqueous coordination chemistry has come the recommendation that aluminum-containing antacids and citrus fruits or juices should not be taken together.[2,84]

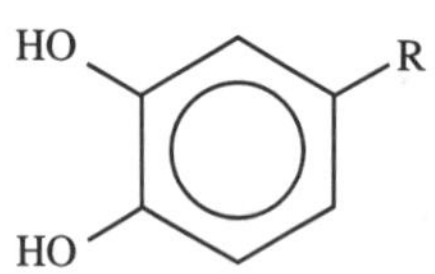

R =	$CH_2CH_2NH_2$	dopamine
	$CH(OH)CH_2NHCH_3$	adrenaline (epinephrine)
	$CH(OH)CH_2NH_2$	noradrenaline (norepinephrine)

In brain tissue, concentrations of citrate (2-200μM) and transferrin are very low. Other small molecules become possible Al^{3+} ion ligands in brain and other tissue. Catechols have been of interest as Al^{3+} chelators for decades.[66,149–153] The catecholamines are neurotransmitters and have been demonstrated to perfuse into tissue, and to bind aluminum strongly. Potentiometric studies have shown that in acidic solution Al^{3+} complexation to catecholamines occurs at the diphenolato catecholate moiety, with the peripheral ammonium group remaining protonated; conditional stability constants for the first complex formation vary from $10^{7.66}$ for catechol to $10^{10.4}$ for tiron (the strongest chelate among them).[149] These conditional constants have the advantage over normal stability constants in that they are independent of conditions and are directly comparable among ligands because pH is specified thereby correcting for proton competition.[2] This work indicates that large excesses of catecholamines would be required to chelate Al^{3+} in the presence of citrate or ATP. Work on aluminum complexation by 2,3-diphosphoglycerate (a possible carrier molecule) has shown that it binds strongly enough for there to be comparable concentrations of Al^{3+} on both sides of a red cell membrane and that the main binder of Al in blood might be ATP.[154]

The binding of Al^{3+} by nucleosides and nucleotides has been of considerable interest in the last decade because of the increasing importance of Al bioinorganic chemistry.[79,84,155–161] Although it has been suggested that Al can bind to both base nitrogens and phosphate oxygens, quantitative work shows the binding to be almost completely to the latter, in agreement with the general trends for ligand binding obvious from the multitude of stability constant determinations.[47] In particular, it appears that ADP and ATP can chelate Al^{3+} to form complexes of greater stability than nucleoside phosphates with only one terminal phosphate such as AMP.[157,160]

The transport chemistry of aluminum is now of considerable concern. After adsorption into the blood it is most likely that Al^{3+} is bound to the ferric transport protein transferrin.[86,162–164] There are significant differences between the two tripositive ions Fe^{3+} and Al^{3+}; however, transferrin has a high affinity for both. Conditional stability constants for the binding of these ions to the two sites of transferrin under blood plasma conditions are $K_1 = 10^{22.7}$ and $K_2 = 10^{22.1}$ for Fe^{3+},[164] and $K_1 = 10^{13.5}$ amd $K_2 = 10^{12.5}$ for Al^{3+}.[163] Ferric ion reacts faster than Al^{3+}; the latter proceeds via a dissociative mechanism and the former via an association mechanism.[39] Comparison of the stability constant of Al^{3+} and Fe^{3+} for ligands of relevance to plasma binding show that the metal-ligand interactions are stronger for ferric ion than for aluminum(III) with all ligands except F^-. Harris and Sheldon[163] and Clevette and Orvig[105] found linear energy relationships relating the thermodynamics of complexation of Al^{3+} and Fe^{3+} with O and N,O ligands and these showed clearly that aluminum complexes are generally weaker than ferric complexes by about 14–20% for O ligands and nearly 40% for N,O ligands.

Since both thermodynamic and kinetic data show that ferric ion is much more thermodynamically stable and kinetically faster than Al^{3+} there should be no competition between the two metal ions for binding sites on the protein and Al^{3+} cannot displace ferric ion. In plasma, however, transferrin is only 30% saturated with iron, so vacant sites can be filled by Al^{3+}.

Dialysis encephalopathy (Al concentration in kidney dialysis patients) and osteomalacia (a bone accumulation of Al) have both been treated using a variety of chelating agents. DFB, NTA, EDTA, HBED, and tiron have all been tested[163,165,166] and DFB has proven to be the best chelating agent for the removal of aluminum from serum proteins. NTA, EDTA, and HBED are all strong non-discriminating chelating agents; they remove not only aluminum, but other trace elements such as Fe, Zn, and Ca as well. Tiron is also good, but not as effective as DFB; the latter is now finding use in the hospital clinic and is the subject of intense scrutiny in the medical community.[162,167–169] DFB has the highest stability constant in Table 3.3 for Al^{3+} and it is also used for Fe^{3+} overload; it presumably chelates a lot of iron while removing aluminum.

This medical interest is a direct result of the aqueous coordination chemistry research which has been undertaken in recent years. It should be obvious from this brief survey that the aqueous coordination chemistry of aluminum will continue to attract considerable interest in the future and that its application to biomedical and environmental problems will persist as both a justification and a nucleus for study. In addition, fundamental research into aluminum chemistry and its applications will serve as synergistic forces for the continuing study of aluminum in water.

Acknowledgments

I am indebted to the Natural Sciences and Engineering Research Council of Canada for their continuing support of our aluminum research, and to Li-Wei Yang, Jane Webb, and Trang Vu for their great help in the preparation of this chapter.

References

1. *Polyhedron Symposium-in-Print Number 10;* Barron, A. R., Ed.; 1990; Vol.9.
2. Martin, R.B. *Metal Ions Biol. Syst.* **1988**, *24*, 1.
3. Jackson, J. A.; Lemons, J. F.; Taube, H. *J. Chem. Phys.* **1960**, *32*, 553.
4. Baldwin, H. W.; Taube, H. *J. Chem. Phys.* **1960**, *33*, 206.
5. Connick, R. E.; Fiat, D. N. *J. Chem. Phys.* **1963**, *39*, 1329.
6. Fiat, D.; Connick, R. E. *J. Am. Chem. Soc.* **1968**, *90*, 608.
7. Baes, C. F. Jr.; Mesmer, R. E. *The Hydrolysis of Cations;* John Wiley: New York, 1976: pp 112–123.
8. Delpuech, J. J. In *NMR of Newly Accessible Nuclei;* Laszlo, P. Ed.; Academic: New York, 1983; Vol.2, pp 153–195.
9. Bertsch, P.M.; Anderson, M. A. *Anal. Chem.* **1989**, *62*, 535.
10. Johansson, G. *Acta Chem. Scand.* **1960**, *14*, 771.
11. Johansson, G. *Acta Chem. Scand.* **1962**, *16,* 403.
12. Bertsch, P. M.; Thomas, G. W.; Barnhisel, R. I. *Soil Sci. Soc. Am. J.* **1986**, *50*, 825.
13. Bottero, J. Y.; Cases, J. M.; Flessinger, F.; Poirier, J. E. *J. Chem. Phys.* **1980,** *84*, 2933.
14. Bradley, S. M.; Kydd, R. A.; Yamdagni, R. *Magn. Reson. Chem.* **1990**, *28*, 746.
15. Bradley, S. M.; Kydd, R. A.; Yamdagni, R. *J Chem. Soc., Dalton Trans.* **1990**, 2653.
16. Brown, P. L.; Sylva, R. N.; Batley, G. E.; Ellis, J. *J. Chem. Soc., Dalton Trans.* **1985**, 1967.
17. Öhman, L.-O.; Forsling, W. *Acta Chem. Scand.* **1981**, *A35*, 795.
18. Thompson, A. R.; Kunwar, A. C.; Gutowsky, H. S.; Oldfield, E. *J. Chem. Soc., Dalton Trans.* **1987**, 2317.
19. Venturini, M.; Berthon, G. *J. Chem. Soc., Dalton Trans.* **1987**, 1145.

20. Akitt, J. W.; Farthing, A. *J. Magn. Reson.* **1978**, *32*, 345.

21. Akitt, J. W.; Mann, B. E. *J. Magn. Reson.* **1981**, *44*, 584.

22. Akitt, J. W.; Farthing, A. *J. Chem. Soc., Dalton Trans.* **1981**, 1606.

23. Akitt, J. W.; Farthing, A.; Howarth, O. W. *J. Chem. Soc.* **1981**, 1609.

24. Akitt, J. W.; Farthing, A. *J. Chem. Soc., Dalton Trans.* **1981**, 1617.

25. Akitt, J. W.; Farthing, A. *J. Chem. Soc., Dalton Trans.* **1981**, 1624.

26. Akitt, J. W.; Milic, N.B. *J. Chem. Soc., Dalton Trans.* **1984**, 981.

27. Akitt, J. W.; Elders, J. M. *J. Chem. Soc., Faraday Trans.* **1985**, *81*, 1923.

28. Akitt, J. W.; Elders, J. M. *J. Chem. Soc., Dalton Trans.* **1988**, 1347.

29. Akitt, J. W.; Elders, J. M.; Fontaine, X. L. R.; Kundu, A. K. *J. Chem. Soc., Dalton Trans.* **1989**, 1889.

30. Akitt, J. W.; Elders, J. M.; Fontaine, X. L. R.; Kundu, A. K. *J. Chem. Soc., Dalton Trans.* **1989**, 1897.

31. Bertsch, P. M. In *The Environmental Chemistry of Aluminum;* Sposito, G. Ed.; CRC: Boca Raton, 1989; pp 87–115.

32. Gennaro, M. C.; Mirti, P.; Casalino, C. *Polyhedron* **1983**, *2*, 13.

33. Evans, D. F.; Iki, N. *J. Chem. Soc., Dalton Trans.* **1990**, 3773.

34. Langford, C. H.; Gray, H. B. *Ligand Substitution Processes;* W. A. Benjamin: Reading, MA, 1986; pp 96–98.

35. Hugi-Cleary, D.; Helm, L.; Merbach, A. E. *Helv. Chim. Acta* **1985**, *68*, 545.

36. Burgess, J. *Metal Ions in Solution*; Ellis Horwood: Chichester, 1978.

37. Garrison, J. M.; Crumbliss, A. L. *Inorg. Chem.* **1987**, *26*, 3660.

38. Garrison, J. M.; Crumbliss, A. L. *Inorg. Chim. Acta* **1987**, *138*, 61.

39. Crumbliss, A. L.; Garrison, J. M. *Comments Inorg. Chem.* **1988**, *8*, 1.

40. Saito, K.; Nagasawa, A. *Polyhedron* **1990**, *9*, 215.

41. Eigen, M.; Tamm, K. *Z. Elektrochem.* **1962**, *66*, 107.

42. Eigen, M.; Wilkins, R. G. In *Mechanisms of Inorganic Reactions*; Gould, R. F. Ed.; American Chemical Society: Washington, DC, 1965; Vol. 49, pp 55–80.

43. Fuoss, R. M. *J. Am. Chem. Soc.* **1958**, *80*, 5059.

44. Driscoll, C. T.; Schecher, W. D. In *Aluminum and Health: A Critical Review*: Gitelman, H. J. Ed.; Marcel Dekker: New York, 1989; pp 27–65.

45. Taylor, M. J. In *Comprehensive Coordination Chemistry;* Wilkinson, G.; Gillard, R. D.; McCleverty, J. A. Ed.; Pergamon: Oxford, 1987; Vol 3, pp 106–126.

46. Nordstrom, D. K.; May, H. M. In *The Environmental Chemistry of Aluminum;* Sposito, G. Ed.; CRC: Boca Raton, 1989; pp 29–53.

47. Martell, A. E.; Smith, R. M. *Critical Stability Constants:* Plenum: New York, 1974–1989; Vol. 1–6.

48. Yoshida, I.; Murase, I.; Motekaitis, R. J.; Martell, A. E. *Can. J. Chem.* **1983**, *61*, 2740.

49. Motekaitis, R. J.; Martell, A. E. *Inorg. Chem.* **1984**, *23*, 18.

50. Martell, A. E.; Motekaitis, R. J.; Smith, R. M. *Polyhedron* **1990**, *9*, 171.

51. Motekaitis, R. J.; Martell, A. E. *Inorg. Chem.* **1989**, *28*, 3499.

52. Evers, A.; Hancock, R. D.; Martell, A. E.; Motekaitis, R. J. *Inorg. Chem.* **1989**, *28*, 2189.

53. Martell, A. E.; Szpoganicz, B. *Inorg. Chem.* **1989**, *28*, 4199.

54. Hedlund, T.; Sjöberg, S.; Öhman, L.-O. *Acta Chem. Scand.* **1987**, *A41*, 197.

55. Hedlund, T.; Bilinski, H.; Horvath, L.; Ingri, N.; Sjöberg, S. *Inorg. Chem.* **1988**, *27*, 1370.

56. Marklund, E.; Sjöberg, S.; Öhman, L.-O. *Acta Chem. Scand.* **1986**, *A40*, 367.

57. Marklund, E.; Öhman, L.-O.; Sjöberg, S. *Acta Chem. Scand.* **1989**, *43*, 641.

58. Marklund, E.; Öhman, L.-O. *J. Chem. Soc., Dalton Trans.* **1990**, 755.

59. Marklund, E.; Öhman, L.-O. *Acta Chem. Scand.* **1990**, *44*, 228.

60. Marklund, E.; Öhman, L.-O. *Acta Chem. Scand.* **1990**, *44*, 353.

61. Öhman, L.-O.; Sjöberg, S. *Acta Chem. Scand.* **1981**, *A35*, 201.

62. Öhman, L.-O.; Sjöberg, S. *Acta Chem. Scand.* **1982**, *A36*, 47.

63. Öhman, L.-O.; Sjöberg, S.; Ingri, N. *Acta Chem. Scand.* **1983**, *A37*, 561.

64. Öhman, L.-O.; Sjöberg, S. *Acta Chem. Scand.* **1983**, *A37*, 875.

65. Öhman, L.-O.; Sjöberg, S. *J. Chem. Soc., Dalton Trans.* **1983**, 2513.

66. Öhman, L.-O.; Sjöberg, S. *Polyhedron* **1983**, *2*, 1329.

67. Öhman, L.-O. *Inorg. Chem.* **1988**, *27,* 2565.

68. Öhman, L.-O. *Inorg. Chem.* **1989**, *28*, 3629.

69. Öhman, L.-O. *Polyhedron*, **1990**, *9*, 199.

70. Öhman, L.-O. *Acta Chem. Scand.* **1990**, *44,* 793.

71. Öhman, L.-O. *Acta Chem. Scand.* **1991**, *45*, 258.

72. Sjöberg, S.; Öhman, L.-O. *J. Chem. Soc., Dalton Trans.* **1985**, 2665.

73. Hedlund, T.; Öhman, L.-O. *Acta Chem. Scand.* **1988**, *A42*, 702.

74. Arp, P. A.; Meyer, W. L. *Can. J. Chem.* **1985**, *63*, 3357.

75. Dayde, S.; Filella, M.; Berthon, G. *J. Inorg. Biochem.* **1990**, *38*, 241.

76. Hsu, P. H. *Adv. Chem.* **1968**, *73*, 115.

77. Jackson, G.E.; Voyi, K. V. V. *S. Afr. J. Chem.* **1988**, *41*, 17.

78. Karlik, S. J.; Elgavish, G. A.; Pillai, R. P.; Eichhorn, G. L. *J. Magn. Reson.* **1982**, *49*, 164.

79. Karlik, S. J.; Elgavish, G. A.; Eichhorn, G. L. *J. Am. Chem. Soc.* **1983**, *105*, 602.

80. Wilson, M. A.; Collin, P. J.; Akitt, J. W. *Anal. Chem.* **1989**, *61,* 1253.

81. Akitt, J. W. *Prog. NMR Spectros.* **1989**, *21*, 1.

82. Finnegan, M.; Lutz, T. G.; Nelson, W. O.; Smith, A.; Orvig, C. *Inorg. Chem.* **1987**, *26*, 2171.

83. Kinrade, S. D.; Swaddle, T. W. *Inorg. Chem.* **1989**, *28*, 1952.

84. Martin, R. B. *J. Inorg. Biochem.* **1986**, *28*, 181.

85. Molitoris, B. A.; Froment, D. A.; MacKenzie, T. A.; Huffer, W. H.; Alfrey, A. C. *Kidney, Int.* **1989**, *36*, 949.

86. van Ginkel, M. F.; van der Voet, G. B.; van Eijk, H. G.; de Wolff, F. A. *J. Clin. Chem. Clin. Biochem.* **1990**, *28*, 459.

87. Venturini, M.; Berthon, G. *J. Inorg. Biochem.* **1989**, *37,* 69.

88. Karlik, S. J.; Tarien, E.; Elgavish, G. A.; Eichhorn, G. L. *Inorg. Chem* **1983**, *22*, 525.

89. Bombi, G. G.; Corain, B.; Sheikh-Osman, A. A.; Valle, G. C. *Inorg. Chim. Acta* **1990**, *171*, 79.

90. Venema, F. R.; Peters, J. A.; van Bekkum, H. *J. Chem. Soc., Dalton Trans.* **1990**, 2137.

91. Jackson, G. E. *S. Afr. J. Chem.* **1982**, *35*, 89.

92. Gregor, J. E.; Powell, H. K. J. *Aust. J. Chem.* **1986**, *39*, 1851.

93. Greenaway, F. T. *Inorg. Chim. Acta* **1986**, *116*, L21.

94. Feng, T. L.; Gurian, P. L.; Healy, M. D.; Barron, A. R. *Inorg. Chem.* **1990**, *29*, 408.

95. Valle, G. C.; Bombi, G. G.; Corain, B.; Favarato, M.; Zatta, P. *J. Chem. Soc., Dalton Trans.* **1989**, 1513.

96. Crapper McLachlan, D. R. *Neurobiol. Aging* **1986**, *7*, 525–532 and following commentaries.

97. Nelson, W. O.; Karpishin, T. B.; Rettig, S. J.; Orvig, C. *Inorg. Chem.* **1988**, *27*, 1045.

98. Clevette, D. J.; Nelson, W. O.; Nordin, A.; Orvig, C.; Sjöberg, S. *Inorg. Chem.* **1989**, *28*, 2079.

99. Nelson, W. O.; Rettig, S. J.; Orvig, C. *Inorg. Chem.* **1989**, *28*, 3153.

100. Zhang, Z.; Rettig, S. J.; Orvig, C. *Inorg. Chem.* **1991**, *30*, 509.

101. Llinas, M.; De Marco, A. *J. Am. Chem. Soc.* **1980**, *102*, 2226.

102. Matsuba, C. A.; Rettig, S. J.; Orvig, C. *Can. J. Chem.* **1988**, *66*, 1809.

103. Jaber, M.; Bertin, F.; Thomas-David, G. *Can. J. Chem.* **1977**, *55*, 3689.

104. Lutz, T. G.; Clevette, D. J.; Rettig, S. J.; Orvig, C. *Inorg. Chem.* **1989**, *28*, 715.

105. Clevette, D. J.; Orvig, C. *Polyhedron* **1990**, *9*, 151.

106. Gansow, O. A.; Holm, R. H. *J. Am. Chem. Soc.* **1969**, *91*, 573.

107. Gansow, O. A.; Holm, R. H. *J. Am. Chem. Soc.* **1969**, *91*, 5984.

108. Abbott, E. H.; Martell, A. E. *J. Am. Chem. Soc.* **1973**, *95*, 5014.

109. Martell, A. E.; Taylor, P. *Inorg. Chem.* **1984**, *23*, 2734.

110. Martell, A. E.; Langohr, M. F.; Tatsumoto, K. *Inorg. Chim. Acta* **1985**, *108*, 105.

111. Szpoganicz, B.; Martell, A. E. *Inorg. Chem.* **1985**, *24*, 2414.

112. Szpoganicz, B.; Martell, A. E. *Inorg. Chem.* **1986**, *25*, 327.

113. Martin, R. B. *Inorg. Chem.* **1987**, *26*, 2197.

114. Taylor, P. A.; Martell, A. E. *Inorg. Chim. Acta* **1988**, *152*, 181–188.

115. Jencks, W. P. *Catalysis in Chemistry and Enzymology*; McGraw-Hill: New York, 1969.

116. Evans, D. F.; Jakubovic, D. A. *Polyhedron* **1988**, *7*, 1881.

117. Evans, D. F.; Jakubovic, D. A. *Polyhedron* **1988**, *7*, 2723.

118. Schwarzenbach, G.; Gutt, R.; Anderegg, G. *Helv. Chim. Acta* **1954**, *37*, 937.

119. Aikens, D. A.; Bahbah, F. J. *Anal. Chem.* **1967**, *39*, 646.

120. Dzugan, S. J.; Goedken, V. L. *Inorg. Chem.* **1986**, *25*, 2858.

121. Gurian, P. L.; Cheatham, L. K.; Ziller, J. W.; Barron, A. R. *J. Chem. Soc., Dalton Trans.* **1991**, 1449.

122. Goedken, V. L.; Ito, H.; Ito, T. *J. Chem. Soc., Chem. Commun.* **1984**, 1453.

123. Ganrot, P. O. *Environ. Health Perspect.* **1986**, *65*, 363.

124. *Aluminum and Its Role in Biology*; Sigel, H.; Sigel, A., Eds.; Marcel Dekker: New York, 1988; Vol. 24.

125. *Aluminum in Food and the Environment*; Massey, R. C.; Taylor, D., Eds.; Royal Society of Chemistry: Cambridge, 1989.

126. *Environmental Chemistry and Toxicology of Aluminum*; Lewis, T. E., Ed.; Lewis: Chelsea, MI, 1989.

127. *Aluminum in Health: A Critical Review;* Gitelman, H. J., Ed.; Marcel Dekker: New York, 1989.

128. *The Environment Chemistry of Aluminum;* Sposito, G., Ed.; CRC: Boca Raton, FL, 1989.

129. Martin, R. B. *Clin. Chem.* **1986**, *32*, 1797.

130. Havas, M.; Jaworski, J. F. *Aluminum in the Canadian Environment*; National Research Council of Canada: Ottawa, 1986; Vol. NRCC 24759.

131. Driscoll, C. T. *Environ. Health Perspect.* **1985**, *63*, 93.

132. Tam, S. C.; Williams, R. J. P. *J. Inorg. Biochem.* **1986**, *26*, 35.

133. Campbell, P. G. C.; Bisson, M.; Bougie, R.; Tessier, A.; Villeneuve, J. -P. *Anal. Chem.* **1983**, *55*, 2246.

134. Campbell, P. G. C.; Tessier, A. *Adv. Chem.* **1987**, *216*, 185.

135. Sorenson, J. R. J.; Campbell, I. R.; Tepper, L. B.; Lingg, R. D. *Environ. Health Perspect.* **1974**, *8*, 3.

136. Stevenson, F. J.; Vance, G. F. In *The Environmental Chemistry of Aluminum;* G. Sposito, Ed.; CRC: Boca Raton, Fl, 1989; pp 117–145.

137. Berg, K.; Bommer, J. C.; Moan, J. *Cancer Lett.* **1989**, *44*, 7.

138. Birchall, J. D.; Chappell, J. S. *Clin. Chem.* **1988**, *34*, 265.

139. Birchall, J. D.; Exley, C.; Chappell, J. S.; Phillips, M. J. *Nature* **1989**, *338*, 146.

140. Crapper McLachlan, D. R.; Lukiw, W. J.; Kruck, T. P. A. *Can. J. Neurol. Sci.* **1989**, *16*, 490.

141. Lione, A. *Food Chem. Toxicol.* **1983**, *21*, 103.

142. Birchall, J. D.; Chappel, J. S. *Lancet* **1989**, *i*, 953.

143. Martyn, C. N.; Osmond, C.; Edwardson, J. A.; Barker, D. J. P.; Harris, E. C.; Lacey, R. F. *Lancet* **1989**, *i*, 59.

144. Slanina, P.; Falkeborn, Y. Y.; Frech, W.; Cedergen, A. *Food. Chem. Toxicol.* **1984**, *22*, 391.

145. Slanina, P.; Frech, W.; Bernhardson A.; Cedergen, A.; Mattson, P. *Acta Pharmacol. Toxicol.* **1985**, *56*, 331.

146. Slanina, P.; Frech, W.; Ekstrom, L.-G; Loof, L.; Slorach, S.; Cedergren, A. *Clin. Chem.* **1986**, *32*, 539.

147. Lopez-Quintela, M. A.; Knoche, W.; Veith, J. *J. Chem. Soc., Faraday Trans.* **1984**, *80*, 2313.

148. Marques, H. M. *J. Inorg. Biochem.* **1991**, *41*, 187.

149. Kiss, T.; Sovago, I.; Martin, R.B. *J. Am. Chem. Soc.* **1989**, *111*, 3611.

150. Kennedy, J. A.; Powell, H. K. J. *Aust. J. Chem.* **1985**, *38*, 879.

151. Kennedy, J. A.; Powell, H. K. J. *Aust. J. Chem.* **1985**, *38*, 659.

152. McBride, M. B.; Sikora, F. J. *J. Inorg. Biochem.* **1990**, *39*, 247.

153. Weinland, R. F.; Denzel, W. *Ber.* **1914**, *47*, 737.

154. Sovago, I.; Kiss, T.; Martin, R. B. *Polyhedron* **1990**, *9*, 189.

155. Bock, J. L.; Ash, D. E. *J. Inorg. Biochem.* **1980**, *13*, 105.

156. Dryssen, D.; Haraldson, C.; Nyberg, E.; Wedborg, M. *J. Inorg. Biochem.* **1987**, *29*, 67.

157. Jackson, G. E.; Voyi, K. V. V. *Polyhedron* **1987**, *6*, 2095.

158. Karlik, S. J.; Eichhorn, G. L.; Lewis, P. N.; Crapper, D. R. *Biochemistry* **1980**, *19*, 5991.

159. Karlik, S. J.; Eichhorn, G. L. *J. Inorg. Biochem.* **1989**, *37*, 259.

160. Kiss, T.; Sovago, I.; Martin, R. B. *Inorg. Chem.* **1991**, *30*, 2130.

161. Laussac. J.-P.; Commenges, G. *Nouv. J. Chim.* **1983**, *7*, 579.

162. Alonsa, J. I. G.; Garcia, A. L.; Parajon, J. P.; Gonzalez, E. B.; Medel, A. S.; Cannata, J. B. *Clin. Chim. Acta* **1990**, *189*, 69.

163. Harris, W. R.; Sheldon, J. *Inorg. Chem.* **1990**, *29*, 119.

164. Martin, R. B.; Savory, J.; Brown, S.; Bertholf, R. L.; Wills, M. R. *Clin. Chem.* **1987**, *33*, 405.

165. Martell, A. E. *Biol. Trace Element Res.* **1989**, *21*, 295–303.

166. Domingo, J. L. *Clin. Toxicol.* **1989**, *27*, 355.

167. McCarthy, J. T.; Milliner, D. S.; Johnson, W. J. *Q. J. Med.* **1990**, *74 (New Ser.)*, 257.

168. Crapper McLachlan, D.R.; Dalton, A. J.; Kruck, T. P. A.; Bell, M. Y.; Smith, W. L.; Kalow, W.; Andrews, D. F. *Lancet* **1991**, *337*, 1304.

169. Yokel, R. A.; Datta, A. K.; Jackson, E. G. *J. Pharmacol. Exp. Ther.* **1991**, *257*, 100.

170. Nitzsche, R.; Hildebrandt, R.; Mizner, R.; Robisch, G. *Monatsh. Chem.* **1988**, *119*, 495.

171. Nitsche, R.; Mitzner, R.; Robisch, G. *Z. Phys. Chem.* (Leipzig) **1988**, *269*, 97.

172. Secco, F.; Venturini, M. *Inorg. Chem.* **1975**, *14*, 1978.

173. Perlmutter-Hayman, B.; Tapuhi, E. *Inorg. Chem.* **1979**, *18*, 875.

174. Plankey, B. J.; Patterson, H. H. *Environ. Sci. Technol.* **1987**, *21*, 595.

175. Perlmutter-Hayman, B.; Tapuhi, E. *Inorg. Chem.* **1977**, *16*, 2742.

176. Rauh, H.; Knoche, W. *Ber. Bunsenges. Phys. Chem.* **1979**, *83*, 518.

177. Plankey, B. J.; Patterson, H. H. *Inorg. Chem.* **1989**, *28*, 4331.

178. Dash, A. C. *Inorg. Chem.* **1983**, *22*, 837.

CHAPTER

4

Low Valent and Paramagnetic Compounds Of Aluminum

Andrew R. Barron

Department of Chemistry, Harvard University
Cambridge, Massachusetts 02138

4.1. Introduction

While the predominant oxidation state of the heavier Group 13 elements is +III, the possible lower oxidation states 0, +I, and +II are known, though the amount of information varies considerably from one element to another. It is well established that thallium(I) is the dominant oxidation state in both solution and the solid state,[1] and the chemistry of indium(I)[2–6] and gallium(I)[7–13] has been an expanding field in recent years. By contrast, the monovalent compounds of aluminum are rare, until recently they were known to exist only in the gas phase or isolated within frozen matrices. The situation with the +II oxidation state is slightly different. Divalent boron compounds are well known,[14] whereas a much smaller number of aluminum and thallium compounds have been prepared. The dihalides of gallium and indium have been the subject of much investigation,[15] and several of their Lewis base adducts have been characterized by X-ray crystallography.[16–21]

If compounds of low valent aluminum only constitute a small fraction of known compounds, why are they of interest? The answer to this question may be found in both traditional and more contemporary areas of organoaluminum chemistry.

Organoaluminum compounds are made almost exclusively by one of four general routes:[22] (a) metathesis between metal alkyls and aluminum compounds (e.g. Equation 4.1), (b) displacement between metal alkyls and aluminum metal (e.g. Equation 4.2), (c) hydroalumination of unsaturated hydrocarbons (e.g. Equations 4.3 and 4.4) and (d) dehalogenation of organoaluminum sesquihalides (e.g. Equations 4.5 and 4.6).

$$3\,RM + AlX_3 \longrightarrow AlR_3 + 3\,MX \qquad (4.1)$$

$$2\,Al + 3\,R_2Hg \longrightarrow 2\,AlR_3 + 3\,Hg \qquad (4.2)$$

$$2\,Al + 3\,H_2 + 6\,CH_2{=}CR_2 \longrightarrow 2\,Al(CH_2CHR_2)_3 \qquad (4.3)$$

$$2\,Al + 3\,H_2 + 4\,AlR_3 \longrightarrow 6\,Al(H)R_2 \xrightarrow{\text{alkene}} 6AlR_3 \qquad (4.4)$$

$$3\,RX + 2\,Al \longrightarrow R_2AlX + RAlX_2 \qquad (4.5)$$

$$3\,R_2AlX \xrightarrow[-NaCl]{Na} 2\,AlR_3 + Al \qquad (4.6)$$

Variants (c) and (d) are the most interest due to their use in the industrial synthesis of $AlEt_3$ and $AlMe_3$ respectively.[13–25] However, it is only in the first of these reactions that the formation of some form of low valent inermediate, most probably surface bound, is not proposed. The undoubted heterogeneous nature of these reactions has prompted researchers to study the reactions of organic molecules with either aluminum surfaces under UHV conditions,[26–28] or with isolated aluminum atoms and clusters in matrix isolation experiments. Since it is only these latter studies that have allowed for the direct spectroscopic observation of low valent organoaluminum compounds, the former is not included in this review.

Since the pioneering work by Mansevit[29] on the use of organometallic compounds for the preparation of thin films by what is now termed metal-organic chemical vapor deposition (MOCVD), there has been increased interest in the gas phase, and surface catalyzed, decomposition of organoaluminum compounds. The pyrolysis[80] and photolysis[31] of a number of organoaluminum compounds has been studied and low valent di- and monoalkyls are proposed as intermediates for either decomposition route (e.g. Equations 4.7–4.9).

$$AlR_3 \xrightarrow[-R]{\Delta} AlR_2 \xrightarrow[-R]{\Delta} AlR \qquad (4.7)$$

$$AlR_3 \xrightarrow[-R]{h\nu} AlR_2 \longrightarrow Al + R{-}R \qquad (4.8)$$

$$AlR_3 \xrightarrow[-2R]{h\nu} AlR \tag{4.9}$$

Over the last thirty years a number of compounds have been reported that appear to contain aluminum in the formal oxidation states of 0, +I, or +II. However, in each case detailed spectroscopic characterization has subsequently indicated otherwise. In most cases low valency was proposed on the basis of the paramagnetic nature of the compounds. It is of course a consequence ot the [Ne]3s°3p° configuration of trivalent aluminum that the vast majority of aluminum compounds are diamagnetic. However, interest in paramagnetic compounds has arisen from mechanistic studies on the reactions of coordinated ligands, in particular organic carbonyls, which have often been proposed to proceed via the formation of ligand centered radical intermediates.[32–36] Since paramagnetic aluminum compounds have historically (although often incorrectly) been linked to the presence of low valent aluminum centers, they are also included in this survey.

4.2. Aluminum–Carbonyl Complexes

The extensive chemistry of zero-valent transition metal carbonyl complexes is undoubtedly due to the π-acidic nature of carbon monoxide, that is its ability to accept electron density from electron-rich zero-valent metal centers via dπ-pπ back donation, **1**.

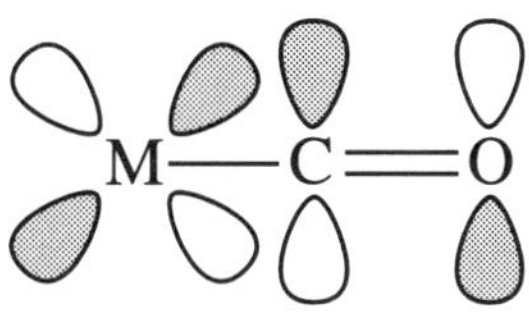

1

Although a wide variety of binary metal–carbonyls have been isolated, others have been identified only at low temperatures by the co-condensation of metal atoms and carbon monoxide in inert gas or hydrocarbon matrices. The isolable nature of the resulting complex has been ascribed to the presence of dπ-pπ back donation, an effect precluded from occurring for pre-transition metals such as aluminum. In view of this it is generally theorized that the only metals which form simple binary carbonyls are those in, or following the first transition metal series. However, aluminum carbonyls have also been generated and spectroscopically characterized.

Ogden and co-workers[37] reported that the co-condensation of aluminum atoms with carbon monoxide in a krypton matrix at 20 K leads to the formation of an aluminum–carbonyl complex. The IR spectra of the deposition products showed, in addition to free CO, two bands at 1988 and 1890 cm^{-1} (Figure 4.1a)

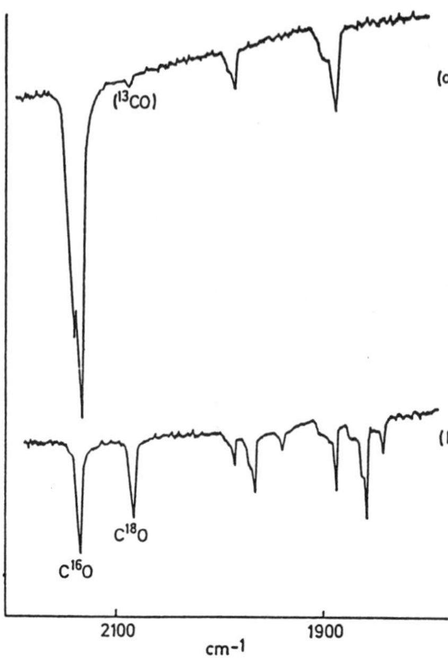

Figure 4.1. IR spectra obtained after condensation of aluminum atoms in a Kr matrix containing (a) 3% $C^{16}O$ and (b) 1.7% $C^{16}O$ and 1.3% $C^{18}O$. Reprinted with permission from ref. 37 Copyright 1972 Royal Society of Chemistry.

assigned to terminal carbonyl vibrations. Isotype studies using $C^{18}O$ gave rise to a triplet of peaks (Figure 4.1b) which was interpreted to be due to the symmetric and antisymmetric vibrations of a bent dicarbonyl moiety, $Al_x(CO)_2$. Based on these studies the number of carbonyl ligands involved could not be ascertained. From observation of the formation of Al_2 in Ar matrices, Ozin and McIntosh[38] suggested that the dicarbonyl should be formulated as $Al_2(CO)_2$, with structures **2** and **3**.

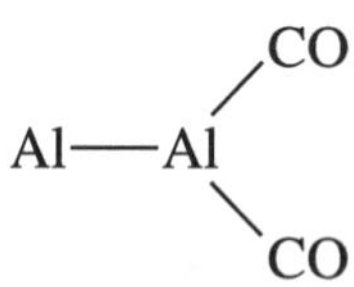

2

3

The correct formulation of $Al(CO)_2$ was unequivocally determined by solution ESR spectral studies of the products obtained from the co-condensation of Al atoms and isotopically labeled CO in solid argon at 4K, and in solid cyclohexane and adamantane at 77K, as well as in adamantane solution at 243K.[39,40] The spectrum at 243K in adamantane solution shown in Figure 4.2a, yielded the isotropic parameters for coupling to a single ^{27}Al ($I=5/2$) center,

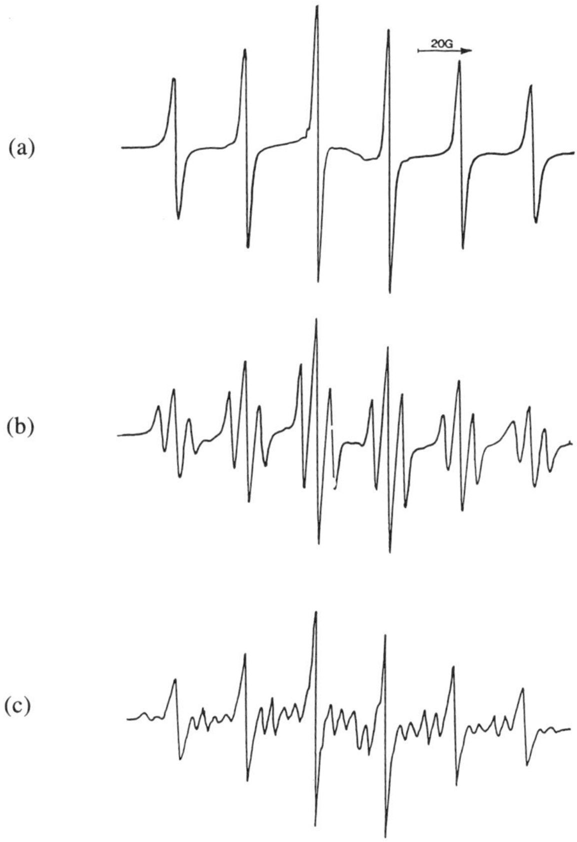

Figure 4.2. The solution ESR spectra of (a) $Al(CO)_2$ (b) $Al(^{13}CO)_2$ and (c) $Al(CO)_2/Al(C^{17}O)_2$ in adamantane at 243K. Reprinted with permission from ref. 40. Copyright 1986 American Chemical Society.

g=2.00, while a sextet of triplets (Figure 4.2b) was obtained with ^{13}CO.[40] These observations proved conclusively that the observed species has one aluminum atom and two carbon monoxide ligands. The use of ^{17}O-enriched CO (Figure 4.2c) allowed the observation of an ^{17}O hyperfine interaction, a first for any metal carbonyl.[40] Based on the frozen (4K and 77K) ESR spectra the bonding in $Al(CO)_2$ was proposed to involve a donation from the two CO ligands into two empty sp^2-hybridized orbitals on the aluminum atom, with its lone-pair electrons in the remaining sp^2 orbital, **4** and the unpaired electron in the remaining p orbital. Such a structure would allow the possibility of π-type back-bonding from the half-filled p_z orbital of Al into the vacant p_z^* orbitals of the two CO ligands, **5**.

The extent of this interaction would seem to be significant since the reduction in the IR(CO) stretching frequency as compared to free CO (2143 cm^{-1}) is comparable to that observed for the d^{10} anion $Co(CO)_4^-$, *ca.* 1890 cm^{-1}.

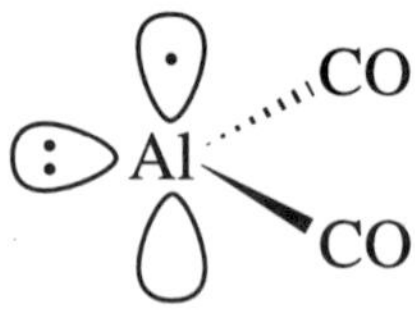

4

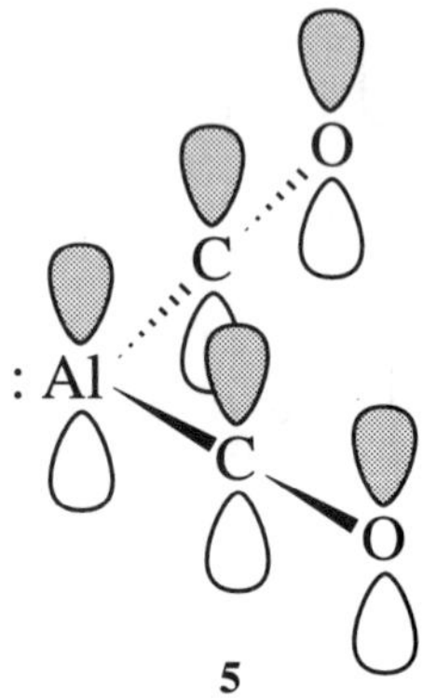

5

The intensity ratio of the symmetric and antisymmetric stretching bands gave the angle between the carbonyl ligands of 110 ± 5° at 77 K.[41] This angle is, within experimental error, the same as that obtained from a comparison of ESR spectra with INDO-MO calculations.[59] There is an apparent anomaly between this value and the C–Al–C angle of 74° found from *ab initio* calculations.[42] However, it should be noted that the IR analysis actually gives the angle between the two CO groups, not the two Al–C bonds. Thus, if the Al–C–O bonds are bent, this angle will differ from the C–Al–C angle. As shown in Figure 4.3, the calculated geometry for each Al–C–O angle does indeed have an outward 13.55° deviation from linearity.[42] The energy for this deformation was estimated to be *ca.* 9.2 kJ/mol. It should be noted that the metal–C–O bonds of many transition metal-carbonyl complexes are also known to be bent.[42,43] There are two factors that could contribute to the small C–Al–C angle in $Al(CO)_2$. First, since the distance between the two carbon atoms is only 2.5 Å, there is the possibility of C–C bonding, **6**, in an analogous manner to that proposed in $L_nM(CO)_2$ complexes.[45] Second, the small angle could result in a more favorable orbital overlap than a larger angle.

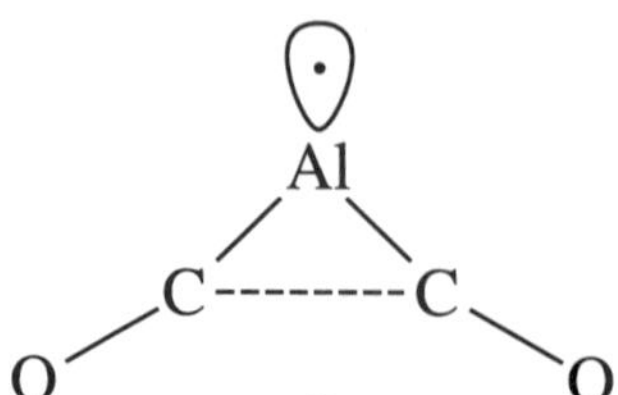

6

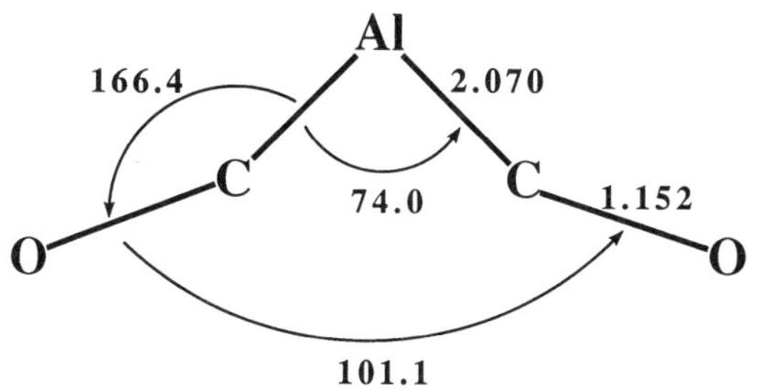

Figure 4.3. 6.311G* [C, O] 6-31G* (2d)[Al] optimized, from ref. 42, geometry for $Al(CO)_2$. Bond lengths are in angstroms, and bond angles in degrees.

Perhaps the most intriguing revelation of the experimental studies is that, whereas the aluminum dicarbonyl is readily formed, aluminum monocarbonyl is not. Aluminum monocarbonyl is a 13-electron system and following the Walsh rule,[46] would have a linear structure, and would thus be a $^2\Pi$ radical and possibly ESR silent. Despite this, no carbonyl bands assigned to Al(CO) have been detected in the IR spectrum. However, one high-resolution IR study[41] in adamantane at 77 K detected, in addition to the bands due to $Al(CO)_2$ (1985 and 1904 cm^{-1}), two broad bands at 2079 cm^{-1} and 1715 cm^{-1}. The former band was observed to grow in intensity on warming above 77 K. It was suggested that this effect results from the diamagnetic dimerization product $Al_2(CO)_4$, **7**. The low frequency band at 1715 cm^{-1} is in the range (1620–1730 cm^{-1}) expected for a triply bridging CO group,[47] and was assigned to Al_3CO, **8**, because of the ease with which Al atoms form Al_3.[48]

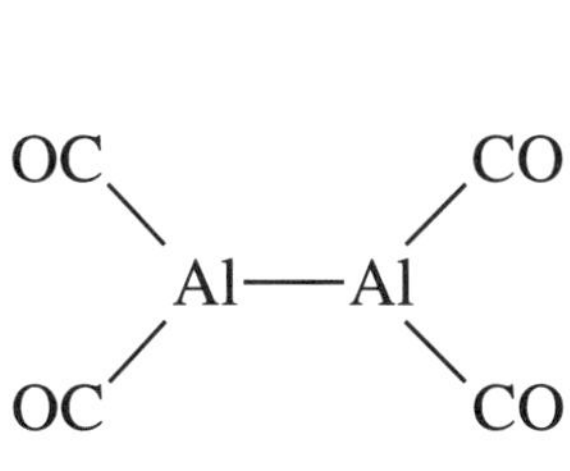

7

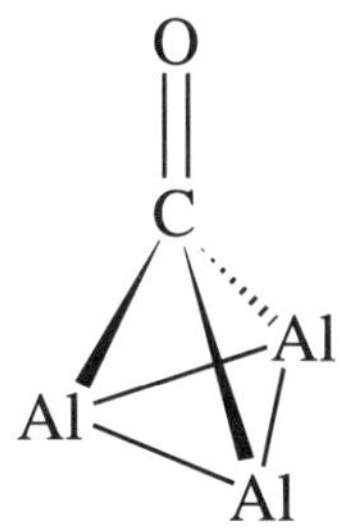

8

4.3. Unsaturated Hydrocarbons

4.3.1. Alkenes

Undoubtedly the most important class of π-complexes are those of compounds with carbon–carbon multiple bonds. The earliest known organotransition metal complex was discovered by Zeise in Copenhagen in about 1845, but the true structure was not determined until the 1950s. Zeise's salt, $K[Pt(C_2H_4)Cl_3]$, **9**, has the C–C axis of the ethene moiety perpendicular to the expected bond direction of the metal.

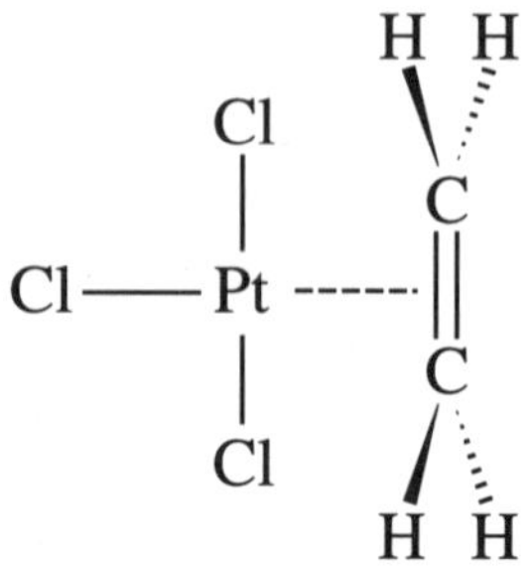

9

The most general description of the bonding was developed by Dewar,[49] Chatt and Duncanson[50] for transition metals. Like CO ligation, the bonding consists of (a) σ-donation from the alkene π orbital to an unfilled metal orbital (**10**), and (b) back bonding from a filled metal d-orbital to the π^* on the carbon atoms (**11**).

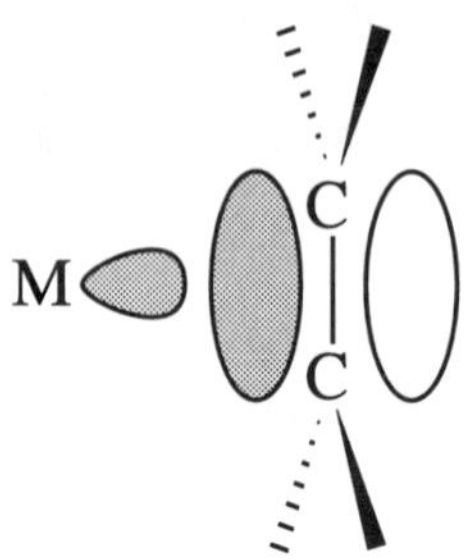

10

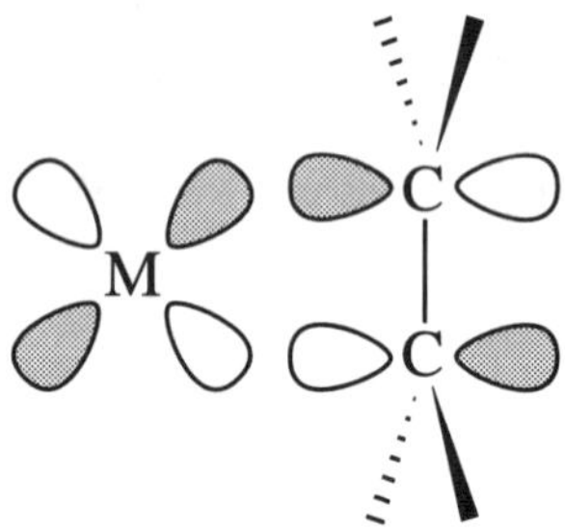

11

The direct reaction of aluminum atoms with alkenes was first reported by Skell and Wolf[51] to yield organoaluminum compounds. Based on the products of deuterium oxide hydrolysis they concluded that the initial product from the reaction was the addition of one aluminum to the alkene double bond; however, no attempt at isolation was made.

The first spectroscopic characterization of an aluminum–alkene adduct was reported by Kasai and McLeod[52] for the reaction of aluminum with ethene in a frozen neon matrix, and latter confirmed in argon[53] and hydrocarbon matrices.[54] When Al atoms were trapped together with ethene at 5K, the matrices appeared green, and an ESR signal centered at about g = 2.00 was observed (Figure 4.4).[52] The prominent sextet feature was attributed to the hyperfine

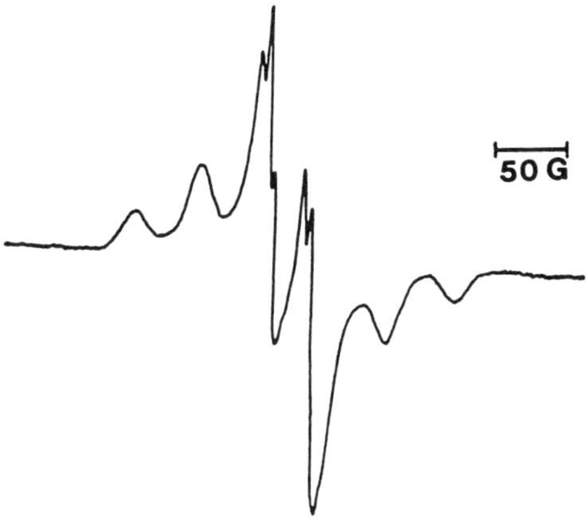

Figure 4.4. ESR spectrum of $Al(C_2H_4)$ in frozen Ne at 5K. Reprinted with permission from ref. 52. Copyright 1975 American Chemical Society.

interaction with the ^{27}Al nucleus, whereas spectra in adamantane at 270K (Figure 4.5)[54] confirmed the presence of hyperfine coupling to four equivalent protons. These results were proposed to be consistent with a π-type interaction between the aluminum and ethene, in which sp hybridization of aluminum allows for the formation of two dative bonds. One results from the migration of electrons from the bonding π-orbital to the vacant sp orbital of Al (**12**) whereas the other consitutes back donation from the semi-filled p_x orbital of Al into the empty ethene π^* orbital (**13**), analogous to the dπ-pπ back donation observed for transition metal complexes.

Despite the high stability of the π-complex as measured from warm-up experiments,[54] and direct determination of the binding energy in the gas phase by time-resolved resonance fluorescence experiments,[55] pseudopotential multi reference Cl calculations indicated that the interaction between the Al and ethene is van der Waals in nature, and predicted a binding energy of only 7.9

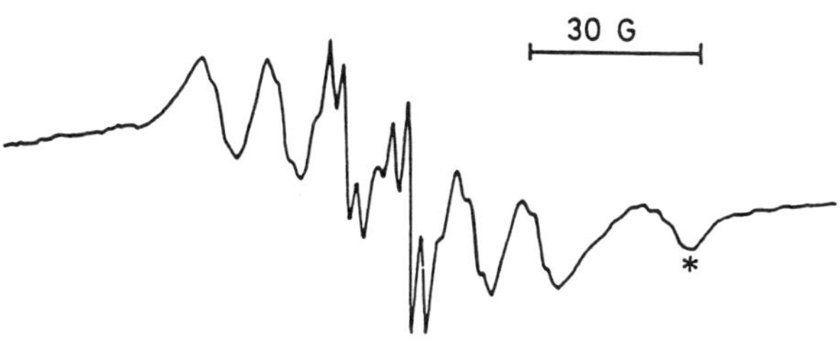

Figure 4.5. ESR spectrum of $Al(C_2H_4)$ in adamantane at 270K. The high-field line marked with an asterisk is due to $Al(CH_2CH_2CH_2CH_2)$. Reprinted with permission from ref. 54. Copyright 1987 Royal Society of Chemistry.

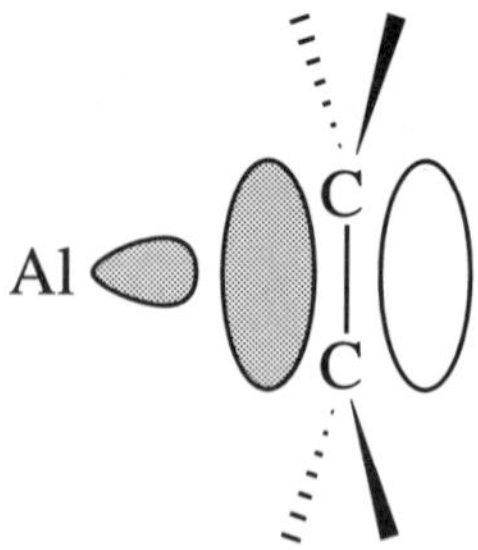

12

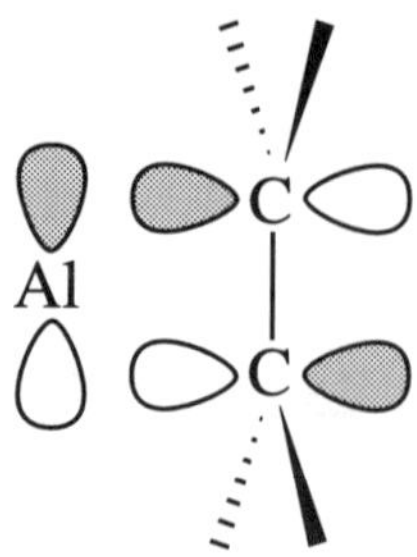

13

kJ/mol.[56] In this study, even though electron correlation treatments were employed, only the C–C and Al–C bond distances were optimized with hydrogen atoms fixed to be coplanar with the carbon atoms. A more recent *ab initio* study[57] at the MP2-DZ* level showed that significant rehybridization at the carbon atoms occurs (Figure 4.6), in addition the calculated binding energy of 51.0 kJ/mol was close to the values found experimentally. The rehybridization at the carbon atoms (∠Al–C–C–H = 100.9°) is consistent with significant π-back donation from the aluminum, and the overall geometry is thus comparable to that observed for transition metal complexes.[58,59] The η^2-coordination was also observed for the interaction with benzene[60] and was implied for the related toluene and *ortho*-xylene systems.[55] The ESR spectrum of Al and benzene in a neon matrix consists of a sextet of triplets due to hyperfine interaction with the ^{27}Al nucleus, and two equivalent protons,[58,59] consistent with a bonding scheme identical to that proposed for $Al(C_2H_4)$. Observation of the η^2– rather than the η^6– mode of coordination, which is commonly observed for transition metals, is consistent with prediction based on the frontier molecular orbitals of benzene.[61]

Irradation of the green $Al/C_2H_4/Ar$ system with visible light ($\lambda > 4000$ Å) results in a gradual but complete change in the ESR spectrum.[53] When the ethene concentration was low a significant fraction of the original spectrum remained, suggesting that a reaction between the Al–monoethene complex and

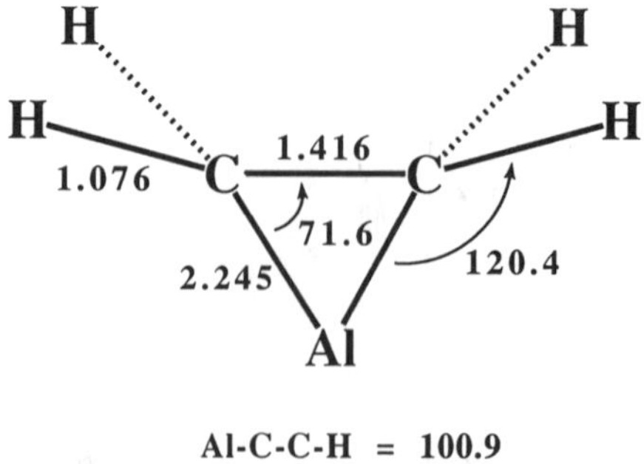

Figure 4.6. MP2-DZ* optimized geometry for $Al(C_2H_4)$ from ref. 57. Bond lengths are in angstroms, and bond angles in degrees.

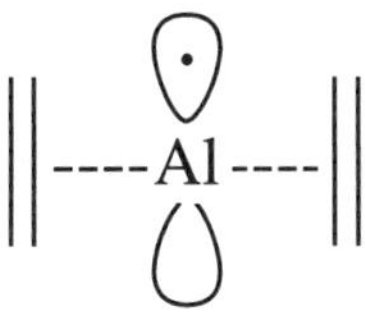

14

a second ethene had occurred. By analogy to studies with silver and copper[62] the product was proposed to be the bis-ethene complex, $Al(C_2H_4)_2$ (**14**).

Photolysis of the aluminum–ethene complex by UV irradiation results in the formation of a radical which does not contain aluminum.[53] The paramagnetic species was postulated, on the basis of its ESR spectral assignment, as the α,β-dimethyl vinyl radical formed *via* the reaction sequence shown in Equation 4.10.

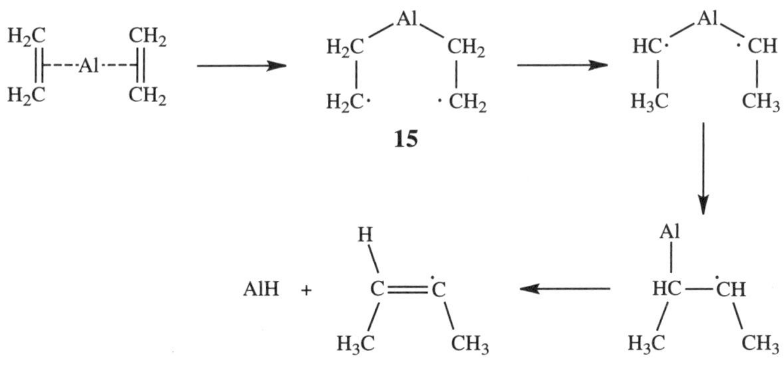

(4.10)

The di-radical **15** was also proposed to be an intermediate in the formation of aluminocyclopentane (**16**)[63] from the oxidative-cyclization[64–66] of two ethene molecules with aluminum (Equation 4.11)

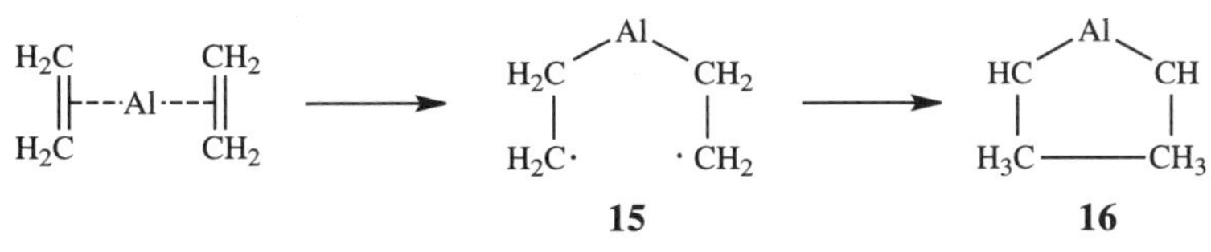

(4.11)

Although not observed directly, an aluminocyclopentane intermediate undoubtedly accounts for the 2,3-dimethylbutane isolated from the hydrolysis of Al/propane matrices.[67]

Oxidative cyclization,[68] and not π-coordination, also occurs when aluminum atoms and buta-1,3-diene are co-condensed in an adamantane matrix (Equation 4.12). The result is the formation of a σ-bonded aluminum cyclopentene complex (**17**).[69]

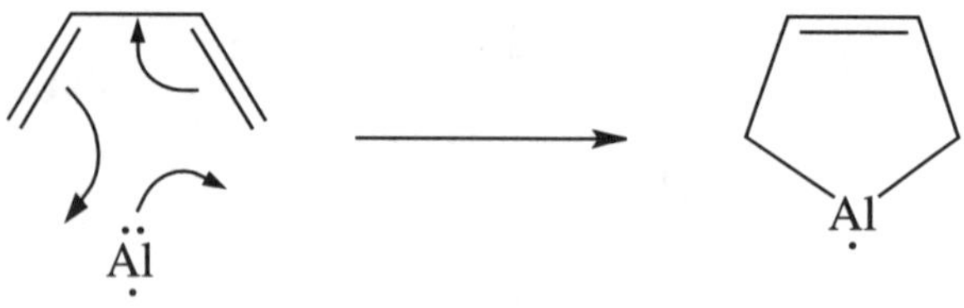

17

(4.12)

The ESR spectra of $\overline{Al(CH_2CH_2CH_2CH_2)}$ and $\overline{Al(CH_2CH{=}CHCH_2)}$ are similar; both indicate an exceptionally large hyperfine interaction with aluminum, consistent with significant Al 3σ orbital contribution to the singly occupied molecular orbital (SOMO). They also point to a hybridization of the Al valence orbitals. Both structures were therefore proposed to contain an approximately sp^2 hybridized aluminum with two of the sp^2 orbitals interacting with the two terminal CH_2 groups, with the unpaired electron on the aluminum residing in an sp^2 orbital pointing away from the metallocyclic framework. This bonding picture was confirmed for $\overline{Al(CH_2CH{=}CHCH_2)}$ by an ab initio study, the results of which are shown in Figure 4.7. The equilibrium geometry adopts a puckered five-membered ring structure resmbling cyclopentene. The Al–C bond distances are similar to the values of monomeric $AlMe_3$,[70] whereas the predicted C–C single and double bonds are reasonable. In the same manner that $\overline{Al(CH_2CH_2CH_2CH_2)}$ is observed as a minor constituent in the formation of $Al(C_2H_4)$ from the reaction of aluminum with ethene,[63] $\overline{Al(CH_2CH{=}CHCH_2)}$ is the minor constituent in the reaction of aluminum with buta-1,3-diene. Based on deuterium labelling experiments, the major species was determined to be the aluminum-substituted alkyl radical **18**.[69]

Oxidative-cyclization of propene to give a dimethyl-substituted aluminocyclopentane was proposed to account for the formation of 2,3-dimethylbutane and 2-methylpentane (Equation 4.13)[51]

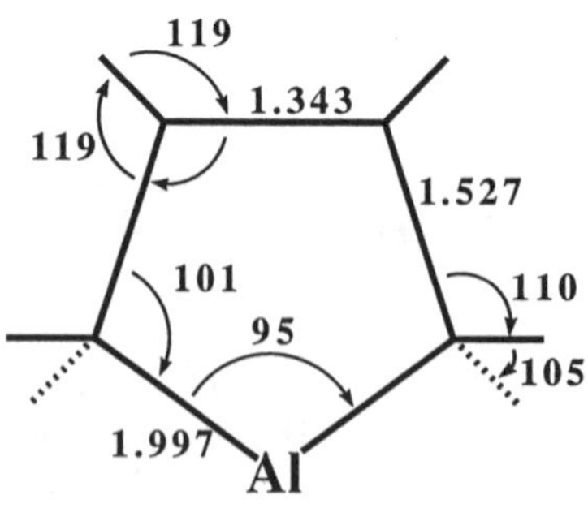

Figure 4.7. DZ Optimized geometry for $\overline{Al(CH_2CH{=}CHCH_2)}$ from ref. 69. Bond lengths are in angstroms, and bond angles in degrees.

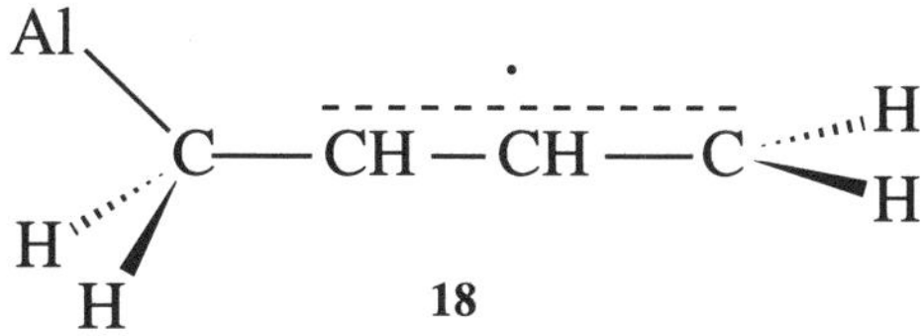

18

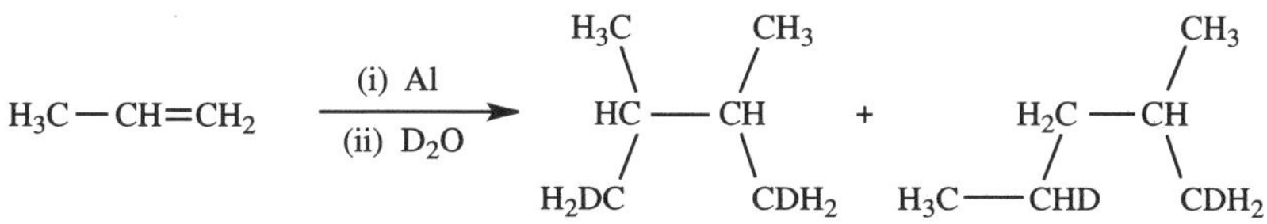

(4.13)

An ESR study of this reaction confirmed that the cyclopentanes are formed along with allyl radicals, and a new, previously unobserved, species.[71] Based on the hyperfine coupling and isotopic labeling studies the new species was proposed to be the aluminum centered radical, $HAl(\eta^3\text{-}C_3H_5)$ (**19**), in which there is considerable hybridization at Al as was found for the aluminocyclopentane[63] and aluminocyclopentene.[69]

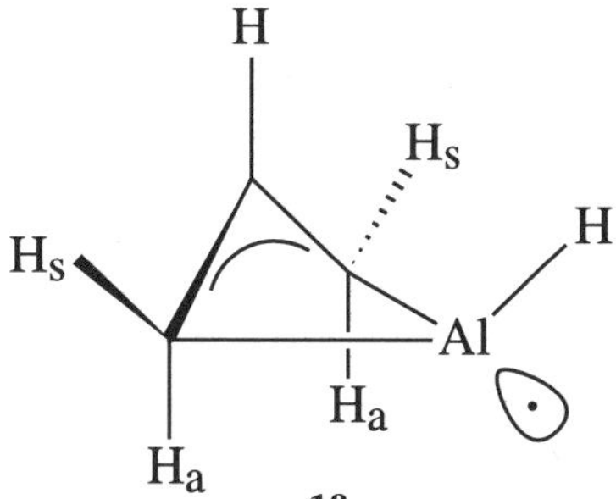

19

It should be noted that the reaction of ground-state Al atoms and propene at 77K gave $HAl(\eta^3\text{-}C_3H_5)$ unlike the photochemically induced oxidative addition of a C–H bond observed for the reaction of Al atoms with methane at 12K.[72] Thus, it is likely the route to the π-allyl complex is via propene complexation, in a manner similar to that observed for the double-bond isomerization of alkenes by coordinatively unsaturated transition metal complexes.[73–75] The alkene forms a π-complex with the transition-metal center which rapidly equilibrates with the π-allyl metal hydride,[76–79] which can give either the original metal hydride or the isomerized alkene π-complex (Equation 4.14).

Supporting evidence for a π-alkene intermeidate in the reaction of Al atoms with propene comes from the perdeutero propene which gave an ESR signal consistent with $Al(\eta^2\text{-}CD_2{=}CDCD_3)$.[71] The isolation of the η^2-complex is a consistent primary isotope effect on the formation of the π-alkyl aluminum hybride via an oxidative addition of a C–H bond.

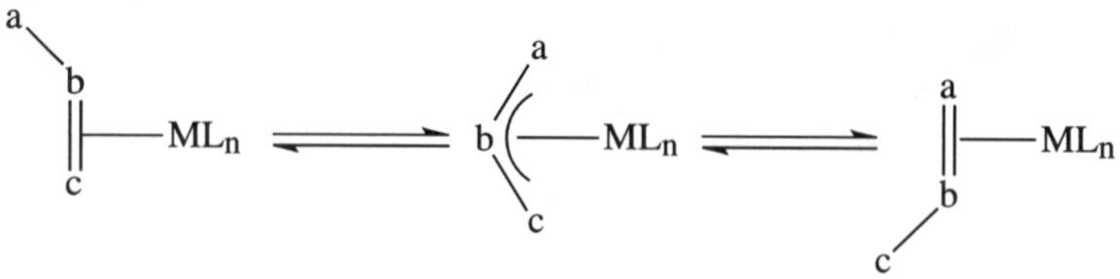

(4.14)

4.3.2. Alkynes

An alkyne, RC≡CR, can, except in rare cases,[80] use only one pair of π-electrons to bond to a metal atom in the same way as an alkene. However, complexes of this type are rare and alkynes are most commonly found as bridging ligands using both pairs of π-electrons (**20**).[81,82] Given the observation of Al_2 as an aggregate in inert matrices of aluminum,[83,84] it is conceivable that species such as **20** should be observed.

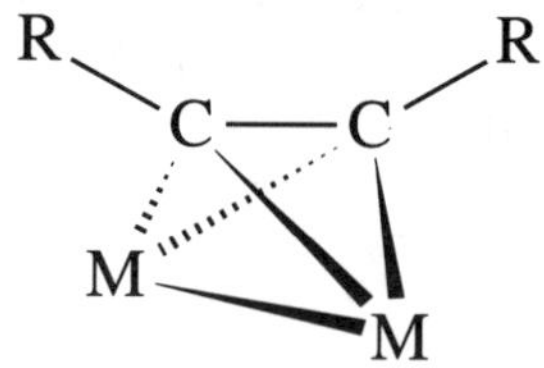

20

The co-deposition of Al atoms and ethyne in a neon matrix resulted in the formation of purple color and the observation of an ESR signal centered at g = 2.00.[85] The spectra observed from neon matrices containing Al and C_2D_2 contains a simple sextet pattern attributed to the essentially isotropic hyperfine structure due to one ^{27}Al nucleus. This sextet is further split in the Al/C_2H_2/Ne system to a doublet of doublets (Figure 4.8) due to two non-equivalent hydrogen atoms. Based on these hyperfine coupling interactions it was proposed that the spectra should be assigned to the $Al(C_2H_2)$ complex possessing the vinyl structure **21**.[53] The larger doublet splitting (see Figure 4.8) *ca.* 55 G was assigned to the β-proton coupling, since a larger hyperfine interaction is expected because of a direct overlap between the 1s orbital on H_b and the non-bonding half-filled orbital at the α-carbon.

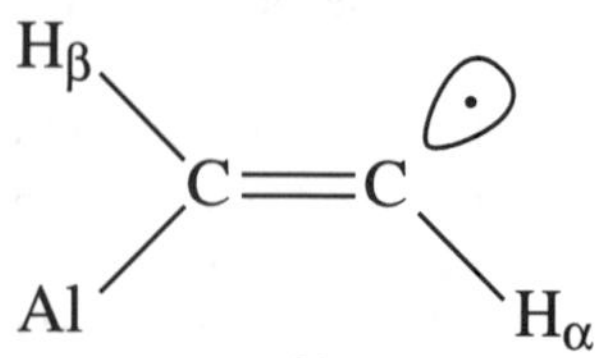

21

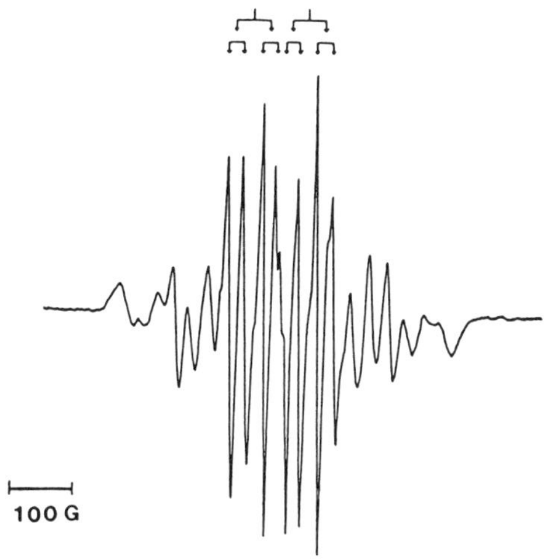

Figure 4.8 ESR spectra of $Al(C_2H_2)$ in frozen Ne matrices at 5K, showing the coupling to two inequivalent hydrogen atoms. Reprinted with permission from ref. 53. Copyright 1982 American Chemical Society.

The formation of a σ– rather than a π-complex for the $Al-CH_2H_2$ system may be rationalized on simple energetic grounds based on the relative stability between the formation of the Al–C bond(s) to the loss of the C–C π interactions. From thermodynamic data, the energy for losing a π bond in ethyne is about 92 kJ/mol less that that for ethene.[86] Thus the energy gained from the formation of a strong Al–C σ bond more than compensates for the loss of a partial π – bond. This, however, may not be energetically feasible for ethene, and thus the π – complex is observed for $Al(C_2H_4)$.

Photolysis of *cis*–$Al(C_2H_2)$ resulted in a gradual change of the ESR spectra assigned to a *cis* to *trans* isomerization (Equation 4.15).[53]

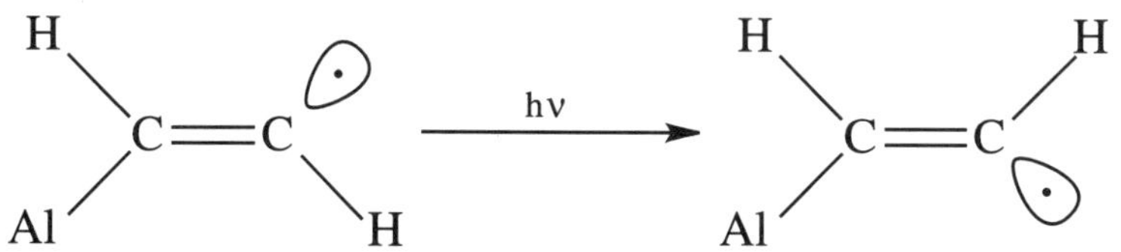

(4.15)

A Huckel MO calculation (Figure 4.9) indicated the *trans* isomer ($\theta \approx -40°$) is more stable than the linear structure by *ca* 16 kJ.[53] The plot of the total energy vs θ curve only showed the possibility of a metastable *cis* configuration at $\theta \approx 40°$. However, a more detailed theoretical treatment[87] indicated the *cis* isomer to be a true local minimum, and 0.8 kJ less stable than the *trans* isomer. Since it has been observed that single metal atoms strongly stabilize carbenes,[88–90] a theoretical study was done of the relative stability of the aluminum–vinylidene complex $Al(C{=}CH_2)$, **22**.[91]

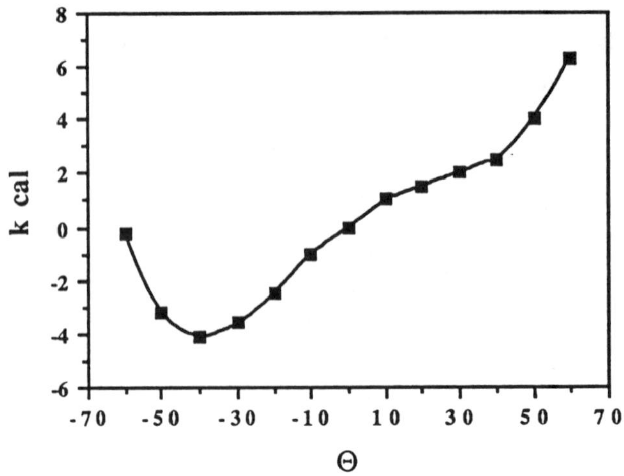

Figure 4.9. Dependency of the total energy of $Al(C_2H_2)$ upon θ (the angle of bend of the $C–C_\alpha–H_\alpha$) determined by MO calculations from ref. 53.

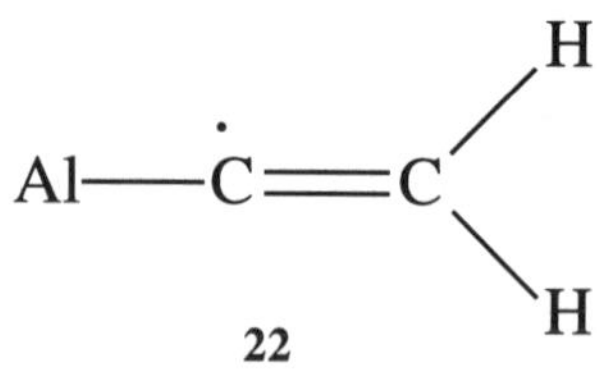

22

At the SCF level of theory,[91] the vinylidene complex was ca. 56 kJ more stable than the σ-complex. However, later studies[92] indicated that the activation barrier for the σ-ethyne to vinylidene complex isomerization (Figure 4.10) was significant ca. 200 kJ, and was consistent with the observed experimental data.

4.4. Hydrides

Although lithium aluminum hydride, $LiAlH_4$, is commonly employed for the reduction of a wide variety of organic compounds, surprisingly little is known about the mechanisms of these reactions. Hydridoaluminates are usually considered to transfer nucleophilic hydrogen heterolytically to the substrate, but seminal work suggests the involvement of free radicals in the reductions of certain organic halides, ketones, and polynuclear arenes.[93–96] These studies subsequently led a number of workers to attempt to isolate, characterize, and investigate reaction chemistry for alane radical anions.

The first ESR spectral observation of the alane radical anion, $\cdot AlH_3^-$, was made by the photolysis of a THF solution of either $LiAlH_4$ or $[n\text{–}Bu_4N][AlH_4]$

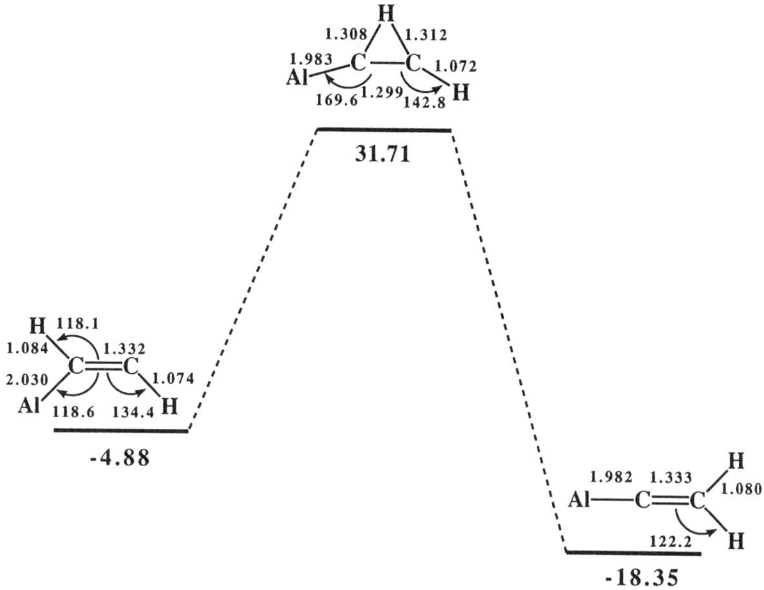

Figure 4.10. Energy diagram and stationary point geometries for the $Al(C_2H_2)$ σ-complex, the transition state and the vinylidne, $Al(CCH_2)$.[92] Bond lengths are in angstroms, bond angles in degrees, and energies in kcal.

in the presence of a radical initiator *t*–BuO–O*t*–Bu at 180–295 K (Figure 4.11).[97] However, subsequent studies showed that the same species is also formed during γ-radiolysis of solid tetrahydroaluminate,[98,99] although an earlier study[100] was shown to be spurious due to possible oxidation and/or hydrolysis of the sample. Based on ESR data it has been proposed that the structure of the alane radical anion clearly resembles $\cdot BH_3^-$ and $\cdot SiH_3$,[101] with near identical s and p orbital populations (*ca.* 31% s). Interestingly, this is in contrast to the result reported for AlR_3^-, SiR_3 and PR_3^{2-} (R = alkyl) which suggests a more marked trend in bond angle, and thus orbital population.[102] However, no clear picture has been developed concerning the structural effects of substituents on alane radicals. Since the value of a (^{27}Al) derived from $\cdot Al(O^tBu)_3^-$ (300.7 G), in which there is the possibility of interaction with the non-bonding electrons on oxygen,[103] is surprisingly close to that obtained for $\cdot AlMe_3^-$ (324 G), it is clear that a greater variety of ligand types must be investigated before these substituent effects are better understood.

The alane radical anion was found to react readily with n-propylhalides to give the n-propyl radical (Equation 4.16),[97]

$$[\cdot AlH_3]^- \quad + \quad n\text{–PrX} \xrightarrow[\text{X=Cl, Br}]{} [AlH_3X]^- \quad + \quad n\text{–Pr}\cdot \qquad (4.16)$$

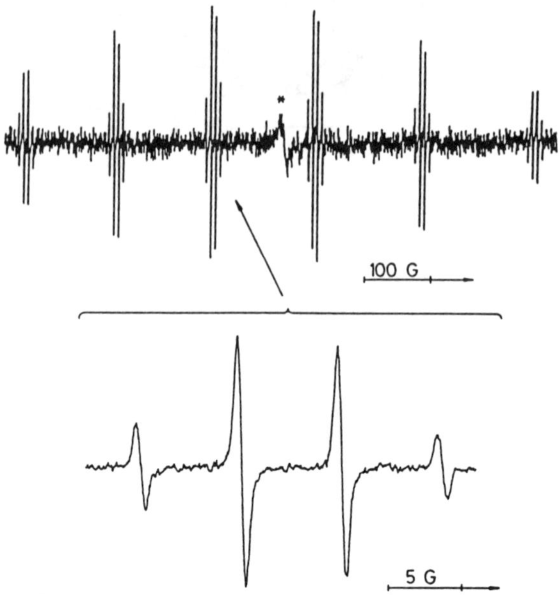

Figure 4.11. ESR spectra of $\cdot AlH_3^-$ obtained from the photolysis of a THF solution of $[N^nBu_4][AlH_4]$ and *t*–BuO–O*t*Bu at 235K. Reprinted with permission from ref. 97. Copyright 1981 Royal Society of Chemistry.

while the reaction of $\cdot AlH_3^-$ with ethene and benzene gives a β-substituted ethyl radical (**23**) and the cyclohexadienyl radical (**24**), respectively.[97]

In contrast to the situation for the alane radical, $\cdot AlH_3^-$, information concerning the alanyl radical, $\cdot AlH_2$, is very scarce. A vibrational spectrum of $\cdot AlH_2$ was observed by Herzberg[104] who found it to be similar to that of $\cdot BH_2$. This was deduced from the geometries of the two lowest-lying electronic states, and the result was supported by a subsequent ab initio calculation.[105] The results of both studies are given in Figure 4.12.

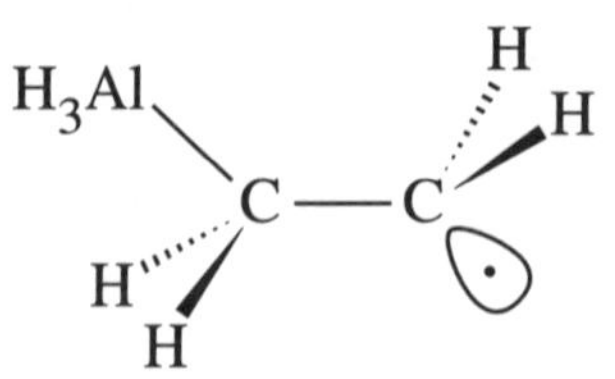

23

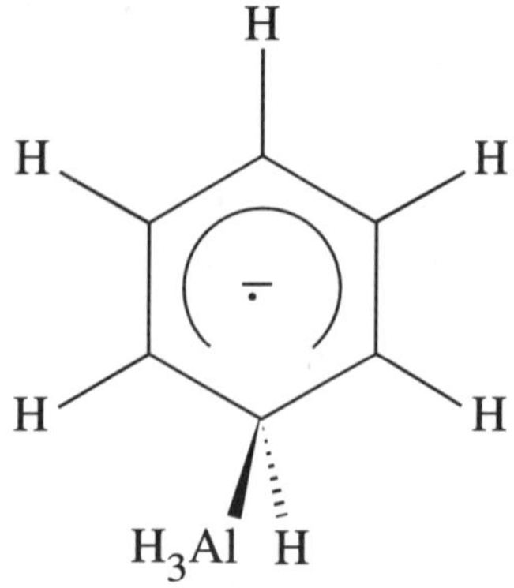

24

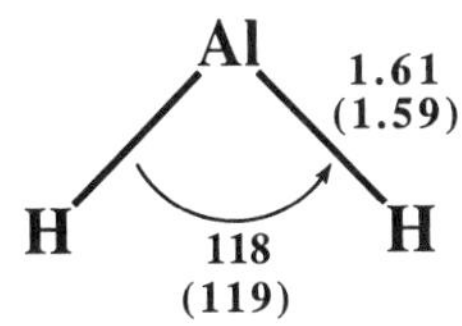

Figure 4.12. MRD Cl optimized geometry of the $\cdot AlH_2$ radical.[105] Bond length and angles in angstroms and degrees respectively. Experimental values are given in parenthesis.[104]

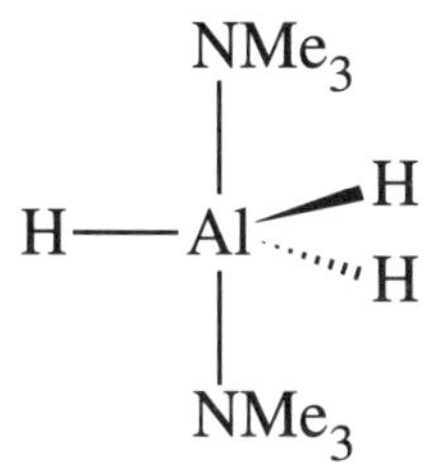

25

Alane forms a bis(trimethylamine) complex, $AlH_3(NMe_3)_2$, which is monomeric in both non-polar solvents and in the crystalline state where it adopts the trigonal bipyramidal (TBP) structure **25**.[106] The alanyl radical complex $\cdot AlH_2(NMe_3)_2$ may be prepared from $AlH_3(NMe_3)_2$ by either its exposure as a solid to γ-rays or UV photolysis of a cyclopropane-toluene solution in the presence of *t*-BuO–O*t*Bu.[107] In both cases the resulting ESR spectrum is centered at g = 2.0018, and is characteristic of axial symmetry (Figure 4.13), and shows hyperfine splitting from one ^{27}Al (255.6 G) and two each of ^{1}H (9.9 G) and ^{14}N (5.7 G). The ESR data also suggest that $\cdot AlH_2(NMe_3)_2$ adopts a

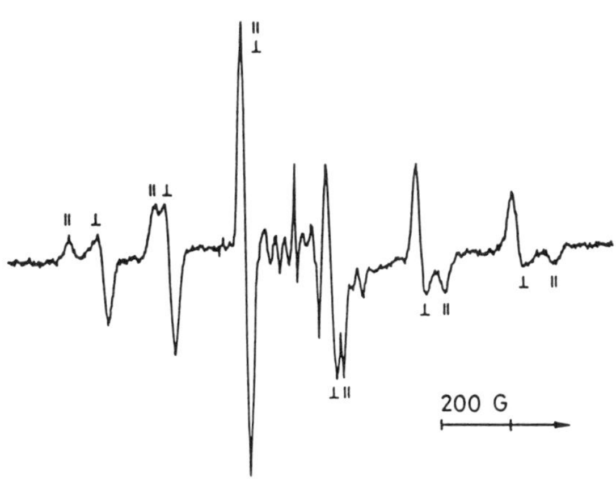

Figure 4.13. ESR spectrum of $\cdot AlH_2(NMe_3)_2$ in polycrystalline $AlH_3(NMe_3)_2$ at 138K. Reprinted with permission from ref. 107. Copyright 1986 Royal Society of Chemistry.

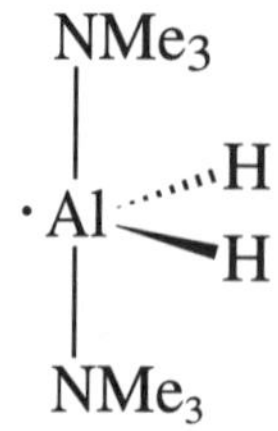

26

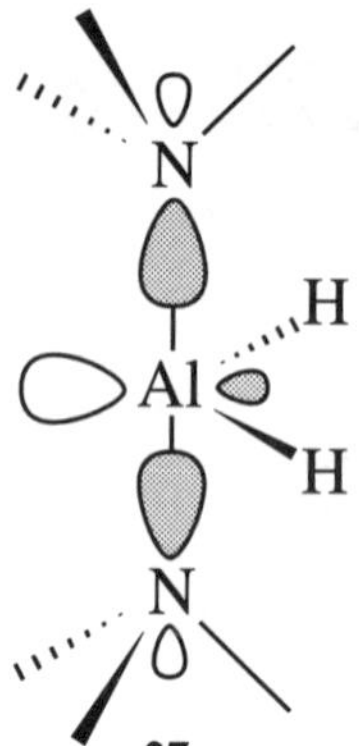

27

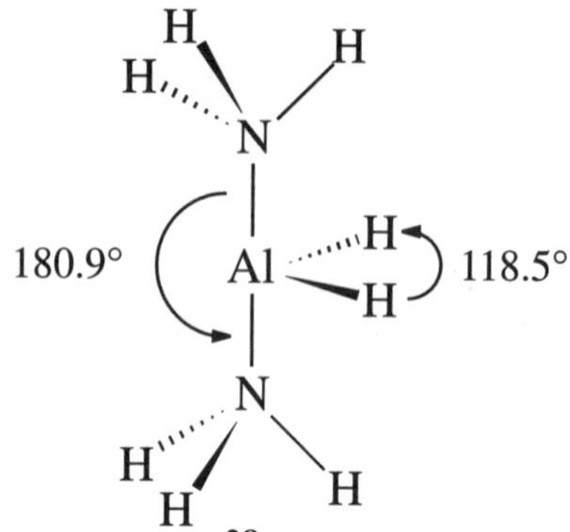

28

quasi-TBP structure (**26**), in which the unpaired electron is distributed mainly between the aluminum and the apical nitrogen atoms in a SOMO (**27**) of composition similar to that established for TBP phosphoranyl radicals.[108,109] Further support for this structure was obtained from ab initio calculations for the model compound $\cdot AlH_2(NH_3)_2$ (**28**).[107]

The chemical reactivity of $\cdot AlH_2(NMe_3)_2$ is similar to that reported for $\cdot AlH_3$ (see earlier). Thus, reaction with an n-propyl halide gives the n-propyl radical (Equation 4.17).[107]

$$\cdot AlH_2(NMe_3)_2 + n\text{–PrX} \xrightarrow[X=Cl,\ Br]{} AlH_2(X)(NMe_3)_2 + n\text{–Pr}\cdot \quad (4.17)$$

In the presence of 1,1-di-*tert*-butylethene, $\cdot AlH_2(NMe_3)_2$ reacts to afford the persistent radical $t\text{-}Bu_2CCH_2AlH_2(NMe_3)_2$, whereas a transient adduct is observed with ethene itself.[107]

4.5. Alkyls

As was indicated in Section 4.1 the recent interest in the monomeric low valent aluminum alkyl derivatives, AlR and AlR_2, has been driven by studies of the thermal and photochemcial decomposition of the trialkyls in chemcial vapor deposition processes. The UV excimer laser-induced photolytic decomposition of $AlMe_3$ was shown to occur by two separate wavelength-dependent mechanisms.[31] Photolysis at wavelengths between 260 and 230nm results in the formation of Al atoms (Equations 4.18 amd 4.19),

$$AlMe_3 + h\nu \xrightarrow{\lambda < 260nm} AlMe_2 + Me \quad (4.18)$$

$$AlMe_2 \longrightarrow Al + C_2H_6 \quad (4.19)$$

whereas photolysis at wavelengths below 230nm results in the additional formation of the monomethyl fragment (Equation 4.20).

$$AlMe_3 + h\nu \xrightarrow{\lambda < 230nm} AlMe + Me + Me \quad (4.20)$$

Since the thermal decomposition of $AlMe_3$ has been reported to occur via a stepwise thermolysis of the aluminum methyl bonds (see Equation 4.7), it is unclear whether in the present case $AlMe_2$ is an intermediate in the formation of AlMe. However, it has been demonstrated that AlMe is not the precursor to Al atoms in the photochemical decomposition.[110,111]

Several examples of monomeric mono- and di-alkyls have been observed spectroscopically, from the reaction of aluminum atoms with hydrocarbons, and these are discussed, along with relevant theoretical data, in Section 4.2.

In the same manner as the $Al(CO)_2$ radical was found to dimerize in frozen matrices to the diamagnetic $Al_2(CO)_4$ (**7**),[41] it is reasonable to expect an analogous reaction to occur for AlR_2 (Equation 4.21).

$$2 \cdot AlR_2 \longrightarrow R_2Al\text{–}AlR_2 \quad (4.21)$$

Although none of these Al–Al bonded species have been observed directly in the reaction of aluminum atoms with hydrocarbons, their presence has been infered from hydrolysis studies.[67]

The first reported isolation of a compound thought to contain an Al–Al bond was as a result of the reaction of $AlMe_3$ with the B–B bonded species $(Me_2N)_2B\text{–}B(NMe_2)_3$.[71] In addition to the previously reported compounds, BMe_3, Me_2BNMe_2, Me_2AlNMe_2, a non-volatile yellow solid was isolated the formulation of which was suggested as $Al_4B(NMe_2)_3Me_6$ based on IR and NMR spectroscopy. The only indirect evidence for the presence of an Al–Al bond was the production of H_2 on hydrolysis. The presence of Al–H or B–H groups, whose hydrolysis would also yield H_2, was discounted on the basis of the lack of suitable IR stretches in the range 2630 – 1430 cm^{-1}.[113–115] The proposed structure for $Al_4B(NMe_2)_3Me_6$ (**29**) is highly circumspect, and is easily discounted based on ring strain considerations. It is likely, therefore, that $Al_4B(NMe_2)_3Me_6$ is in fact a mixture of species. Similar concern may be raised concerning the proposed structure of $Al_3(NMe_2)_2Me_3$, (**30**).[116]

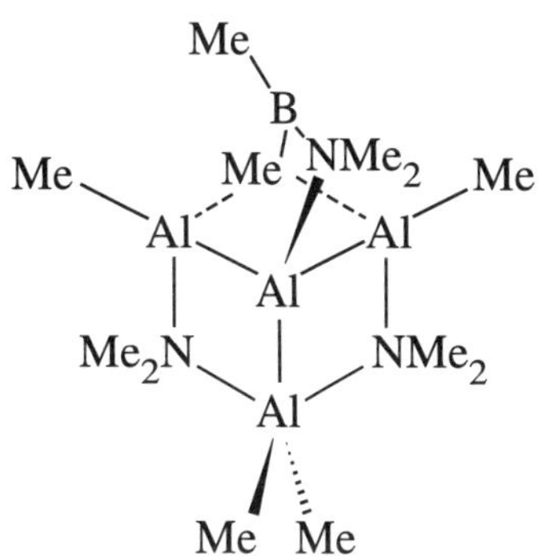

29

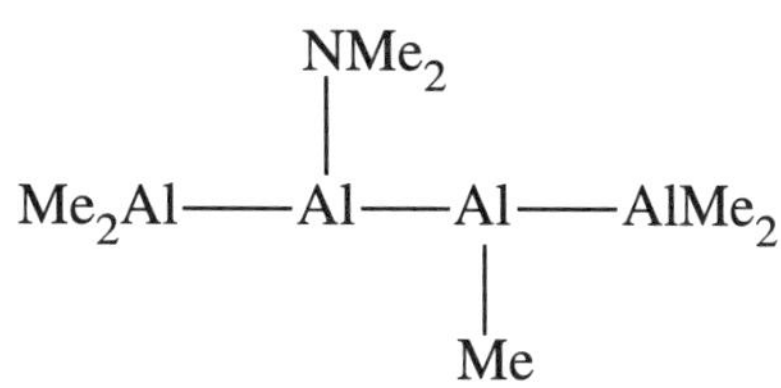

30

The first confirmed synthesis of isolated compounds containing Al–Al bonds,[117] involved the dehalogenation of *i*-Bu(X)Al–Cl (X = *i*-Bu, O-*i*-Bu, NMe_2) with alkali metals (Equation 4.22)

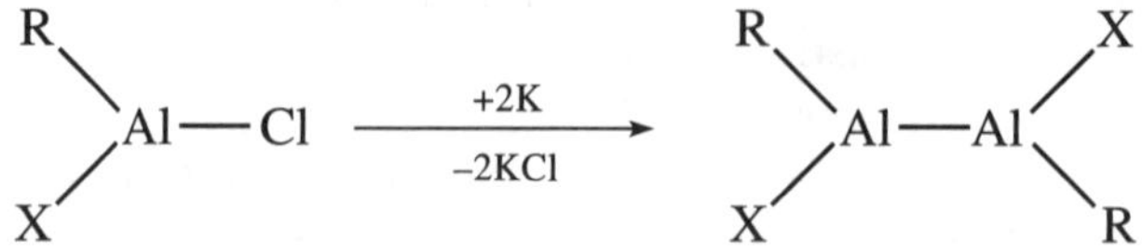

(4.22)

31

In the absence of sterically demanding alkyl groups this reaction resulted in the quantitative deposition of aluminum metal (Equation 4.23).[117]

$$3\ R_2AlCl + 3K \longrightarrow 2\ AlR_3 + Al + 3\ KCl \qquad (4.23)$$

It was thus proposed that steric effects were at least as important as electronic stabilization.[117] Confirming evidence was not obtained until twelve years later when Uhl reported the synthesis and molecular structure of an Al–Al bonded species stabilized by the sterically demanding *bis*(trimethylsilyl)methyl ligand, $CH(SiMe_3)_2$.[118]

The reaction of $[(Me_3Si)_2CH]_2AlCl$ with potassium metal yielded $[(Me_3Si)_2CH]_2Al{-}Al[CH(SiMe_3)_2]_2$ in moderate yield (*ca.* 60%).[118] The

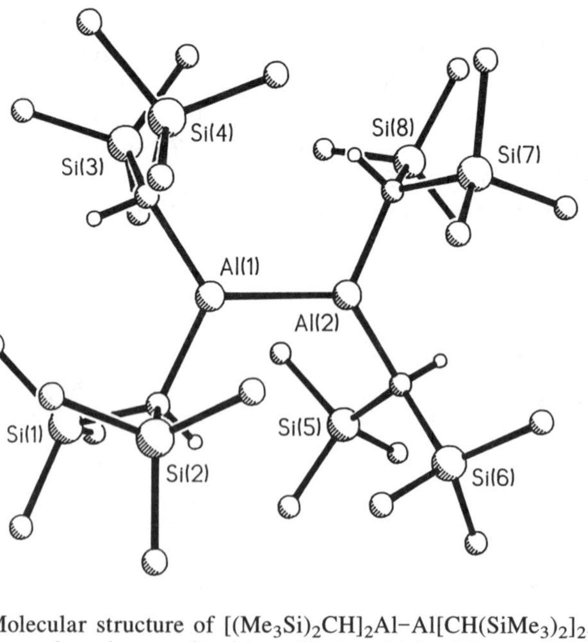

Figure 4.14. Molecular structure of $[(Me_3Si)_2CH]_2Al{-}Al[CH(SiMe_3)_2]_2$. For clarity the methyl hydrogens have been omitted (coordinates obtained from ref. 118).

molecular structure of $[(Me_3Si)_2CH]_2Al{-}Al[CH(SiMe_3)_2]_2$ is shown in Figure 4.14. The core of the molecule is almost planar ($\angle$C–Al–Al–C = *ca* 4.0°) which, given the lack of any possible classical pπ – pπ multiple bond character, is somewhat surprising. The Al–Al distance of 2.660(1) Å is comparable to that found in Al_7Te_{10} (2.60(1) Å),[119] and has become the bench-mark value for an Al–Al single bond. Two interesting spectroscopic results are also worthy of note. First, a band in the Raman spectrum at 373 cm^{-1} was assigned to the Al–Al stretch.[118] Second, the ^{13}C NMR resonance for the carbon α to the aluminum is shifted downfield (δ 17.7 ppm), well out of the range usually expected (δ 0 to −30 ppm).[118] The former observation should allow for other $R_2Al{-}AlR_2$ compounds whose X-ray structure have not been determined to be more fully characterized. The latter result indicates a strongly electron-withdrawing aluminum center, and the ^{13}C NMR of other $R_2Al{-}AlR_2$ compounds should corroborate this effect.

Although structural evidence for compounds containing Al–Al bonds has only recently been obtained,[118] the reactivity of the Al–Al bonds has been studied extensively. As mentioned above,[117] the observation of dihydrogen liberation from the hydrolysis of $i\text{-}Bu_2Al{-}Al\,i\text{-}Bu_2$ was taken as evidence for the formation of an Al–Al bond, by the reaction sequence in Equations 4.24 and 4.25.

$$R_2Al{-}AlR_2 + H_2O \longrightarrow R_2AlH + R_2AlOH \qquad (4.24)$$

$$R_2AlH + H_2O \longrightarrow R_2AlOH + H_2 \qquad (4.25)$$

Thus, one molar equivalent of hydrogen should be evolved per Al–Al bond, as is observed.

Reaction of $i\text{-}Bu_2Al{-}Al\,i\text{-}Bu_2$ with excess H_2O (pH 1) results in 100% cleavage of both Al–Al and Al–C bonds, whereas the Al–Al bonds exhibit greater reactivity towards gaseous HCl (Equations 4.26 – 4.28).[120]

$$i\text{-}Bu_2Al\text{-}Al\,i\text{-}Bu_2 + HCl \longrightarrow i\text{-}Bu_2AlH + i\text{-}Bu_2AlCl \qquad (4.26)$$

$$i\text{-}Bu_2AlH + HCl \longrightarrow i\text{-}Bu_2AlCl + H_2 \qquad (4.27)$$

$$i\text{-}Bu_2AlCl + HCl \longrightarrow i\text{-}BuAlCl_2 + i\text{-}BuH \qquad (4.28)$$

Analogous reactivity is observed with dimethylamine (Equation 4.29 and 4.30).[121]

$$i\text{-}Bu_2Al{-}Al\,i\text{-}Bu_2 + HNMe_2 \longrightarrow i\text{-}BuAlH + i\text{-}Bu_2AlNMe_2 \qquad (4.29)$$

$$i\text{-}Bu_2AlH + HNMe_2 \longrightarrow i\text{-}Bu_2AlNMe_2 + H_2 \qquad (4.30)$$

However, in the latter case, some prior disproportionation was proposed based on the low hydrogen yield; by contrast no reaction is observed with the weakly acidic complex $HMn(CO)_6$.[120]

A highly reactive Al–Al bond was also observed for the sterically hindered $[(Me_3Si)_2CH]_2Al–Al[CH(SiMe_3)_2]_2$, which reacts with CS_2 to give the sulfido bridged species (**32**), whose structure has been confirmed by X-ray crystallography (W. Uhl, personal communication).

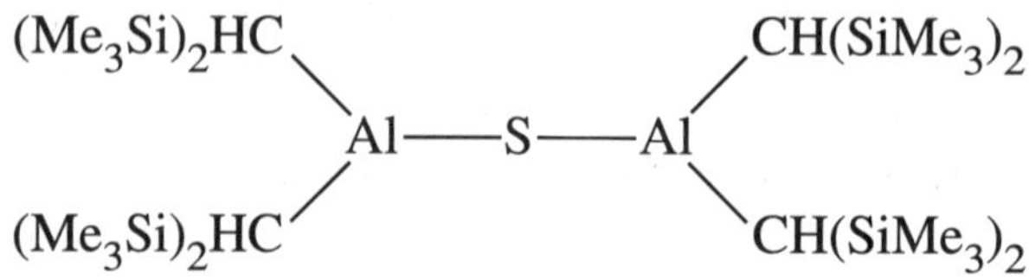

32

The decomposition of i-$Bu_2Al–Al i$-Bu_2 to give aluminum metal (Equation 4.31) was observed to be dependent on solvent, temperature, and possibly concentration.[121]

$$3\ i\text{-}Bu_2Al–Al i\text{-}Bu_2 \longrightarrow 4\ Al i\text{-}Bu_3 + 2\ Al^0 \qquad (4.31)$$

In addition, Lewis basic solvents such as Et_2O and THF were found to catalyze the dissociation, whereas γ-picoline formed the stable Lewis acid-base adduct **33**.[121] It should be noted that analogous gallium halide complexes have been structurally characterized.[16–21]

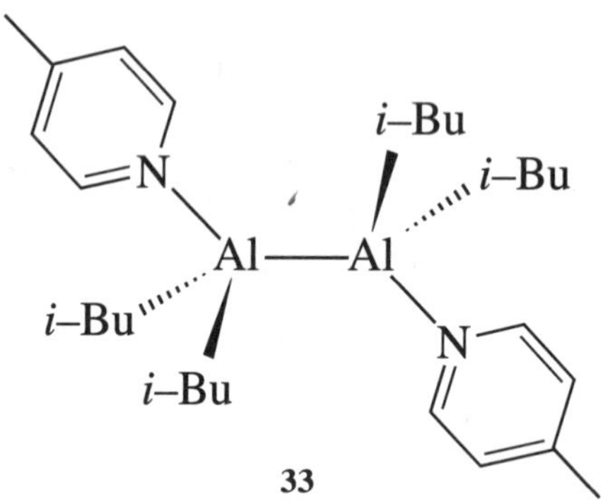

33

Thermolysis of i-$Bu_2Al–Al i$-Bu_2 at 125° C for 10h affords isobutylene and H_2 in a 2:1 molar ratio, consistent with a β-hybride elimination process (Equation 4.32).[121]

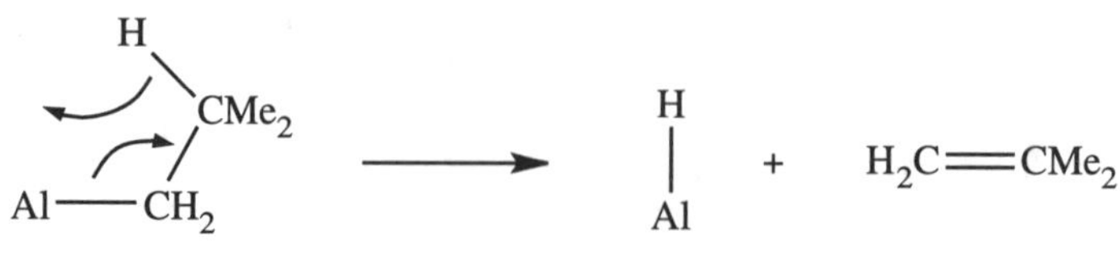

(4.32)

Al–Al bonds have also been proposed but not structurally confirmed for the dipotassium salts $K_2[Al_2\textit{i}\text{-}Bu_6]$,[122] and $K_2[Al_2\textit{i}\text{-}Bu_4H_2]$ (**34**)[123]

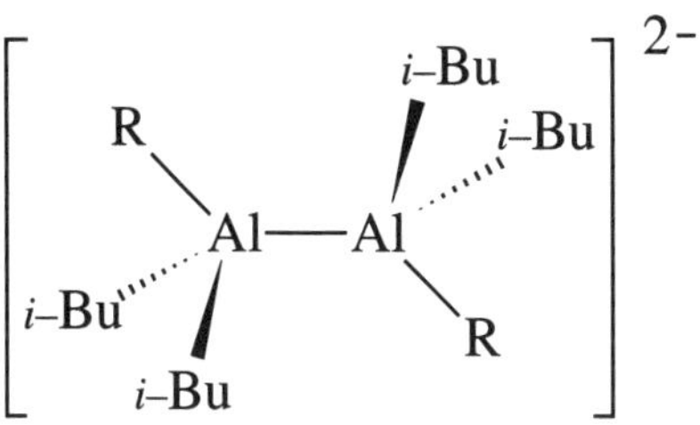

34

R = H, *i*-Bu

4.6. Clusters

Although electron-deficient polyhedral boranes containing B–B cluster bonding have been known since 1912,[124] there have until recently been no examples isolated of the aluminum analogs. There has, however, been spectroscopic evidence for aluminum cluster formation in both frozen matrices and the gas phase.[125]

Condensation of Al atoms in adamantane or cyclohexane matrices at 77 K resulted in the observation of an ESR signal at g = 2, which consisted of 16 lines compatible with three equivalent aluminum atoms, and was thus assigned to the neutral triatomic cluster Al_3 (Figure 4.15).[48] Based on ESR data and unrestricted Hartree–Fock calculations, the electronic structure of Al_3 was proposed to be a quartet $^4A_2$ ground state[126,127] with the unpaired electrons

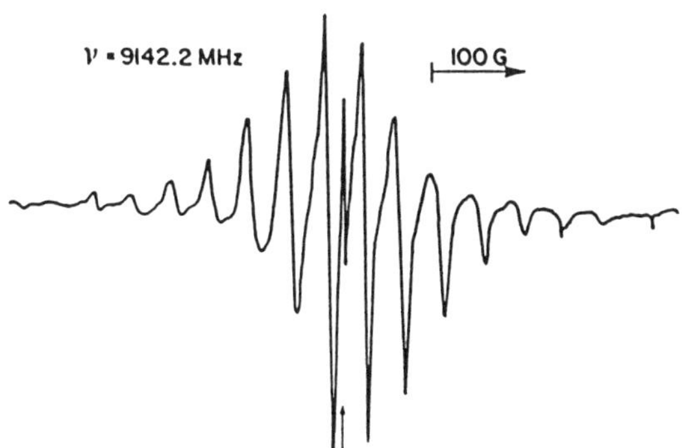

Figure 4.15. Al_3 in adamantane at 77K. Reprinted with permission from ref. 48 Copyright 1985 American Chemical Society.

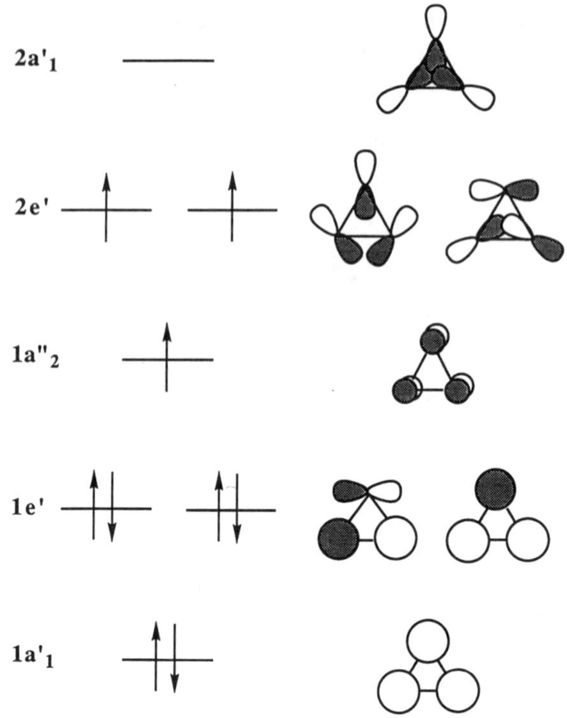

Figure 4.16. MO orbitals for equilateral triangular Al_3.[48]

populating the $1a_2''$, 2a', and one component of the degenerate 2e orbitals (see Figure 4.16). The Al–Al bond length and Al–Al–Al bond angle were calculated to be 2.709 Å and 72.6 Å respectively. The former value is close to the theoretical Al–Al bond distance of 2.779 Å for Al_2 in the 3Π state,[128] and the experimentally determined value for $[(Me_3Si)_2CH]_2Al{-}Al[CH(SiMe_3)_2]_2$ of 2.660(11) Å (see earlier).[118]

It is interesting to note that although the dimer Al_2 is predicted to have a triplet ground state, no ESR spectral evidence has been obtained. However, Al_2 was found to be one of the more reactive aluminum clusters in the gas phase. using the pulsed cluster beam flow reactor technique[129,130] Al_2, Al_6, Al_{10}, and Al_n ($n \geq 16$) were determined as the most reactive clusters towards D_2, D_2O, CO, CH_3OH, and O_2. This result was in line with theoretical calculations of the chemisorption of molecular hydrogen on small aluminum clusters ($n \geq 2$–6), which showed that only Al_6 has an exothermic heat of absorption.[131]

As has been discussed earlier, the synthesis of tetraalkyldialuminum ($R_2Al{-}AlR_2$) compounds has been accomplished by the alkali metal reduction of dialkylaluminum chlorides. If the reduction of $i\text{-}Bu_2AlCl$ is carried out in hexane, a brown precipitate is formed (primarily KCl) from which can be extracted a low but reproducible yield of the toluene-soluble dark red

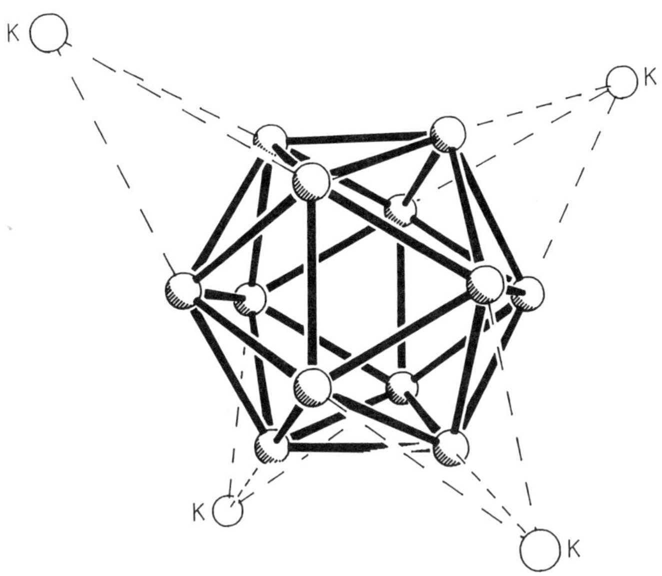

Figure 4.17. Molecular structure of the $K_2[Al_{12}i–Bu_{12}]$ (coordinates obtained from reference 132), iso-butyl substituents have been omitted for clarity.

crystalline salt $K_2[Al_{12}i\text{-}Bu_{12}]$.[132] $K_2[Al_{12}i\text{-}Bu_{12}]$ is surprisingly air stable; oxidation and hydrolysis to $Al(OH)_3$ occurs over several hours, and thermal decomposition to aluminum metal only above 150°C.

The solid-state structure of $K_2[Al_{12}i\text{-}Bu_{12}]$ was determined by X-ray crystallography[132] and consists of an Al_{12} icosahedral core (Figure 4.17), in which each aluminum is bonded to one terminal ^{i}Bu group and five adjacent aluminum atoms. The Al–Al bond distances range from 2.679 (5) Å to 2.696 (5) Å, and are slightly shorter than that predicted for Al_3.[48] The potassium atoms lie on a three-fold axis above each of four triangular faces of the icosahedron and serve as a bridge between two adjacent icosahedra. The structural parameters associated with the bridging iso-butyl groups are consistent with the presence of an interaction between the potassium and the C–H bonds.

The Al_{12} core of $K_2[Al_{12}i\text{-}Bu_{12}]$ is isostructural to $[B_{12}H_{12}]^{2-}$ [133] and, as would be expected, obeys Wade's rules for a closo structure. It is interesting to speculate as to plausible reasons why compound $K_2[Al_{12}i\text{-}Bu_{12}]$ is red, whereas the boron analog, $[B_{12}H_{12}]^{2-}$, is colorless. The red shift of the ultraviolet absorption edge is indicative of a decrease in the HOMO–LUMO energy gap for $[Al_{12}i\text{-}Bu_{12}]^{2-}$, which would in turn indicate weaker intercluster bonding for aluminum than for boron.

Molecules of the type $E_nX_n (n > 2)$ are well established in boron chemistry.[134–136] However, unlike the boranes, these closed Δ-faced polyhedra are electron-hyperdeficient and violate Wade's rules. Little has been reported of the analogous molecules of the heavier group 13 elements, although the monochloride, AlCl, has long been known as a gas phase species.[137–139] More recently,[140] a solution of AlCl has been identified which is metastable at −78°C in a mixture of toluene and Et_2O. Addition of $Mg(Cp^*)_2$ ($Cp^* = C_5Me_5$) to this solution allows for the isolation of the first stable monovalent compound of aluminum, $[AlCp^*]_4$ (Equation 4.33).[141]

$$4\,AlCl(Et_2O)_x + 2\,MgCp_2^* \longrightarrow [AlCp^*]_4 + 2\,MgCl_2(Et_2O)_2 + (4x\text{-}4)\,Et_2O \quad (4.33)$$

The tetrameric nature of $[AlCp^*]_4$ was confirmed by X-ray crystallography[141] (Figure 4.18). The four aluminum atoms form a regular tetrahedron, while each Cp* ligand is η^5-coordinated to an aluminum atom. The average Al-C distance (2.34 Å) is comparable to that found for $Mg(Cp^*)_2$.[142] The average Al−Al bond distance in $[AlCp^*]_4$ (2.769 Å) is longer than the Al−Al single bond in $[Me_3Si)_2CH]_2Al{-}Al[CH(SiMe_3)_2]_2$ (2.660 Å) but it is close to the values calculated for the naked clusters Al_2 and Al_3 (2.779 Å and 2.709 Å respectively). Extended basis set SCF calculations on the model compounds AlX and $[AlX]_4$ [X = H, F, Cl, Cp (C_5H_5)] indicate that in the case of Cp,

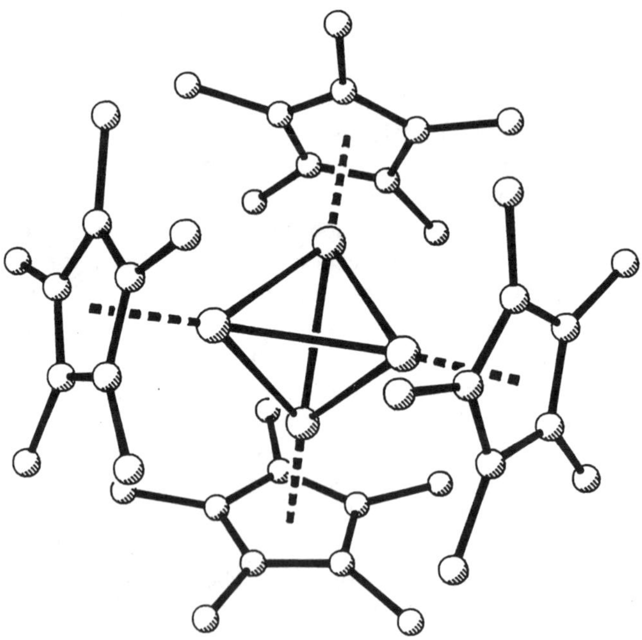

Figure 4.18. Molecular structure of $[AlCp^*]_4$ with hydrogen atoms omitted for clarity (coordinates obtained from reference 141). Selected bond lengths Å and angles (°): Al−Al 2.773(6) – 2.767(5) Å, Al−Cp (center) 2.032 – 1.997 Å; Al−Al−Al 60.1(1) – 59.9(1)°.

significant weakening in the Al–Al bonding occurs due to the π-back donation from the Cp ligand. This effect is readily apparent from the calculated Al–Al distances in $[AlX]_4$; X = H, 2.617 Å; F = 2.645 Å; Cl = 2.639 Å and Cp = 2.796 Å. Conversely, the AlCp monomer was found to be stabilized by the presence of π-back donation from Cp to Al.[143] Perhaps the most interesting characteristic reported for $[AlCp^*]_4$ is the ^{27}Al NMR spectrum, which shows a sharp signal at $\delta = -80.8$ ppm ($W_{1/2}$ = 170Hz). This exceptionally high chemical shift [Al(III) compounds are commonly observed in the range 0 to +300 ppm] is in direct contrast to the large downfield shift ($\delta = 1639$ ppm) reported for aluminum metal.[144]

Although not structurally characterized it would appear that the monovalent tri-*tert*-butylsilylaluminum also exists as a tetramer $[Al(Si\text{-}t\text{-}Bu_3)]_4$, based on mass and NMR spectral studies (H. Schnöckel, personal communication).

4.7. Paramagnetic Compounds of Al(III)

The first report of an isolated complex with an apparently low valent aluminum center was that of the neutral complex $Al(bipy)_3$ (**35**) (bipy = 2,2'-bipyridyl)[145] which can be prepared by the reduction of $AlCl_3$ with Li(bipy) (Equation 4.34) or the reaction of $LiAlH_4$ with the neutral ligand (Equation 4.35).

$$AlCl_3 + 3\,Li(bipy) \xrightarrow[THF]{} Al(bipy)_3 + 3LiCl \tag{4.34}$$

$$LiAlH_4 + 3\,bipy \xrightarrow[THF]{} Al(bipy)_3 + LiH + 3/2\,H_2 \tag{4.35}$$

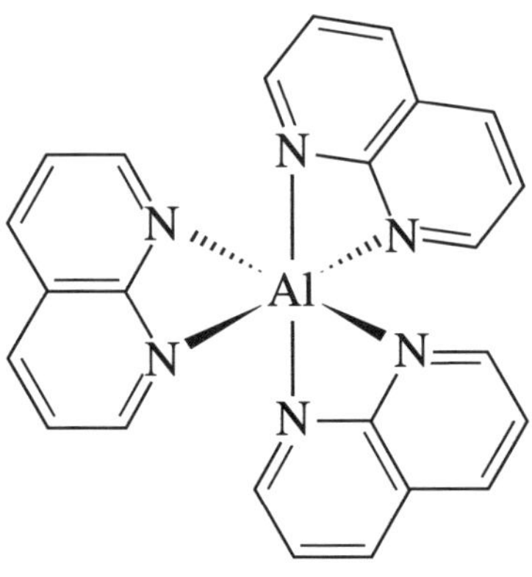

35

Subsequent magnetic measurements showed that although $Al(bipy)_3$ is paramagnetic, its susceptibility, $\mu = 2.32$ Bohr magnetons, can best be explained by postulating an Al^{3+} cation, with the unpaired electrons extensively delocalised on the ligands, giving rise to three bipy radical anions. Further evidence for

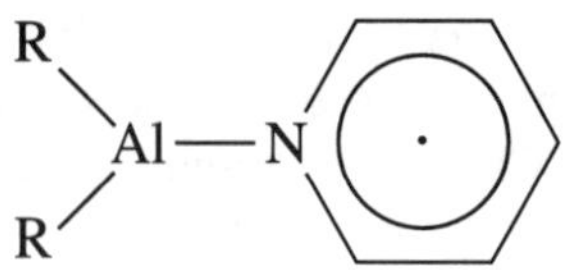

36

37

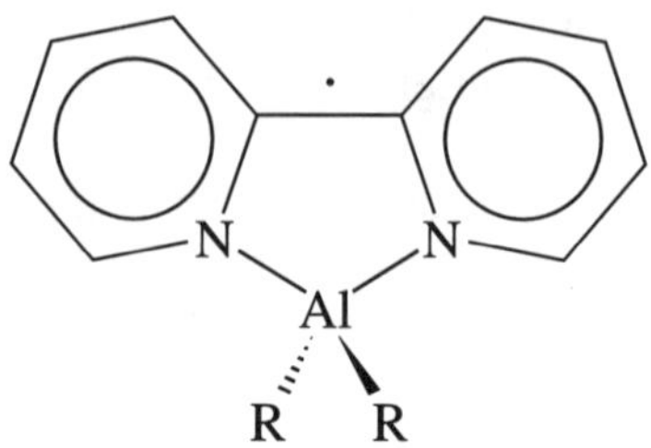

38

this picture was gained by investigations of the intermolecular interactions of $Al(bipy)_3$ with tetracyanoethylene.[146]

Following the work with $Al(bipy)_3$, a number of related organoaluminum radicals, $[R_2AlL]$ (**36, 37** and **38**) were isolated. In each case the aluminum is in the formal oxidation state of +II. However, ESR measurements and HMO calculations indicated that all of these compounds are best considered as being complexes of the dialkyl aluminum cations, R_2Al^+, with the radical anions of the pyridine, pyrazine or bipyridine ligands.[147–151]

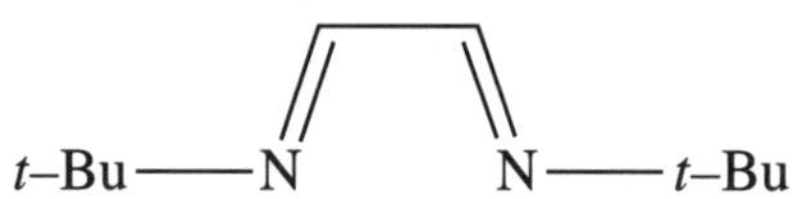

39

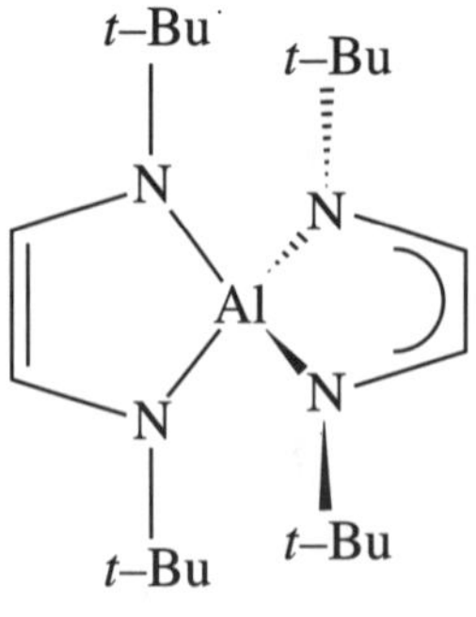

40

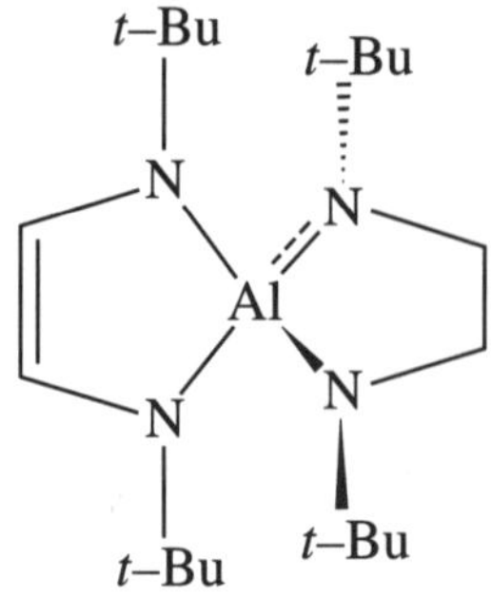

41

Recent attempts to prepare low-valent aluminum complexes have centered on the reaction of 1,4-di-*tert*-butyl-1,4-diazabutadiene (dbdab) (**39**) with either aluminum metal or alanes.[152] Thus, co-condensation of aluminum vapor with dbdab, or treating dbdab with lithium aluminum hydride-aluminum powder mixture in hexanes, both lead to the homoleptic complex $Al(dbdab)_2$ (**40**). Reaction of the trimethylamine adduct of alane, $AlH_3(NMe_3)$ with dbdab gives the partically hydrogenated complex $Al(dbdab)[N(t\text{-}Bu)CH_2]_2$ (**41**). ESR spectra of both compounds (Figure 4.19) show hyperfine coupling. However,

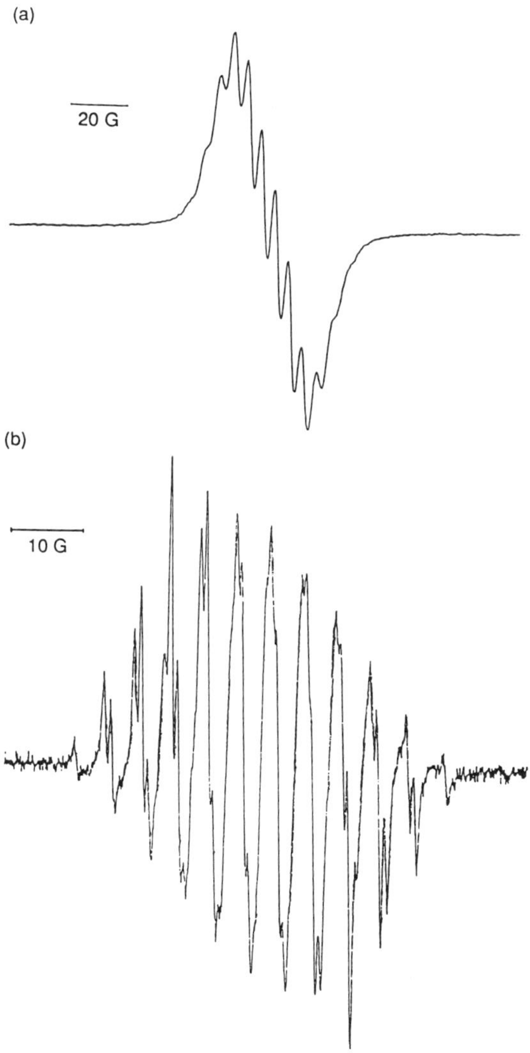

Figure 4.19. ESR spectra of (a) $Al(dbdab)_2$ and (b) $Al(dbdab)\{[N(t\text{–}Bu)CH_2]_2\}$. Reprinted with permission from ref. 152. Copyright 1990 Royal Society of Chemistry.

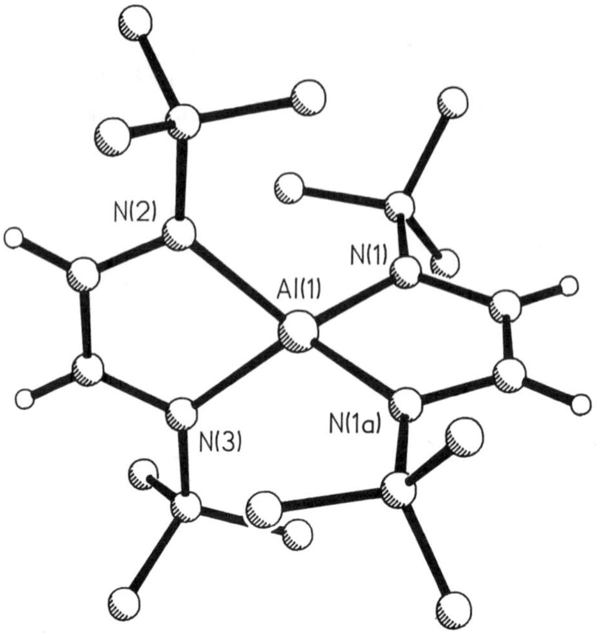

Figure 4.20. Molecular structure of $Al(dbdab)_2$, for clarity the *tert*-butyl hydrogen are omitted. (Coordinates obtained from ref. 152.) Selected bond lengths and angles: Al(1)–N(1) 1.823(2) Å, AL(1)–N(2) 1.916 (3) Å, Al(1)–N(3) 1.924 (3) Å, N(1)–C(1) 1.415 (3) Å, N(2)–C(6) 1.346 (5) Å, N(3)–C(7) 1.331 (5) Å, C(1)–C(1a) 1.343(5) Å, C(6)–C(7) 1.395 (6) Å, N(1)–Al(1)–N(1a) 92.3(1)°, N(2)–Al(1)–N(3) 86.1(1)°.

neither exhibits the six line pattern expected for coupling to an ^{27}Al nucleus. This suggests that the unpaired electron is primarily ligand centered, with the aluminum in a trivalent state. The molecular structure of both compounds has been determined by X-ray crystallography.[152] $Al(dbdab)_2$ (Figure 4.20) has two crystallographically independent ligands with different geometries, indicative of the localization of the unpaired electrons on one of the dbdab ligands (**42**). The bond distances found within the ligands, and photoelectron spectral data,[153] are entirely consistent with this picture. However, it should be noted that the ESR spectrum (Figure 4.19a) shows coupling to four equivalent nitrogens, suggesting that an electron hops from one ligand to another on a time scale comparable to that of the ESR experiment.

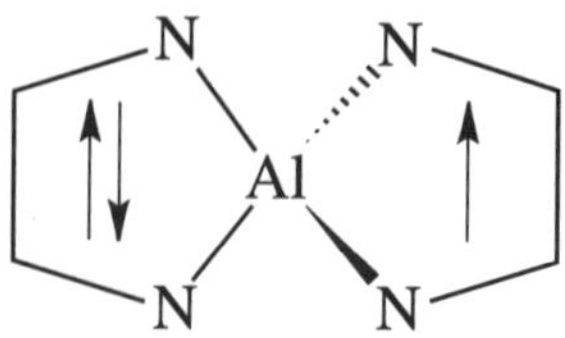

42

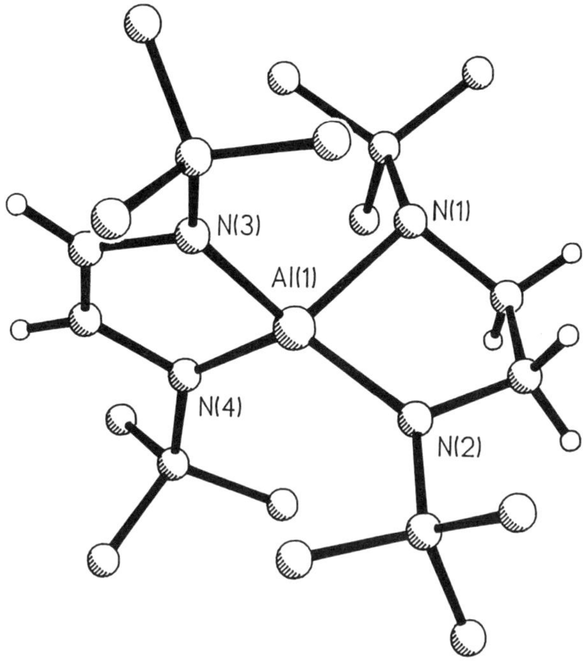

Figure 4.21. Molecular structure of Al(dbdab) $[N(t\text{-}Bu)CH_2]_2$, for clarity the *tert*-butyl hydrogen atoms are omitted (Coordinates obtained from ref. 152). Selected bond lengths and angles: Al(1)–N(1) 2.057 (4) Å, Al(1)–N(2) 1.813 (3) Å, Al(1)–N(3) 1.839(4) Å, Al(1)–N(4) 1.831 (4) Å, N(1)–C(1) 1.787 (6) Å, N(2)–C(2) 1.443(6) Å, C(1)–C(2) 1.470(6) Å, N(3)–C(11) 1.408(5) Å, N(4)–C(12) 1.419(6) Å, C(11)–C(12) 1.311(8) Å, N(1)-Al(1)-N(2) 86.7(1)°, N(3)-Al(1)-N(4) 92.4(2)°, N(1)-Al(1)-N(3) 108.1(2), N(1)-Al(1)-N(4) 117.4(2)°, N(2)-Al(1)-N(3) 127.5(2)°, N(2)-Al(1)-N(4) 125.6(2)°.

The molecular structure of Al(dbdab)$[N(t\text{-}Bu)CH_2]_2$ is shown in Figure 4.21. The dbdab ligand is nearly symmetrical; by contrast the hydrogenated ligand is highly unsymmetrical with a long [Al(1)–N(1) = 2.057 (4) Å] and a short [Al(1)–N(2) = 1.813 (3) Å] Al–N distance. This suggests some π-interaction between the four coordinate aluminum and the lone pair on one of the nitrogens, N(2). In support of this proposal N(2) is close to planar whereas N(1) is pyramidal. Since, the aluminum is approximately tetrahedral with sp^3 hybridization, any π-type interaction must be between the nitrogen lone pair and the Al(1)–N(4), Al(1)–N(3) σ antibonding orbitals, that is **43**. Similar interactions have been proposed for aluminum aryloxide complexes.[154–159]

Although not crystallographically confirmed, a ligand-centered radical was also observed for the monomeric dialkylaluminum species Al[2-$C(SiMe_3)_2C_5H_4N]_2$ (**44**), which was prepared via the alkali metal reduction of the analogous chloride (Equation 4.36).[160]

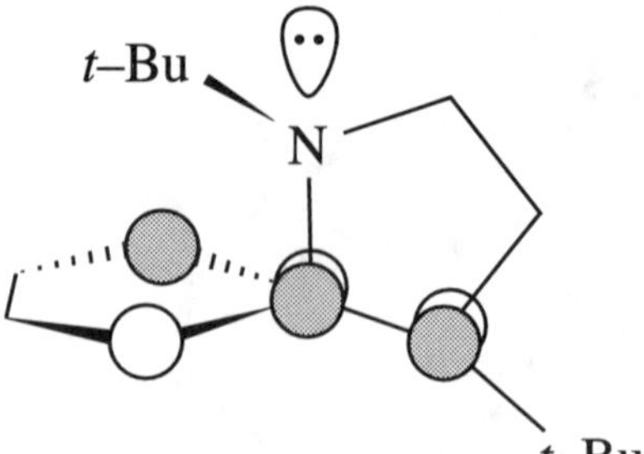

43

Although $Al[2\text{-}C(SiMe_3)_2C_5H_4N]_2$ is monomeric in THF or benzene, removal of the solvent in vacuo gave a diamagnetic dimeric product which was assumed to be the 4,4'-dipyridyl derivative on the basis of subsequent hydrolysis. The reductive coupling of the pyridyl groups[161] is taken as further evidence for the ligand-centered nature of the radical.

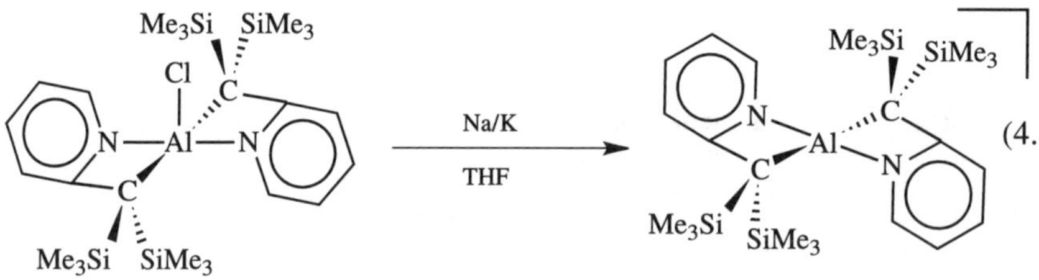

(4.36)

44

Despite the proposal [153] that $Al(dbdab)_2$ should be readily oxidized to give the cation, $[Al(dbdab)_2]^+$, where both ligands are singly reduced, it is only recently that any attempt has been made to prepare a homologous series from the neutral complex to either the trication (Equation 4.37) or the trianion (Equation 4.38).

$$AlL_x \xrightarrow{-e^-} [AlL_x]^+ \xrightarrow{-e^-} [AlL_x]^{2+} \xrightarrow{-e^-} [AlL_x]^{3+} \qquad (4.37)$$

$$AlL_x \xrightarrow{+e^-} [AlL_x]^- \xrightarrow{+e^-} [AlL_x]^{2-} \xrightarrow{+e^-} [AlL_x]^{3-} \qquad (4.38)$$

The 1,3-diphenyltriazenido (dpt) ligand, $[PhNNNPh]^-$, has been shown to be resistant to fragmentation[162], and is also found to stabilize a number of paramagnetic transition metal complexes.[163–165] The tris-complexes of aluminum, gallium and indium, $M(dpt)_3$, (**45**) are readily prepared from the triazene (dpt-H) and the appropriate groups 13 trimethyl compound (Equation 4.39).[166]

$$MMe_3 + 3\ \text{dpt–H} \xrightarrow[M = Al,\ Ga,\ In]{} M(dpt)_3 + 3\ MeH \qquad (4.39)$$

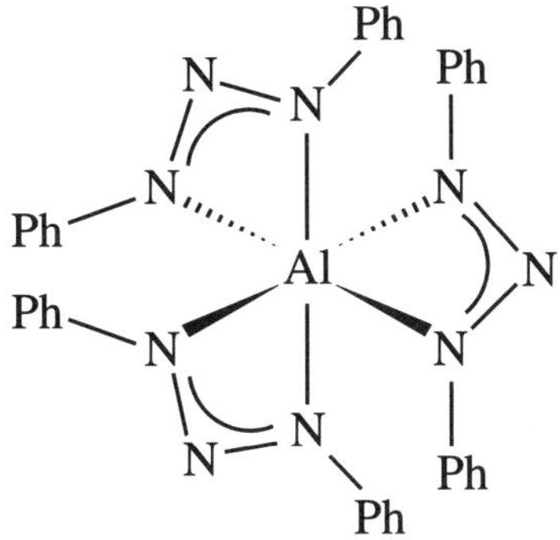

45

Electrochemical studies of $Al(dpt)_3$ by cyclic voltammetry in THF (Figure 4.22)[167] revealed three successive pseudo-reversible reduction waves, with nearly identical interwave spacings, indicating that chemical reduction of $Al(dpt)_3$ should result in the formation of a series of radical anion species by successive reduction of each of the triazenide ligands.[167] Thus, the stoichiometric reduction of $Al(dpt)_3$ with Na metal in THF affords the corresponding radical anion complexes $Na_x[Al(PhNNNPh)_3]$ according to Equation 4.40. These complexes, which are isolated as dark purple–black solids, are stable at room temperature, both in solution and in the solid state under an inert atmosphere.

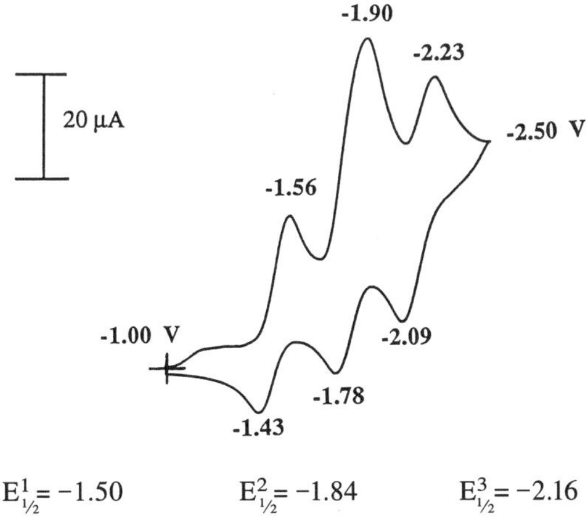

Figure 4.22. Cyclic-voltamograms of $Al(dpt)_3$ in the THF. $E^1 = -1.50$ V, $\Delta E^1 = 130$ mV; $E^2 = -1.84$ V, $\Delta E^2 = 120$ mV; $E^3 = 2.16$ V, $\Delta E^3 = 120$ mV.

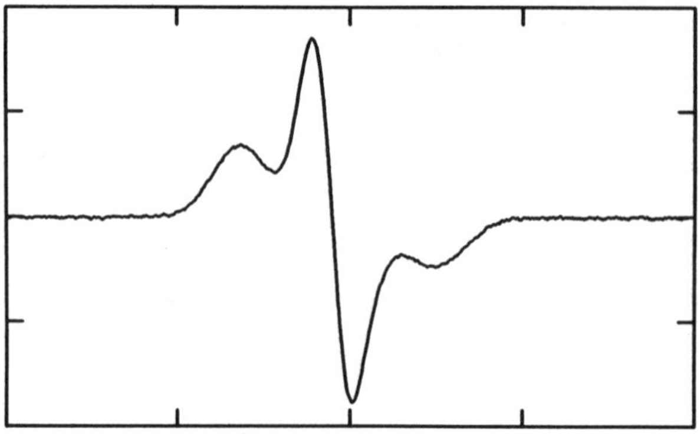

Figure 4.23. ESR spectrum of the radical anion $[Al(dpt)_3]^-$ in frozen THF (77K).

$$Al(dpt)_3 + x\,Na \xrightarrow[THF]{} Na_x[Al(dpt)_3] \quad x = 1, 2, 3 \tag{4.40}$$

Ab initio calculations [3–21G] performed on the model system $Al[HNNNH]_3$ (W. M. Cleaver and A. R. Barron, unpublished results) indicated that the LUMO is the triazenide N_3 π^* (**46**), which is localized on the central nitrogen. This supports the ESR spectra (77 K) which show hyperfine coupling to a single nitrogen (Figure 4.23)[167] and not to aluminum.

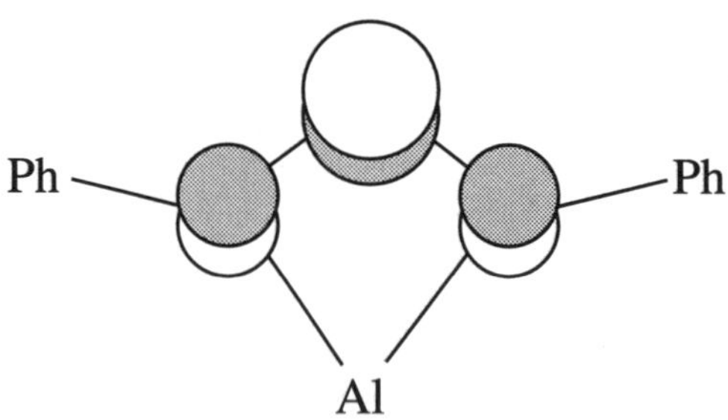

46

The UV–visible spectrum for $Al(dpt)_3$ (Figure 4.24a)[167] shows three principle absorption bands with the most intense band at 354 nm (ε = 65000) attributable to the $\pi \rightarrow \pi^*$ transition for the triazenide's unsaturated N_3 framework.[168] Upon reduction with three equivalents of Na (i.e. $Na_3[Al(dpt)_3]$) this band is bathochromically shifted to lower energy (412 nm) (Figure 4.24b). In addition a new band is observed (520 nm ε = 5500) which assigned to a $\pi^* \rightarrow$

(a)

(b)

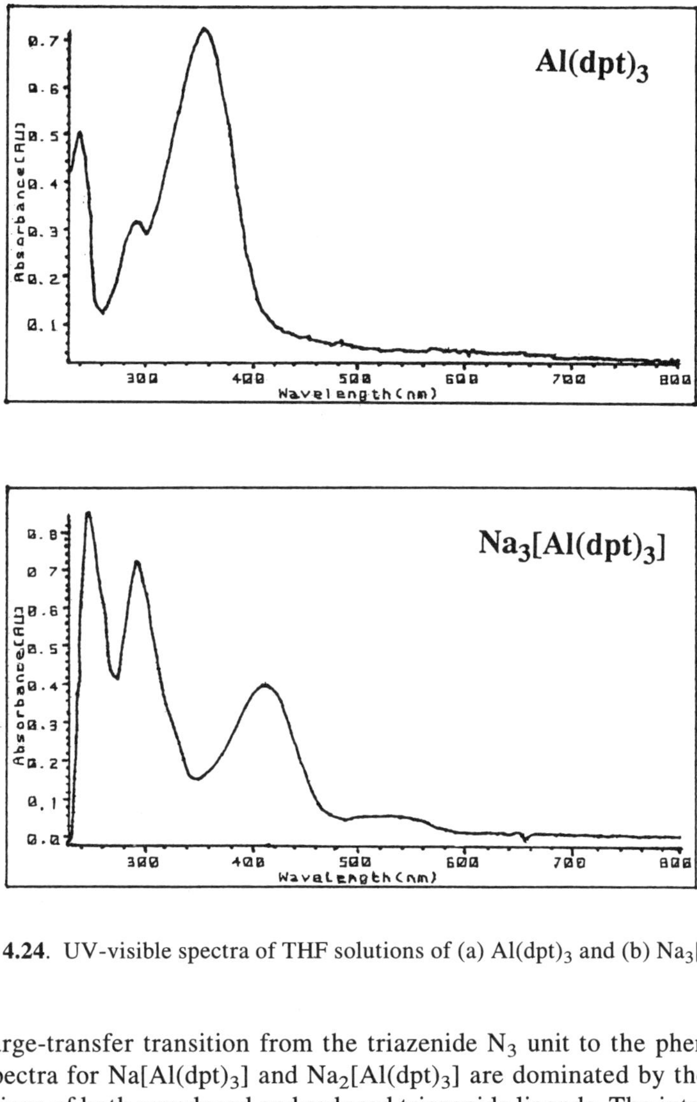

Figure 4.24. UV-visible spectra of THF solutions of (a) $Al(dpt)_3$ and (b) $Na_3[Al(dpt)_3]$.

π^* charge-transfer transition from the triazenide N_3 unit to the phenyl rings. The spectra for $Na[Al(dpt)_3]$ and $Na_2[Al(dpt)_3]$ are dominated by the $\pi \rightarrow \pi^*$ transitions of both unreduced and reduced triazenide ligands. The intensities of these bands change in a continuous manner on increasing reduction of the $Al(dpt)_3$ complex. These spectra are consistent with the successive and independent reduction of each of the triazenide ligands. A metathesis reaction of $Na[Al(PhNNNPh)_3]$ with (PPN)Cl [PPN = bis(triphenylphosphine)iminium] allowed for the isolation of the crystalline complex, $PPN[Al(PhNNNPh)_3]$.[167] The structure of this compound was determined by X-ray crystallography (Figure 4.25), and revealed two equivalent and one inequivalent triazene ligands consistent with the localization of the electron on one of the ligands in the solid state. The unique triazenido ligand shows a lengthening in the N–N bond distances and concomitant decrease in the N–N–N bond angle as

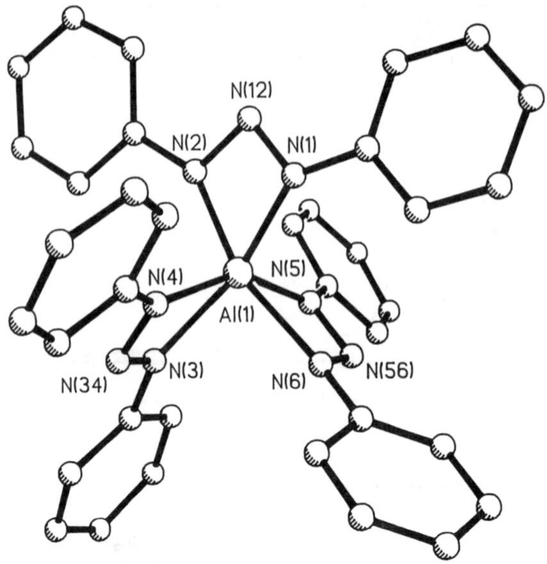

Figure 4.25. Molecular structure of the radical anion $[Al(dpt)_3]^-$. Hydrogen atoms omitted for clarity.

compared to the two equivalent triazenides, and those in the unreduced complex $Al(dpt)_3$. These distortions in the reduced ligand are as expected from the population of the $N_3\pi^*$ orbital (see earlier). Although in the solid state the unpaired electron is clearly localized on one of the dpt ligands, in solution at room temperature the ESR spectra indicate coupling to three equivalent nitrogen atoms consistent with the electron hopping from one triazene ligand to another on a time scale comparable to that of the ESR experiment.

Acknowledgments

The author is grateful to the Aluminum Research Board, and Arthur D. Little for financial support during this exercise; to Drs W. Uhl, C. Raston and H. Schnockel for their assistance; my research group for useful discussion, immoral support and libations; Ms Meg Chute, my secretary, for typing this manuscript; and lastly Peg for suffering in good grace.

References

1. Lee, A. G. *Coord. Chem. Rev.* **1972,** *8*, 289.
2. Greeland, B. H.; Tuck, D. G. *Inorg. Chem.* **1976,** *15*, 475.

3. Habeeb, J. J.; Tuck, D. G. *J. Chem. Soc., Dalton Trans.* **1975**, 1815.

4. Habeeb, J. J.; Tuck, D. G. *J. Chem. Soc., Dalton Trans.* **1976**, 866.

5. Beachley, O. T., Jr.; Hallock, R. B.; Zhang, H. M.; Atwood, J. L. *Organometallics* **1985**, *4*, 1675.

6. Beachley, O. T., Jr.; Hallock, R. B. *Organometallics* **1987**, *6*, 170.

7. Schmidbaur, H.; Thewalt, U.; Zafiropoulos, T. *Organometallics* **1983**, *2*, 1550.

8. Usøn-Finkenzeller, M.; Bublark, W.; Huber, B.; Müller, G. *Z. Naturforsch.* **1986**, *41B*, 346.

9. Beck, J.; Strähle, J. *Z. Naturforsch.* **1986**, *41B*, 1381.

10. Strausz, S. H.; Noirot, M. D.; Anderson, O. P. *Inorg. Chem.* **1986**, *25*, 3850.

11. Frank, W.; Weber, J.; Fuchs, E. *Angew. Chem. Int. Edn. Engl.* **1987**, *26,* 74.

12. Schmidbaur, H.; Nowak, R.; Huber, B.; Müller, G. *Z. Naturforsch.* **1988**, *43B*, 1447.

13. Schmidbaur, H.; Nowak, R.; Huber, B.; Müller, G. *Polyhedron* **1990**, *9,* 283.

14. *Comprehensive Organometallic Chemistry;* Wilkinson, G.; Stone, F. G. A.; Abel, E. W., Eds.; Pergamon; Oxford, **1986**; Vol. 1, pp 523-626.

15. Tuck, D. G. *Polyhedron* **1990**, *9*, 377, and references therein.

16. Beamish, J. C.; Small, R. W. H.; Worrall, I. J. *Inorg. Chem.* **1979**, *18*, 220.

17. Small, R. W. H.; Worrall, I. J. *Acta Crystallogr., Sect. B* **1982**, *38*, 250.

18. Small, R. W. H.; Worrall, I. J. *Acta Crystallogr., Sect. B* **1982**, *38*, 86.

19. Cumming, H. J.; Hall, D.; Wright, C. E. *Cryst. Struct. Commun.* **1974**, *3*, 107.

20. Khan, M. A.; Taylor, M. J.; Tuck, D. G.; Roger, D. A. *J. Cryst. Spectr. Res.* **1986**, *16*, 895.

21. Khan M. A.; Peppe, C.; Tuck, D. G. *Can. J. Chem.* **1984**, *62*, 601.

22. Eisch, J. J. *Comprehensive Organometallic Chemistry,* Wilkinson, G.; Stone, F. G. A.; Abel, E. W., Eds.; Pergammon: Oxford, **1986**; Vol. 1. pp 523-626.

23. Lehmkuhl, H.; Ziegler, K. *Methoden der Organischen Chemie;* Mueller, E., Ed.; Thieme: Stuttgart, **1970**; Vol. 13, p 48.

24. Ziegler, K. *Organometallic Chemistry;* Zeiss, H. Ed.; American Chemical Society Monograph. 147; Reinhold: New York, **1960**, p 194.

25. Sittig, M. *Organometallics 1966.* Chemical Process Monograph No. 20; Noyes Development: Park Ridge, NJ.

26. Chen, J. G.; Beebe, T. P., Jr.; Crowell, J. E.; Yates, J. T., Jr. *J. Am. Chem. Soc.* **1987**, *109*, 1726.

27. Domen, K.; Chuang, T. J. *J. Am. Chem. Soc.* **1987**, *109*, 5288.

28. Hara, M.; Domen, K.; Kato, M.; Onishi, T.; Nozoye, H. *J. Chem. Soc., Chem Commun.* **1990**, 1717.

29. Mansevit, H. M. *Appl. Phys. Lett.* **1968**, *12*, 156.

30. Leys, M. R. *Chemtronics* **1988**, *3,* 179, and references therein.

31. Bouermann, T.; Stuke, M. *Chemtronics,* **1989**, *4*, 189, and references therein.

32. Boone, J. R.; Ashby, E. C. *Top. Stereochem.* **1979**, *11*, 53.

33. Haubenstock, H. *Top. Stereochem.* **1983**, *14,* 231.

34. Noth, H. Z. *Z. Naturforsch. B* **1980**, *35*, 119.

35. Ashby, E. C.; DePriest, R. N.; Pham, T. N. *Tetrahedron Lett.* **1983**, *24*, 2825.

36. Power, M. B.; Bott, S. G.; Atwood, J. L.; Barron, A. R. *J. Am. Chem. Soc.* **1990**, *112*, 2446.

37. Hinchcliffe, A. J.; Ogden, J. S.; Oswald, D. D. *J. Chem. Soc., Chem. Commun.* **1972**, 338.

38. McIntosh, D.; Ozin, G. A. *J. Am. Chem. Soc.* **1976**, *98*, 3167.

39. Kasai, P. H.; Jones, P. M. *J. Am. Chem. Soc.* **1984**, *106*, 8018.

40. Chenier, J. H. B.; Hampson, C. A.; Harvard, J. A.; Mile, B.; Sutcliffe, R. *J. Phys. Chem.* **1986**, *90*, 1524.

41. Chenier, J. H. B.; Hampson, C. A.; Howard, J. A.; Mile, B. *J. Chem. Soc., Chem. Commun.* **1986**, 730.

42. Balaji, V.; Sunil, K. K.; Jordan, K. D. *Chem. Phys. Lett.* **1987**, *136*, 309.

43. Kettle, S. F. A. *Inorg. Chem.* **1965**, *4*, 1661.

44. Abel, E. W.; Stone, F. G. A. *Quant. Rev. Chem. Soc.* **1969**, *23*, 325.

45. Hoffman, R.; Wilkes, C. N.; Lippard, S. J.; Templeton, J. L.; Brower, D. C. *J. Am. Chem. Soc.* **1983**, *105*, 146.

46. Walsh, A. D. *J. Chem. Soc.* **1953**, 2266.

47. Cotton, F. A.; Wilkinson, G. *Advanced Inorganic Chemistry,* 5th Ed.; Wiley: Chichester, **1986**.

48. Howard, J. A.; Sutcliffe, R.; Tse, J. S.; Dahmane, H.; Mile, B. *J. Phys. Chem.* **1985**, *89*, 3595.

49. Dewar, M. J. S. *Bull. Soc. Chim. Fr.* **1951**, *C71*, 18.

50. Chatt, J.; Duncanson, L. A. *J. Chem. Soc.* **1953**, 2939.

51. Skell, P. S.; Wolf, L. R. *J. Am. Chem. Soc.* **1972**, *94*, 7919.

52. Kasai, P. H.; McLeod, D., Jr. *J. Am. Chem. Soc.* **1975**, *97*, 5609.

53. Kasai, P. H. *J. Am. Chem. Soc.* **1982**, *104*, 1165.

54. Howard, J. A.; Mile, B.; Tse, J. S.; Morris, H. *J. Chem. Soc., Faraday Trans. I* **1987**, *83*, 3701.

55. Mitchell, S. A.; Simard, B.; Rayner, D. M.; Hackett, P. A. *J. Phys. Chem.* **1988**, *92*, 1655.

56. Miralles-Sabater, J.; Merchan, M.; Nebot-Gil, I. *Chem. Phys. Lett.* **1987**, *142*, 136.

57. Tse, J. S.; Morris, H. *J. Chem. Soc., Chem. Commun.* **1989**, 78.

58. Ittel, S. D.; Ibers, J. A. *Adv. Organomet. Chem.* **1976**, *14*, 33.

59. Mingos, D. P. *Adv. Organoment. Chem.* **1977**, *15*, 1.

60. Kasai, P. H.; McLeod, D., Jr. *J. Am. Chem. Soc.* **1979**, *101*, 5860.

61. Barrett, J. *Introduction to Atomic and Molecular Structure,* Wiley: Chichester, **1979**, pp 228-234.

62. Kasai, P. H.; McLeod, D., Jr.; Watanabe, T. *J. Am. Chem. Soc.* **1980**, *102*, 179.

63. Chenier, J. H. B.; Howard, J. A.; Mile, B. *J. Am. Chem. Soc.* **1987**, *109*, 4109.

64. Albright, T. A.; Burdett, J. K.; Whangbo, M. H. *Orbital Interactions in Chemistry;* Wiley, New York, 1985; pp 339-352.

65. Pillai, S. M.; Ravindvanathan, M.; Sivaram, S. *Chem. Rev.* **1986**, *86*, 353.

66. Stockis, A.; Hoffman, R. *J. Am. Chem. Soc.* **1980**, *102,* 2952.

67. Skell, P. S.; McGlinchey, M. J. *Angew. Chem. Int. Ed. Engl.* **1975**, *14*, 195.

68. Yasuda, H.; Tatsumi, K.; Nakamura, A. *Acc. Chem. Res.* **1985**, *18*, 120.

69. Chenier, J. H. B.; Howard, J. A.; Tse, J. S.; Mile, B. *J. Am. Chem. Soc.* **1985**, *107*, 7290.

70. Amenningen, A.; Halvorsen, S.; Haaland, A. *Acta Chem. Scand.* **1971**, *25*, 1937.

71. Histed, M.; Howard, J. A.; Morris, H.; Mile, B. *J. Am. Chem. Soc.* **1988**, *110*, 5290.

72. Parnis, J. M.; Ozin, G. A. *J. Am. Chem. Soc.* **1986**, *108*, 1699.

73. Ochiai, E. *Coord. Chem. Rev.* **1968**, *3*, 49.

74. Masters, C. *Homogeneous Transition Metal Catalysts – A Gentle Art;* Chapman Hall: London, **1981**.

75. Bönnemann, H. *Angew. Chem. Int. Ed. Engl.* **1970**, *9*, 736.

76. Byrne, J. W.; Blarer, H. V.; Osborn, J. A. *J. Am. Chem. Soc.* **1975**, *97*, 3871.

77. McGee, W. D.; Bergman, R. G. *J. Am. Chem. Soc.* **1988**, *110*, 4246.

78. Batchelor, R. J.; Einstein, F. W. B.; Jones, R. H.; Zhuang, J. -M.; Sutton, D. *J. Am. Chem. Soc.* **1989**, *111*, 3468.

79. Barron, A. R.; Anderson, S. J.; McGregor, K.; Wilkinson, G. *Polyhedron,* **1989**, *8*, 2599.

80. Richard, L.; Weiss, R.; Newton, W. E.; Chen, G. J. -J.; McDonald, J. W. *J. Am. Chem. Soc.* **1978**, *100*, 1318.

81. Bailey, W. I., Jr.; Chisholm, M. H.; Cotton, F. A.; Rankel, L. A. *J. Am. Chem. Soc.* **1978**, *100*, 5764.

82. Day, V. W.; Abdel-Meguid, S. S.; Dabestani, S. *J. Am. Chem. Soc.* **1976**, *98*, 8289.

83. Abe, H.; Kolb, D. M. *Ber. Bunsenges. Phys. Chem.* **1983**, *87*, 523.

84. Douglas, M. A.; Hauge, R. H.; Margrave, J. L. *J. Phys. Chem.* **1983**, *87*, 2945.

85. Kasai, P. H.; McLeod, D. Jr.; Watanabe, T. *J. Am. Chem. Soc.* **1977**, *99*, 3521.

86. *Handbook of Chemistry and Physics,* 56th ed.; Weast, R. C., Ed.; CRC; Boca Raton, **1975**.

87. Scheiner, A. C.; Schaefer, H. F., III *J. Am. Chem. Soc.* **1985**, *107*, 4451.

88. Brooks, B. R.; Schaefer, H. F., III *Mol. Phys.* **1977**, *34*, 193.

89. Schaefer, H. F., III *Acc. Chem. Res.* **1977**, *10*, 287.

90. Rappe, A. K.; Goddard, W. A. *J. Am. Chem. Soc.* **1977**, *99*, 3966.

91. Trenary, M.; Casida, M. E.; Brooks, B. R.; Schaefer, H. F., III *J. Am. Chem. Soc.* **1979**, *101*, 1638.

92. Sakai, S.; Morokuma, K. *J. Phys. Chem.* **1987**, *91*, 3661.

93. Ashby, E. C.; Goel, A. B.; DePriest, R. N. *J. Am. Chem. Soc.* **1980**, *102*, 7779.

94. Ashby, E. C.; Goel, A. B.; DePriest, R. N.; Prasad, H. S. *J. Am. Chem. Soc.* **1981**, *103*, 973.

95. Ashby, E. C.; DePriest, R. N.; Goel, A. B. *Tetrahedron Lett.* **1981**, *22*, 1763.

96. Ashby, E. C.; Goel, A. B. *Tetrahedron Lett.* **1981**, *22*, 1879.

97. Giles, J. R. M.; Roberts, B. P. *J. Chem. Soc., Chem. Commun.* **1981**, 1167.

98. Symons, M. C. R.; Harris, L. *J. Chem. Soc. Faraday Trans.* **1982**, *78*, 3109.

99. Brand, J. C.; Roberts, B. P. *J. Chem. Soc., Chem. Commun.* **1984**, 109.

100. Catton, R. C.; Symons, M. C. R. *J. Chem. Soc. A* **1969**, 2001.

101. Giles, J. R. M.; Roberts, B. P. *J. Chem. Soc., Chem. Commun.* **1981**, 360.

102. Begum, A.; Lyons, A. R.; Symons, M. C. R. *J. Chem. Soc. A* **1971**, 2290.

103. Krusic, P. J.; Bingham, R. C. *J. Am. Chem. Soc.* **1976**, *98*, 230.

104. Herzberg, G. *Molecular Spectra and Molecular Structure*, Vol. 3. *Polyatomic Molecules;* Van Nostrand: Princeton, 1966.

105. Nestmann, B.; Peric, M. *Chem. Phys.* **1984**, *89*, 257.

106. Heitsch, C. W.; Nordman, C. E.; Parry, R. W. *Inorg. Chem.* **1963**, *2*, 508.

107. Hudson, R. L.; Roberts, B. P. *J. Chem. Soc., Chem. Commun.* **1986**, 1194.

108. Roberts, B. P. *Adv. Free Radical Chem.* **1980**, *6*, 225.

109. Roberts, B. P. *Tetrahedron Lett.* **1983**, *24*, 3377.

110. Zhang, Y.; Stuke, M. *Jpn. J. Appl. Phys.* **1988**, *27*, L1349.

111. Zhang, Y.; Beuermann, Th.; Stuke, M. *Appl. Phys. B* **1989**, *48*, 97.

112. Schram, E. P. *Inorg. Chem.* **1966**, *5*, 1291.

113. Lord, R. C.; Nielson, E. *J. Chem. Phys.* **1951**, *19*, 1.

114. Schomburg, G.; Hoffman, E. G. *Z. Elektrochem.* **1957**, *61*, 1110.

115. Ruff, J. K. *Inorg. Chem.* **1963**, *2*, 515.

116. Schram, E. P.; Hall, R. E.; Glore, J. D. *J. Am. Chem. Soc.* **1969**, *91*, 6643.

117. Hoberg, H.; Krause, S. *Angew. Chem., Int. Ed. Engl.* **1976**, *15*, 694.

118. Uhl, W. *Z. Naturforsch.* **1988**, *436*, 1113.

119. Nesper, R.; Curda, J. *Z. Naturforsch*, **1987**, *426*, 557.

120. Miller, M. S.; Schram, E. P. *Organometallics* **1985**, *4*, 1362.

121. Schram, E. P.; Miller, M. M. *Inorg. Chim. Acta* **1986**, *113*, 131.

122. Hoberg, H.; Krause, S. *Angew. Chem., Int. Ed. Engl.* **1978**, *17*, 949.

123. Gavrilenko, V. V.; Chekulaeva, L. A. *Zh. Obshch. Khim.* **1962**, *32*, 689.

124. Stock, A. *Hybrides of Boron and Silicon.* Cornell University Press: Ithaca, NY, 1933.

125. Cox, D. M.; Trevor, D. J.; Whetton, R. L.; Kaldor, *J. Phys. Chem.* **1988**, *92*, 421.

126. Pacchioni, G.; Plavsic, D.; Koutecky, J. *Bev. Bunsenges. Phys. Chem.* **1983**, *87*, 503.

127. Pacchioni, G.; Koutecky, J. *Bev. Bunsenges Phys. Chem.* **1984**, *88,* 242.

128. Basch, H.; Stevens, W. J.; Krauss, M. *Chem. Phys. Lett.* **1984**, *109*, 212.

129. Whetton, R. L.; Cox, D. M.; Trevor, D. J.; Kaldor, A. *Phys. Rev. Lett.* **1985**, *54*, 1494.

130. Cox, D. M.; Whetton, R. L.; Zakin, M. R.; Trevor, D. J.; Reichmann, K. C.; Kaldor, A. *Adv. Laser Sci. I.* AIP Conf. Proc. No. 146; Nov. 1985.

131. Upton, T. H. *Phys. Rev. Lett.* **1986**, *56*, 2168.

132. Hiller, W.; Klinkhammer, K. -W.; Uhl, W.; Wagner, J. *Angew. Chem. Int. Ed. Engl.* **1991**, *30*, 179.

133. Wunderlich, J. A.; Lipscomb, W. N. *J. Am. Chem. Soc.* **1960**, *82*, 4427.

134. Morrison, J. A. In *Advances in Boron and the Boranes*; Liebman, J. F.; Greenberg, A.; Williams, R. E.; Eds.; VCH, Weinheim, 1988. p151.

135. Mannekes, T.; Paetzold, P.; Boese, R.; Blaser, D. *Angew. Chem. Int. Ed. Engl.* **1991**, *30*, 173.

136. Nöth, H.; Pommerening, H. *Angew. Chem. Int. Ed. Engl.* **1980**, *19*, 482.

137. Klemm, W.; Oss, E.; Geiersberger, K. *Z. Anorg. Allg. Chem.* **1948**, *256*, 15.

138. Schnöckel, H. *Z. Naturforsch* **1976**, *B31*, 1291.

139. Schnöckel, H. *J. Mol. Struct.* **1978**, *50*, 275.

140. Tacke, M.; Schnöckel, H. *Inorg. Chem.* **1989**, *28*, 2895.

141. Dohmeier, C.; Robl, C.; Tacke, M.; Schnöckel, H. *Angew. Chem. Int. Ed. Engl.* **1991**, *30*, 564.

142. Anderson, R. A.; Boncella, J. M.; Burns, C. J.; Blom, R.; Haaland, A.; Volden, H. V. *J. Organomet. Chem.* **1986**, *312*, C49.

143. Ahlrichs, R.; Ehrig, M.; Horn, H. *Chem. Phys. Lett.* **1991**, *183*, 227.

144. Kellberg, L.; Bildsøe, H.; Jakobsen, H. J. *J. Chem. Soc. Chem. Commun.* **1990**, 1294.

145. Herzog, S.; Geisler, K.; Präkel, H.; *Angew. Chem. Int. Ed. Engl.* **1963**, *2*, 47.

146. Flamini, A.; Poli, N. *Inorg. Chim. Acta.* **1988**, *150,* 149.

147. Köster, R.; Benedikt, G.; Schrötter, H. W. *Angew. Chem. Int. Ed. Engl.* **1964**, *3*, 514.

148. Lehmkuhl, H.; Fuchs, G.; Köster, R. *Tetrahedron Lett.* **1965**, *29*, 2511.

149. Kaim, W. *J. Organometal. Chem.* **1980**, *201*, C5.

150. Kaim, W. *Chem. Ber.* **1981**, *114*, 3789.

151. Kaim, W. *Z. Naturforsch* **1982**, *37B*, 783.

152. Cloke, F. G. N.; Dalby, C. I.; Henderson, M. J.; Hitchcock, P. B.; Kennard, C. H. L.; Lamb, R. N.; Raston, C. L. *J. Chem. Soc., Chem. Commun.* **1990**, 1394.

153. Cloke, F. G.; Dalby, C. I.; Daff, P. J.; Green, J. C. *J. Chem. Soc., Dalton Trans.* **1991**, 181.

154. Healy, M. D.; Wierda, D. A.; Barron, A. R. *Organometallics* **1988, 7**, 2543.

155. Healy, M. D.; Ziller, J. W.; Barron, A. R. *J. Am. Chem. Soc.* **1990**, *112*, 2949.

156. Lichtenberger, D. L.; Hogan, R. H.; Healy, M. D.; Barron, A. R. *J. Am. Chem. Soc.* **1990**, *112*, 3369.

157. Healy, M. D.; Ziller, J. W.; Barron, A. R. *Organometallics* **1991**, *10*, 597.

158. Lichtenberger, D. L.; Hogan, R. H.; Healy, M. D.; Barron, A. R. *Organometallics* **1991**, *10*, 609.

159. Barron, A. R.; Dobbs, K. D.; Francl, M. M. *J. Am. Chem. Soc.* **1991**, *113*, 39.

160. Kynast, U.; Skelton, B. W.; White, A. H.; Henderson, M. J.; Raston, C. L. *J. Organomet. Chem.* **1990**, *384*, C1.

161. Ziegler, E.; Fuchs, G.; Lehmkuhl, H. Z. *Anorg. Allg. Chem.* **1967**, *355*, 145.

162. Moore, D. S.; Robinson, S. D. *Adv. Inorg. Chem. Radiochem.* **1986**, *30*, 1.

163. Pfeiffer, E.; Kokkes, M. W.; Vrieze, K. *Trans. Met. Chem.* **1979**, *4*, 393.

164. Vanderlinden, J. G. M.; Dix, A. H.; Pfeiffer, E. *Inorg. Chim. Acta* **1980**, *39*, 271.

165. Connelly, N. G.; Finn, C. J.; Freeman, M. J.; Orpen, A. G.; Stirling, J. *J. Chem. Soc., Chem. Comm.* **1984**, 1025.

166. Leman, J. T.; Barron, A. R.; Ziller, J. W.; Kren, R. M. *Polyhedron* **1989**, *8*, 1909.

167. Braddock-Wilking, J.; Leman, J. T.; Barron, A. R. *Abstracts of Papers.* North East Regional Meeting of the American Chemical Society, Amherst, MA: American Chemical Society: Washington, DC, 1991.

168. Zaitseu, B. E.; Zaitseza, V. A.; Molodkin, A. K.; Lisitsina, E. S. *Russ. J. Inorg. Chem.* **1977**, *22*, 504.

CHAPTER

5

The Chemistry of Alkoxides, Thiolates, and the Heavier Group 16 Derivatives of Aluminum and Gallium

John P. Oliver, Rajesh Kumar and Majid Taghiof

Department of Chemistry, Wayne State University Detroit, Michigan, 48202

5.1 Introduction

The chemistry of organoaluminum and -gallium derivatives bound to either a Group V (Group 15) or VI (Group 16) element has been under investigation for many years. The earlier work has been extensively reviewed.[1–7] There has been extensive discussion of exchange processes, including those of the Group III (Group 13) elements.[8,9] The Group V containing derivatives of aluminum and the heavier Group III metals have come under intensive investigation for the past several years because of the interest in III–V semiconductor materials.[10–26] A number of laboratories have been pursuing these studies with those of Beachley, Cowley and Jones and Wells having contributed substantial studies on these systems over a number of years. The III–VI compounds, on the other hand, have received only modest attention. The recent recognition that the III–V compounds may serve as precursors for ceramic materials, as models for aluminum oxide catalysts, and in applications for regio- and stereoselective organic syntheses has led to a renewed interest in the synthesis and characterization of these derivatives.

This chapter, which will focus on recent developments of the chemistry in this area, is not intended to be exhaustive but will point to some of the more interesting results and some of the potential areas for further investigation. It will limit coverage of areas that have been recently reviewed by this author and others. The chapter is divided into three major sections: the first providing a brief overview of synthetic methods; the second dealing with simple monofunctional ligands; the third briefly introducing the chemistry of polyfunctional ligands with organometallic derivatives. In the second and third sections, the syntheses, structures, and dynamic processes as determined by NMR are discussed. Emphasis is placed on the alkylaluminum thiolates which have not been discussed in detail previously.

5.2. Syntheses

5.2.1. General Synthetic Procedures

A number of synthetic methods have been used for the preparation of organoaluminum alkoxides, thiolates and heavier Group III–VI compounds. These include salt elimination, silyl halide elimination, proportionation reactions followed by alkyl transfer and alkane elimination. All these methods have been used for the synthesis of Group III–V and III–VI derivatives; however, the alkane and the salt elimination methods are the simplest and most direct and have been widely used. The reactions for the formation of the alkoxides have been extensively described by others and will not be discussed in detail, but, in general, these follow the same paths described for the formation of the thiolates and heavier derivatives.

5.2.2. Synthesis of Group III Thiolates, Selenolates and Tellurolates

The synthetic procedures for the preparation of Group III compounds containing sulfur are well established; those for the preparation of the selenium and tellurium compounds have not been studied as extensively, but the limited data available suggest that the techniques used for preparation of the thiolates should be effective for the heavier Group III–VI compounds. In general, the same approaches can be taken for all these derivatives, keeping in mind the different oxidation states available and the ease of oxidation-reduction of the

Group III and VI elements. The most common methods used for the preparation of these derivatives are described below.

5.2.2.1 Method 1: Elimination Reactions

The most common method for the formation of Group III thiolates, selenolates, and tellurolates is the reaction of the corresponding thiol, selenol or tellurol with the appropriate Group III organometallic compound with the elimination of an organic molecule. This is depicted in Equations 5.1 – 5.3 and is thought to proceed with the initial formation of the addition compound followed by elimination of alkane. The reaction may proceed in a step-wise fashion leading to the *bis* and *tris* derivatives indicated in Equations 5.2 and 5.3 (E = S). Extensive reviews of the earlier literature have appeared.[1,2,27,28]

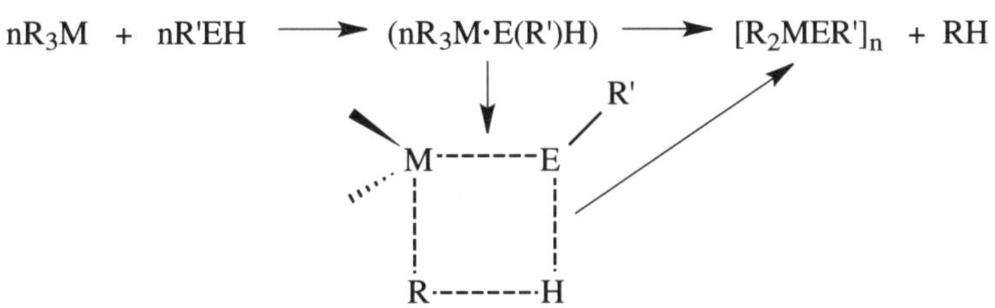

(5.1)

$$[R_2MER']_n + nR'EH \longrightarrow [RM(ER')_2]_n + nRH \quad (5.2)$$

$$[RMER'_2]_n + nR'EH \longrightarrow [M(ER')_3]_n + nRH \quad (5.3)$$

Recently a number of groups have used this approach for the preparation of a variety of these derivatives. For example, we have prepared $(R_2AlSC_6F_5)_2$,[29] $[Me_2Al(SC_6H_4\text{-}2\text{-}i\text{-}Pr)]_3$,[30] $\{i\text{-}Bu_2Al[SC_6H_2\text{-}2,4,6\text{-}(i\text{-}Pr)_3]\}_3$,[30] and $[Me_2Al(SC_6H_3Me_2\text{-}2,6)]_4$.[31] Power and Barron[32] and Cowley et al.[33] have examined the reaction of $(t\text{-}Bu)_3Ga$ with H_2S and obtained two products, initially the species $[(t\text{-}Bu)_2GaSH]_2$ and, after subsequent reaction, $(t\text{-}BuGaS)_4$, a compound with a cubane structure. It should be noted that Cowley et al.[33] obtained the same product by method **4** (Section 5.2.2.4). Hoffmann and Burschka[34] have used method **1** for the preparation of $(Ph_2GaSEt)_2$ and other alkyl- and arylgallium thiolates. In addition, by using dithiols with indium alkyls and halides the formation of polymeric chains with thiol bridges has been observed.[35] Using the alkane elimination (Equation 5.1), Nomura et al.,[36,37] have studied the reaction of a series of thiols and dithiols with R_3In derivatives and have characterized the resulting complexes by analysis, physical measurements, and ^1H NMR spectroscopy.

Reaction of Me_3Al with $2,6\text{-}Me_2C_6H_3SH$, in either a 1:1 or a 1:3 ratio, yields only a single new product, $Al(SC_6H_3Me_2\text{-}2,6)_3$.[31] This reaction is illustrated in Equation 5.4 and proceeds to completion as indicated.

$$Me_3Al + 2,6\text{-}Me_2C_6H_3SH \rightleftharpoons Al(SC_6H_3Me_2\text{-}2,6)_3 \quad (5.4)$$

The formation of the *tris* thiolate most likely results from the disproportion of the derivative R_2AlSR to the products. When small ligands such as Me, Et or Ph are attached to the metal center, the reaction can be stopped along the way with only one or two alkyl or aryl units replaced, leading to the mixed species, $[R_2M(ER')]_n$. When basic solvents (ethers, amines) are present, the mixed products can also be obtained with the formation of $Mes_2RSAl \cdot B$ addition compounds, as noted above. These observations indicate that the steric interactions between the substituent groups may determine the course of this reaction.

Another elimination process that has been explored[35] is the reaction of Me_3SiSR with a metal halide as shown in Equation 5.5, which offers another route for the preparation of III–VI derivatives.

$$MX_3 + nMe_3SiSR \longrightarrow M(SR)_nX_{3\text{-}n} + nMe_3SiX \quad (5.5)$$

5.2.2.2. Method 2: Salt Elimination

The reaction of a Group III organometallic halide with a metal thiolate, selenolate, or tellurolate also provides a direct and effective means for the formation of the desired products and may prove particularly useful where the parent hydride is unstable or where the starting organometallic derivative is either difficult to prepare or to handle. This is illustrated in Equation 5.6 (E = Se).

$$R_{3\text{-}n}MX_n + nM'ER' \longrightarrow R_{3\text{-}n}M(ER')n + nM'X \quad (5.6)$$

Use of this technique has been illustrated for the preparation of $Al[S(2,4,6\text{-}t\text{-}Bu_3C_6H_2)]_3$, the isomorphous gallium derivative,[38] and, subsequently the selenium derivative, $Ga[Se(2,4,6\text{-}t\text{-}Bu_3C_6H_2)]_3$.[39] This group has also prepared other derivatives related to these compounds but full details of this work, which typically leads to the *tris* derivative and not to the mixed organometallic Group III–VI species, have not been published. Starting with $(Me_3CCH_2)_2GaCl$, $[(Me_3CCH_2)_2GaTePh]_2$ has been prepared using this approach,[40] thus indicating that this procedure can be extended to the synthesis of a variety of organometallic chalcogenides.

5.2.2.3. Method 3: Cleavage-Insertion of Dichalcogenides

The reaction of a disulfide, -selenide or -telluride with a Group III alkyl may provide a means for the preparation of the corresponding thiolate, selenolate or tellurolate. As illustrated in Equation 5.7 this leads to mixed products (E = S, Se, or Te).

$$R_3M + R'EER' \longrightarrow R_2MER' + RER' \quad (5.7)$$

We have used this technique very effectively for the preparation of organoaluminum thiolates and selenolates. Alternatively, at least for the indium compounds, oxidative addition as shown in Equation 5.8 may lead to formation of thiolates, selenolates and tellurolates.

$$R'EER' + M \longrightarrow M(ER')_3 \tag{5.8}$$

In this case only the *tris* ER compounds are formed.

5.2.2.4. Method 4: Direct Addition-Insertion of the Elements

The simple Group VI elements insert into carbon-metal bonds in many instances leading to the formation of III–VI derivatives (Equation 5.9).

$$R_3M + nE \longrightarrow R_{3-n}M(ER)_n \tag{5.9}$$

This approach has been used very effectively for the formation of indium compounds. This method offers the advantage of not requiring the preparation or handling of the Group VI hydride, metallate or dielement derivative. It may have the disadvantage of yielding only the *tris* derivatives; alternatively, it may yield more complex products of interest such as the cubane structures or the chalcogen bridged compounds.

5.2.2.5. Method 5: Electrochemical Synthesis

Green et al.[41] have reported synthesis of In(I) and In(III) thiolates by electrochemical synthesis. This procedure does not appear to offer a general method of synthesis but does offer an alternative approach to the synthesis of the $In(SR)_3$ derivatives and provides a method for the preparation of InSR species which could prove very useful for oxidative addition reactions leading to other compounds.

5.2.2.6. Method 6: Exchange-Redistribution

Exchange/redistribution reactions that occur on mixing the R_3M derivative with the metal trichalcogenide, $M(ER')_3$, provide another means for the formation of the mixed organometallic Group VI derivatives. These proportionation/disproportionation reactions are also of concern in the formation of all mixed products, $R_nM(ER')_{3-n}$, since the isolation of a specific product will depend on its thermodynamic and/or kinetic stability. For these reasons it is essential that the stability/equilibrium position of the reaction shown in Equation 5.10 is known. Therefore, exchange studies will be carried out to determine the stabilities of the mixed products toward redistribution.

$$R_3M + M(ER')_3 \rightleftharpoons R_2MER' + RM(ER')_2 \tag{5.10}$$

5.3. Structure and Bonding

5.3.1. Monodentate Ligands

The aluminum center in $(R_2AlER')_n$ derivatives has been shown to exist with various degrees of aggregation ranging from one to four and different coordination numbers ranging from three to six. There are two major classes of derivatives to be discussed: three-coordinate species and bridged four-coordinate derivatives of the general form $(R_2Al{-}\mu{-}ER')_n$. It is also important to recognize that simple four-coordinate addition compounds, $R_n(ER)_{3-n}Al{\cdot}B$ (n = 0-3), may be formed either with the monomers or by disruption of the bridge-bonded oligomeric species.

The structure of these R_2AlER' derivatives depends on the steric and electronic properties of the R group and ER' moiety and include monomeric species, dimeric, trimeric and tetrameric rings, and in some instances, more complex derivatives containing metals in different coordination environments in the same molecule. The tendency to form bridged structures is characteristic of the Group VI derivatives of these metals because of the affinity of the metal atoms for the oxygen or sulfur. Much less is known about the heavier chalcogens. We have reviewed much of this information recently and note that the type of complex or aggregate formed is dependent on several variables, including the organic group bound to the metal and the character of the substituent on the non-metallic element.[7]

5.3.2. Three-coordinate Derivatives

The formation of electron-deficient three-coordinate organoaluminum compounds has been invoked in many cases to account for the reactivity of these derivatives. Until recently, the only structural data for the three-coordinate derivatives was that obtained for Me_3Al in the gas phase,[9] for mesityl species,[42] and tris[bis(trimethylsilyl)amide]aluminum. Uhl[44] has shown that species can be stabilized in this configuration by use of suitable bulky ligands such as $-CH(SiMe_3)_2$. The heavier Group 13 derivatives are frequently found with the metal atom in planar three-coordinate environments when alkyl substituents are present, but with higher coordination numbers of five or six when halide or chalcogenides are bound to the metal.

For aluminum to form stable three-coordinate derivatives, bulky substituents such as BHT (2,6-di-*t*-butyl-4-methylphenoxide) are required. The first structure of this type was reported for $MeAl(BHT)_2$, **1**.[45] Subsequently Benn et al.,[46] showed the *tris*-BHT derivative also to be monomeric.

In the three-coordinate compounds the Al–O–C angle is large and the Al–O bond distance is short in comparison with Al–O distances observed in other compounds. These changes support the proposed interaction between the π-electrons on the oxygen and the vacant p-orbital of aluminum (Table 5.1). With the BHT and *bis*(trimethylsilyl)amide derivative, delocalization of electron

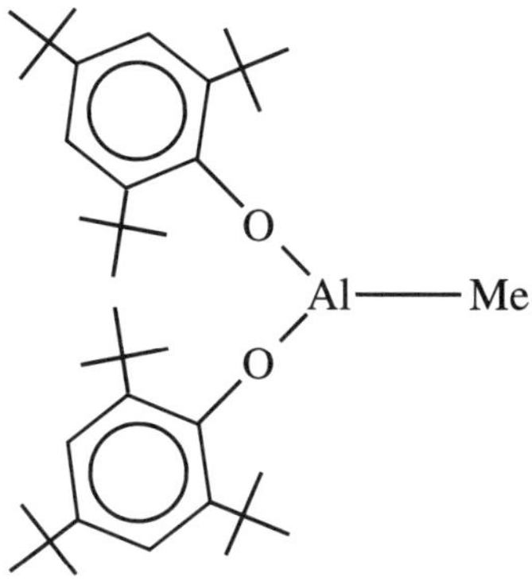

1

density from the oxygen or nitrogen might be expected. Similarly, the heavier chalcogens could interact in this fashion and stabilize the Al–E bond. The extent of the delocalization will depend on several factors, including the separation of the atoms, the relative energies of the orbitals involved in bond formation, and the orientation of the p-orbitals. These mesomeric effects in the alkoxide, **2**, should be reflected in two ways; first, the Al–E bond should decrease in length as the electron delocalization increases, and second, the formation of addition compounds should become less favorable with the increasing electron delocalization. This has been suggested by many authors for the aluminum alkoxides.

Ligands that involve a heavier chalcogen atom attached to the metal generally require much larger substituents to effect the same degree of crowding. Further, the orbital match may be less favorable, reducing the back donation of electrons suggested for the alkoxides. For these reasons and because of the strong electrophilic character of the metal atoms, there are very few low coordinate (three or less) neutral derivatives of the Group III metals bound to a heavier chalcogen. Exceptions to this are the *tris*(thiolato) derivatives of

Table 5.1. Bond distances and angles in three-coordinate organoaluminum alkoxides and thiolates.

Compound	Al–C(Å)	Al–E(Å)	C–Al–E(°)	E–Al–E(°)	Al–E–C(°)	Ref.
$MeAl(BHT)_2$	1.927	1.686	123.7	111.9	143.6	45
i-$BuAl(OC_6H_3$-2,6$Me_2)_2$	1.952	1.692	124.9	110.1	146.0	46
t-Bu_2AlBHT	1.980	1.71	117.3	–	129.4	47
n-BuAl(S-2,4,6-t-$Bu_3C_6H_2)_2$	–	2.19	–	–	–	38
Al(S-2,4,6-t-$Bu_3C_6H_2)_3$	–	2.185	–	115.1-125.9	100.4	38
Ga(Se-2,4,6-t-$Bu_3C_6H_2)_3$	–	2.324	–	117.7-122.5	93.9	39
Ga(S-2,4,6-t-$Bu_3C_6H_2)_3$	–	2.205	–	115.6-125.5	100.4	38

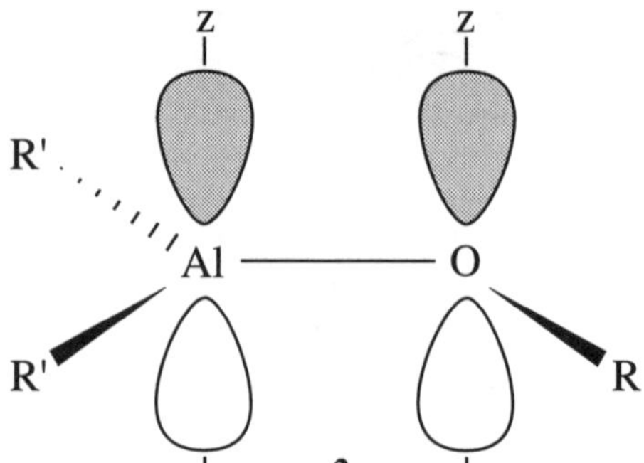

2

aluminum and gallium, $Al(S\text{-}2,4,6\text{-}t\text{-}Bu_3C_6H_2)_3$ and $Ga(S\text{-}2,4,6\text{-}t\text{-}Bu_3C_6H_2)_3$, and the gallium selenolate, $Ga(Se\text{-}2,4,6\text{-}t\text{-}Bu_3C_6H_2)_3$, which have been prepared and their structures determined. These compounds are all planar about the metal center. The average Al–S bond length in $(n\text{-}Bu)(SMes^*)_2Al$ and $(Mes^*S)_3Al$ ($Mes^* = 2,4,6\text{-}t\text{-}Bu_3C_6H_2$) are 2.19 and 2.185 Å, respectively which is substantially shorter than the related bond lengths found in other organoaluminum thiolato complexes in higher coordination numbers.[38] This shortening of Al–S bond length was, however, not interpreted in terms of a π-interaction between the empty metal p-orbital and the lone pairs of the sulfur atom but in terms of increased ionic character in the Al–S bond. The spectroscopic and struc-tural data were interpreted in support of this hypothesis. Compounds with three-coordinate structures of Group III chalcogenides are listed in Table 5.1. The monomeric alkoxide species, $Me(OR)_2Al$, are highly regio- and stereoselective in organic synthesis. There are a number of studies on BHT and related organoaluminum derivatives giving details of the use of this and similar reagents as synthons for sterically-controlled reactions.[48–54] These bulky reagents have proven particularly useful in this area. To date, the three-coordinate derivatives, $R_2(OR')Al$ (R = Me, Et or *i*-Bu; OR' = 2,4,6-tri-*t*-butyl-phenoxide, 2,6-diphenylphenoxide, 2,6-di-*t*-butyl-4-bromophenoxide, 2-*t*-butyl-6-(1,1-diethylpropyl)-4-methylphenoxide, (R)-(+)-3,3'-bis (triphenyl)binaphthol), have been shown to exist in solution but never isolated and characterized. In one specific example the stereoselective, equitorial (anti-Cram) alkylation of a substituted cyclohexane was rationalized

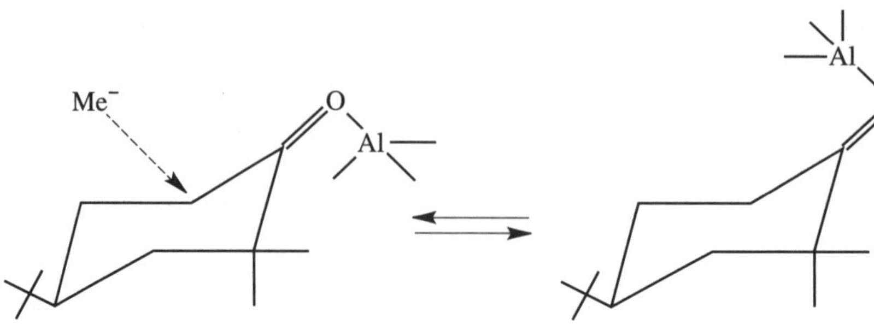

(5.11)

by postulating preferential complexation of $Me(OR)_2Al$ (OR = 2,6-di-*t*-butyl-4-methyl-phenoxide) to the equitorial diastereotopic π-face of the ketone as shown in Equation 5.11.[55]

A second class of reactions that is of importance and is related to the stereoselectivity of these systems is their interaction with bases. When small bases are used or with loss of one of the BHT groups, the fourth coordination site is available. A number of addition compounds of this type have been characterized[56–59] and the influence of the bulky substituents on the reactivity of the $MeAl(BHT)_2$ species has been discussed. It should also be noted that these three-coordinate derivatives, $R_2(OR')Al$, undergo disproportionation reactions to give monomeric $RAl(OR')_2$ species. Only one significant NMR study has been reported which describes the behavior of these compounds.[45] In this study it was demonstrated that there is an equilibrium between *mono* and *di* BHT species in solution. This equilibrium is depicted in Equation 5.12 with all four species in the equilibrium mixture observed. Further, it was noted that the simple alkylaluminum dimers such as Al_2Me_6 undergo fast exchange with the R_2AlBHT derivative, but that the exchange is relatively slow for the $RAl(BHT)_2$ species on the NMR time scale. It appears that this may be accounted for in terms of the ease of bridge formation with the BHT-containing compound. With only one BHT group present, alkyl bridged dimers can form, providing a path for exchange, but with two BHT moieties present, formation of the bridged intermediate becomes far more difficult and slows the exchange process. No similar studies have appeared for the heavier chalcogen derivatives, but it seems likely that as these reagents become available they will prove useful in selected synthesis and will be shown to form addition compounds and to undergo chemical exchange.

$$Al_2R_6 + RAl(BHT)_2 \rightleftharpoons R_2Al(BHT) + R_5Al_2(BHT) \qquad (5.12)$$

5.3.3. Four-coordinate Bridged Alkoxides, Thiolates and Tellurolates

The predominant structure of Group III (Al and Ga) chalcogenides are strongly associated through the bridging oxygen, sulfur, or tellurium and typically exist as dimers, trimers or tetramers with the position of the equilibrium dependent on the substituents on both the metal and non-metal atoms. Tables 5.2 – 5.5 provide a listing of the known structures for the III–VI chalcogen derivatives. This equilibrium between monomer and aggregate is shown in Equation 5.13 and typically lies to the aggregate with the formation of cyclic dimers, trimers and tetramers.

$$nR_2MER' \rightleftharpoons [R_2M(\mu\text{-}ER')]_n \qquad (5.13)$$

The exceptions to this have been noted earlier. The degree of aggregation, *n*, is dependent on the nature of E and on the groups R and R'. Several studies have

Table 5.2. Selected bond distances and angles in dimeric four-coordinate organoaluminum alkoxides.

Compound	Al–O(Å)	Al–C(Å)	Al–Al(Å)	C–Al–C(°)	O–Al–O(°)	Al–O–Al(°)	Ref.
$[Me_2Al(l\text{-borneolate})]_2$	1.841	1.956	2.779	122.3	82.0	98.0	62
$[Me_2Al(l\text{-mentholate})]_2$	1.841	1.963	2.807	118.7	80.6	99.4	62
$[(i\text{-Bu})_2Al(l\text{-mentholate})]_2$							
Molecule 1	1.839	1.956	2.823	123.3	79.0	100.2	62
Molecule 2	1.842	1.942	2.840	120.7	79.1	100.9	62
$[Me_2Al(OC_6H_3\text{-2-allyl-6-Me})]_2$	1.861	1.946	2.866	115.4	79.2	100.8	63
$[Et_2Al(OC_6H_3\text{-2-allyl-6-Me})]_2$	1.862	1.951	2.865	117.0	79.3	100.7	61
$[AsMe_4]_2[Me_2AlOAlMe_3]_2$	1.79	–	–	–	–	96.0	64
$[Al_2(OSiMe_3)_4(Acac)_2]$	1.838	–	–	–	86.8	97.4	65
$[Me_2AlOMe]_3$ (gas phase)	1.851	1.957	–	117.3	103.2	125.8	66
$[Me_2Al(OSiMe_2Ph)]_2$	1.853	1.948	2.760	121.3	83.7	96.3	67
$[Me_2Al(OSiMe_3)]_2$	1.843	1.940	–	–	–	–	68
$[Al(OSiMe_3)_3]_2$	1.813 (bridge)	–	2.692	–	95.4	–	69
$[Me_2Al(OC_6F_5)]_2$	1.895	1.920	2.951	127.5	77.7	102.3	70
$[(i\text{-Bu})_2Al(OC_6H_3\text{-2,6-Me}_2)]_2$	1.873	1.971	2.899	115.6	78.5	101.5	46
$[Al(OC_6H_3\text{-2,6-Me}_2)_3]_2$	1.834	–	2.815	–	79.7	100.3	46
	1.686 (terminal alkoxide)				120.2 (terminal alkoxide)		
$[Al(O\text{-}t\text{-Bu})_3]_2$	1.827	–	2.779	–	81.1	98.9	71
	1.690 (terminal alkoxide)				115.1 (terminal alkoxide)		
$[Me_2AlO(Me)ZrCp_2]_2$	1.844	1.956	2.667	116.5	87.4	92.6	72
$[(Me_2Al)(Me)\{OCMe_2)(OC)ZrCp_2\}_2]$	1.808	1.961	–	114.4	102.4	–	73
$[Me_2AlOCH_2C_6H_2\text{-2,4,6-}(t\text{-Bu})_3]_2$	1.850	1.949	2.840	114.8	79.8	100.2	74

addressed the structure of this class of compounds and the factors that determine the relative stability of the dimeric, trimeric, or tetrameric forms. These studies have largely been qualitative in nature and have generally applied thermodynamic arguments to rationalize the results of solid-state molecular structure determinations. The current reasoning, based on enthalpic changes and entropy arguments, suggests that the stability of the simple Al–E–Al bridge bond is quite high with values for the bond energy for the oxygen bridged derivatives typically in the range of 80 –150 kJ/mol, thus making it difficult to disrupt this bond. Earlier studies established a lower limit of 131.5 kJ/mol for the Al–O–Al bridge bond in $[Al(O\text{-}i\text{-}Pr)_3]_4$,[60] which is in mid-range for these bridge bond energies. The Al–S–Al bridge bonds are thought to be lower in energy but good quantitative information has not appeared for these systems. The stability of the alkoxide bridge bonds inhibits the interchange between bridge and terminal sites, and the equilibrium between dimers, trimers and higher aggregates. The degree of aggregation will be determined by the relative importance of the entropy terms and the enthalpy for the reactions. The latter will be determined by steric interactions involving the substituent groups, by angle strain and by the change in interaction energy between the oxygen and metal as a function of orbital overlap. This would be enhanced by the planarity of the oxygen atom which places the substituent in the ring plane. A four-membered Al_2O_2 ring (dimer) will have greater ring bond angle strain but reduced steric interference. Thus, larger groups should favor the dimer over the trimer for oxygen bridged systems. Entropy also favors the dimeric form over trimeric, as more independent molecules are present when the substance is dimeric. Further, the formation of a larger ring system is not common since little additional energy of stabilization is gained, and the entropy change for the formation of larger aggregates is unfavorable. An examination of the X-ray crystallographic data presented in Table 5.2 supports the concepts cited above. Thus, the Al–O–Al bridge bond angle in the dimer ranges from 92.6–101.5° whereas in the trimer, though limited data are available, this angle is significantly wider with values in the 120–125° range.

The derivatives commonly observed are typically dimeric for simple alkoxides as shown in **3**, although $(Me_2AlOMe)_3$ forms a six-membered ring in a

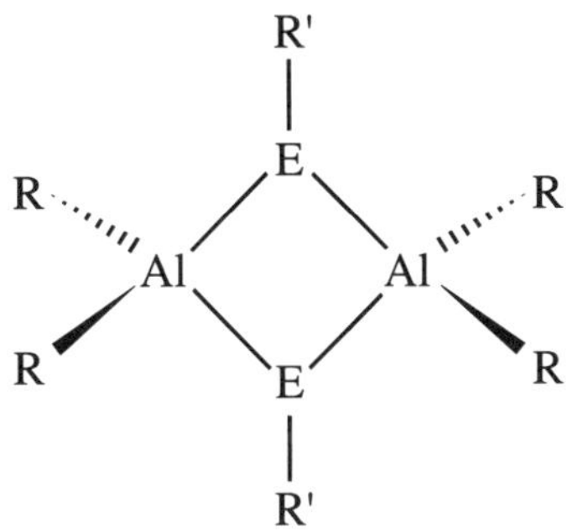

3

Table 5.3. Reported structures of aluminum thiolates, selenolates, tellurolates and related compounds.

Compound	Comments	Al–E distance (Å)	E–Al–E(°)	Al–E–Al(°)	Sum of angles around E	Ref.
$(Me_2AlSMe)_n$	-(Al–S)- infinite chain (crystal)	2.348	100.1	103.0	310.3	75
$(Me_2AlSMe)_2$	planar, dimer (gas phase)	2.370	85.5	94.5	–	76
$(Me_2AlSSiPh_3)_2$	nearly planar, dimer	2.36	87.51	91.85	330.21	30
			87.68	92.54	330.72	
$(Me_2AlSC_6F_5)_2$	dimer, butterfly	2.402	89.41	86.88	300.98	29
			89.61	87.31	308.01	
$[Me_2Al(SC_6H_4\text{-}2\text{-}t\text{-}Bu)]_3$	trimer, chair	2.35	100.6	115.1	323.9	29
			88.8	123.2	338.5	
			99.1	114.6	324.6	
$[Me_2Al(SC_6H_4\text{-}2\text{-}i\text{-}Pr)]_3$	trimer, twisted chair	2.36	101.53	115.83	324.53	30
			99.88	114.23	337.1	
			90.22	122.5	320.03	
$[(i\text{-}Bu)_2Al][SC_6H_2\text{-}2,4,6\text{-}(i\text{-}Pr)_3]_3$	trimer, twisted boat	2.37	89.1	127.6	360	30
			93.2	126.8	357.4	
			96.1	131.5	356.3	
$(Me_2AlSC_6H_3Me_2\text{-}2,6)_4$	tetramer, extended chair	2.36	99.46	126.92	342.8	31
			94.60	130.68	348.3	
$[Me_2Al(\mu\text{-}PyS)]_2$	dimer, head-to-head	2.338	97.34	–	–	77
$Al_4I_4(SMe)_4S_2$	tetrameric cage, Al–S–Al bridged	2.18	–	–	–	78
$Al(S_2CNMe_2)_3$	octahedral, six-coordinates/S-bonded	2.395	75	–	–	79
			164			
$Al[S(2,4,6\text{-}t\text{-}Bu_3C_6H_2)]_3$	three-coordinate monomer, nearly planar	2.185	119.8	–	–	38
$K[Me_3AlSCNAlMe_3]$	ionic, Al–S and Al–N bonded	2.498	–	–	–	80
$K[MesS(AlMe_3)_3]\cdot 2C_6H_6$	ionic, tetrahedral Se	2.578	–	114.6	–	81
$(Me_3Se)(MeAlCl_3)$	ionic, secondary interaction	–	–	–	–	82
$[Me_3Te][Me_2AlCl_2]$	ionic, secondary interaction gives six-coordinate Te	–	–	–	–	83
$[(t\text{-}Bu)_2AlTe(t\text{-}Bu)]_2$	planar dimer	2.732	86.1	93.9	322.6	33
$[i\text{-}Bu_2Al(\mu\text{-}OC_6H_4\text{-}2\text{-}SMe)]_2$	dimer, O-bridged, weak S-Al interaction	2.778	–	–	301.1	84
$(Me_3Al)(12)aneS_4$	addition compound	2.718	–	–	–	85
$(Me_3Al)_4(14)aneS_4$	addition compound	2.522	–	–	–	86

$(Me_3Al)_4(12)aneS_4$	addition compound	2.515	–	–	–	87
$Cl_3Al\cdot SPPh_3$	addition compound	2.297	–	–	–	88
$Cl_3Al\cdot SePPh_3$	addition compound	2.452	–	–	–	88
$Me_3Al\cdot SMe_2$ (gas phase)	addition compound, gas phase	2.55	–	–	–	89
$\{(i\text{-}BuAl)_2[(C_6H_5)_2P(S)CP(C_6H_5)_2\text{-}(S)2][Al(i\text{-}Bu)_2]_2\}$		2.304		83.7	–	90
$[Me_2Al][Ph_2P(S)CHP(S)Ph_2AlMe_3$	Al-S adducts	2.388 2.460	–	–	–	91
$[Me_3Al]_2[Ph_2P(S)CH_2CH_2P(S)Ph_2]$	Al-S adducts	2.506	–	–	–	91

Table 5.4. Reported structures of gallium thiolates, selenolates, tellurolates and related compounds.

Compound	Comments	Ga–E distance (Å)	E–Ga–E(°)	Ga–E–Ga(°)	Sum of angles around E	Ref.
$(Ph_2GaSEt)_2$	dimer, planar	2.378	93.58	86.42	294.92	34
$(t\text{-}Bu_2GaSH)_2$	dimer, planar	2.432	90.5	89.5	282.5	32
$(Me_2GaSC_6F_5)_2$	dimer, butterfly	2.449	89.45 88.59	87.75 87.99	301.75 304.89	70
$[Me_2Ga(\mu\text{-}SC_6H_3Me_2\text{-}2,6)]_4$	tetramer, extended chair	2.397	93.07	126.21	340.71 346.0	31
$[I_2GaSMe]_2$	dimer	2.379	94.5	85.5	–	92
$[I_2GaS(i\text{-}Pr)]_2$	dimer, butterfly	2.329	89.1	84.7	–	93
$(t\text{-}BuGaSe)_4$	tetramer, cubane	2.481	99.3	79.9	–	32
$(t\text{-}BuGaS)_4$	tetramer, cubane	2.359	97.3	82.1	–	32,33
$Ga[Se(2,4,6\text{-}t\text{-}Bu_3C_6H_2)]_3$	monomer, planar	2.324	119.9	–	–	39
$[(Me_3CCH_2)_2GaTePh]_2$	dimer, butterfly	2.755	92.01	83.77	289.92	40
$Ga[S(2,4,6\text{-}t\text{-}Bu_3C_6H_2)]_3$	monomer, nearly planar	2.205	119.8	–	–	38
$(n\text{-}Pr)_4N[Ga(SEt)_4]$	ionic, tetrahedral GaS_4 unit	2.262	109.5	–	–	94
$(Et_4N)_4[Ga(SPh)_4]$	ionic, tetrahedral GaS_4 unit	2.257	109.6	–	–	94
$(Et_4N)_2[Ga_2(SPh)_4]$	dimer, sulfide bridges	2.276	109.5	80.9	–	94

Table 5.5. Reported structures of indium thiolates, selenolates, tellurolates, and related compounds.

Compound	Comments	In–E distance (Å)	E–In–E(°)	Ref.
$In(SePh)_3 \cdot PPh_3$	monomer, four-coordinate	2.546	115.01	95
$[In(SePh)_3]_n$	linear, one-dimensional polymer	2.776	90.02	96
$In(SPh)_3 \cdot 2Py$	monomer, five-coordinate In	2.458	119.9	96
$[(n\text{-Bu})_4N]_2\{[S_2C_2(CN)_2]_2 InCl\}$	five-coordinate, square pyramidal	2.537	107.2	97

chair conformation.[61] The Al_2O_2 ring system is normally planar, with this plane perpendicular to the plane described by the four R groups and the two Al atoms. The ring containing sulfur or the heavier chalcogens may be planar or in some instances has been shown to be puckered. The configuration of the ring is likely dependent on the steric interactions between the substituents and the low energy required for angular deformation. The most important distinction between oxygen and the heavier chalcogens is the geometry around the bridging element E. In the aluminum alkoxides, the oxygen atom is in a trigonal planar environment. Typically the Al–O–Al angle is approximately 100° and the Al–O–C angles are in the range of 125 – 135°. When sulfur or heavier chalcogen is in the bridging atom, the angles around it are acute and the sulfur is in a pyramidal environment. This leads to different steric requirements and stabilities for the aggregates. In the alkoxides this is manifested by strong Al–O–Al bridge bonds, transfer of spin-coupling across the bridge and almost certainly by diminished reactivity of the compounds. In the thiolates the bonds are weaker, and the geometry is altered by the configuration around the sulfur atom.

The chemistry of thiolato derivatives of aluminum and gallium compounds of the class $[R_2M(\mu\text{-ER}')]_n$ has not received significant attention. They also form cyclic dimers, trimers and tetramers with the degree of aggregation, *n*, dependent on the same factors cited for the oxygen derivatives, but the change in bond energy and the change in bond distance alter the relative importance of the steric interactions and angle strain. Thus, the nature of the groups R and R' will differ in these systems. The first structural study established that $[Me_2Al(\mu\text{-SMe})]_n$ is a linear polymer.[75] In the gas phase this compound is reported to be primarily a cyclic dimer with a planar $(AlS)_2$ ring with the methyl groups oriented in the *anti* conformation.[76] The few studies reported show that the four-membered Al_2S_2 ring (dimer) appears to be formed with thiolato ligands which have reduced steric interference as observed in $[Me_2Al(\mu\text{-SC}_6F_5]_2$[29] and may have a butterfly conformation shown in **4**. However, when ligands with larger steric interference are used, the trimers, $[Me_2Al(\mu\text{-SC}_6H_4\text{–}2\text{-}i\text{-Pr})]_3$, **5**,[30] $[Me_2Al(\mu\text{-SC}_6H_4\text{–}2\text{-}t\text{-Bu})]_3$, **6**,[30] and the tetramer, $[Me_2Al(\mu\text{-SC}_6H_3\text{–}2,6\text{-Me}_2]_4$, **8**,[31] are formed.

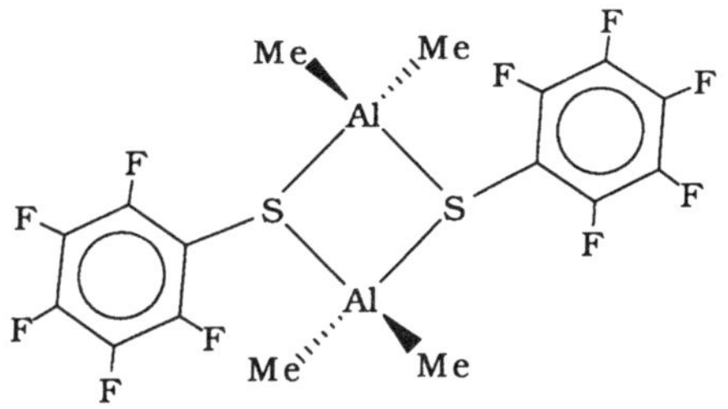

$[Me_2Al(\mu\text{-}SC_6F_5)]_2$
4

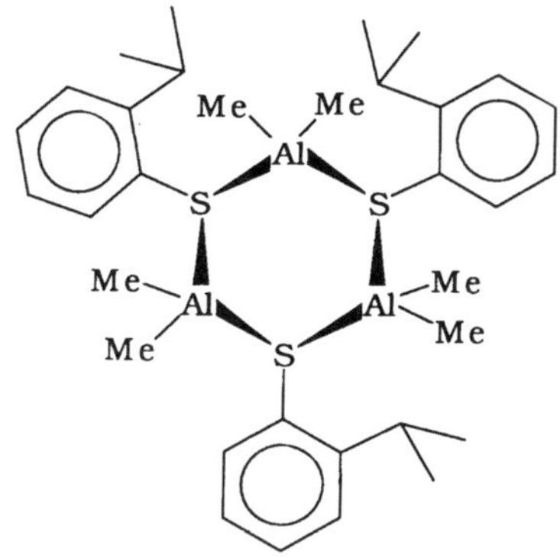

$[Me_2Al(\mu\text{-}SC_6H_4\text{-}2\text{-}i\text{-}Pr)]_3$
5

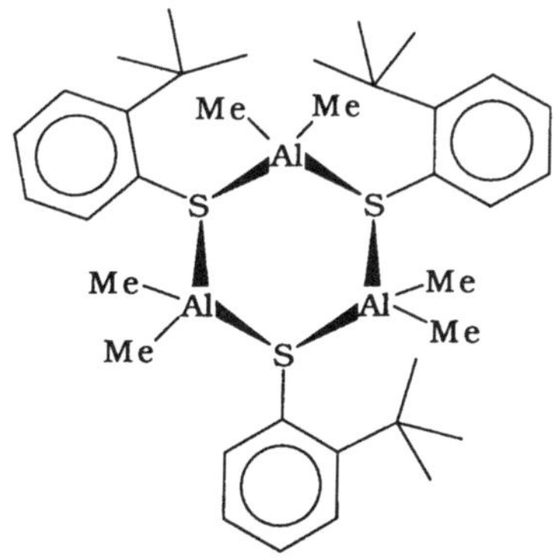

$[Me_2Al(\mu\text{-}SC_6H_4\text{-}2\text{-}t\text{-}Bu)]_3$
6

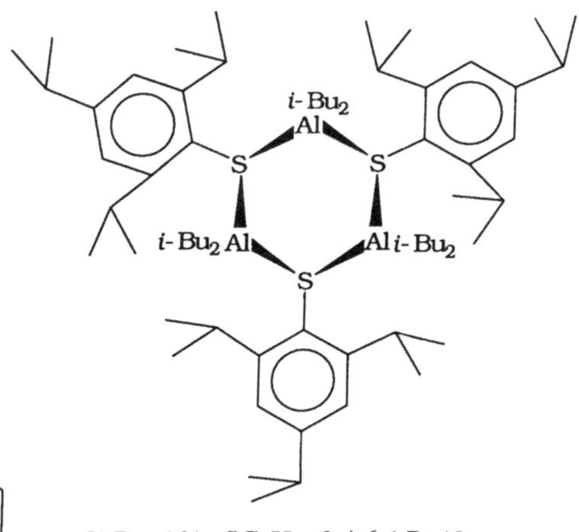

$[i\text{-}Bu_2Al(\mu\text{-}SC_6H_2\text{-}2,4,6\text{-}i\text{-}Pr_3)]_3$
7

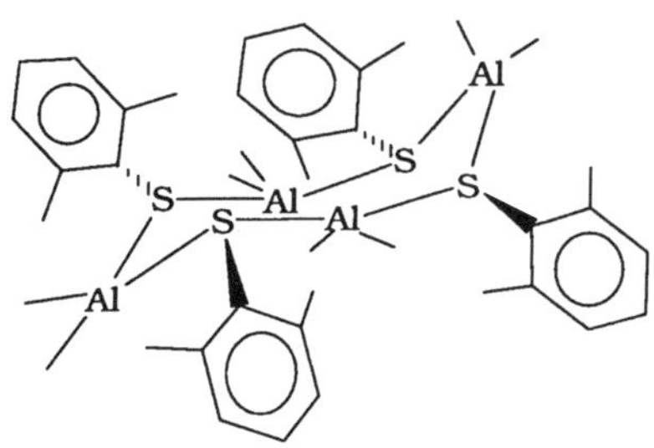

$[Me_2Al(\mu\text{-}SC_6H_3\text{-}2,6\text{-}Me_2)]_4$
8

With the more crowded molecule, $[i\text{-}Bu_2Al(\mu\text{-}SC_6H_2\text{-}2,4,6\text{-}i\text{-}Pr_3)]_3$, **7**,[30] the central ring adopts a twisted-chair conformation. In the tetramer, $[Me_2Al(\mu\text{-}SC_6H_3\text{-}2,6\text{-}Me_2)]_4$, **8**, the central $(AlS)_4$ ring is in an extended chair. The same conformation is present in the isomorphous $[Me_2Ga(\mu\text{-}SC_6H_3\text{-}2,6\text{-}Me_2)]_4$ derivative.

The stabilities of the six- and eight-membered rings was unanticipated since the more sterically-hindered systems have been assumed to give smaller aggregates. The ultimate product obtained with very bulky substituents is the monomer as noted earlier. The X-ray crystallographic data presented in Table 5.3 show that the sulfur is pyramidal in the four-membered $(AlS)_2$ ring system in all cases. The ring may be puckered as in $[Me_2Al(\mu\text{-}SC_6F_5)]_2$ or planar as in $[Me_2Al(\mu\text{-}SMe)]_2$ (gas phase) or in $[Me_2Al(\mu\text{-}SSiPh_3)]_2$. There is no reason other than steric interactions for the thiolate derivative to assume the puckered configuration. In the trimeric six-membered ring systems, $[Me_2Al(\mu\text{-}SC_6H_4\text{-}2\text{-}i\text{-}Pr)]_3$ and $[Me_2Al(\mu\text{-}SC_6H_4\text{-}2\text{-}t\text{-}Bu)]_3$, the rings are in the chair conformation with the substituents on the sulfur atoms in the seat of the chair in equatorial positions while the substituent on the back of the chair is in the axial position. In the sterically hindered trimer, $[i\text{-}Bu_2Al(\mu\text{-}SC_6H_2\text{-}2,4,6\text{-}i\text{-}Pr_3)]_3$, the ring is in a twisted boat conformation. The two substituents on sulfur remain in the equatorial positions while the third, located above the central plane, is in the axial position. In the tetrameric eight-membered rings, $[Me_2Al(\mu\text{-}SC_6H_3\text{-}2,6\text{-}Me_2)]_4$ and $[Me_2Ga(\mu\text{-}SC_6H_3\text{-}2,6\text{-}Me)_2]_4$, the molecules are in extended chair conformations with all the sulfur atoms in the seat of the chair and all the substituents bound to sulfur in equatorial positions. The M–S–M angles observed are similar to those found in the trimers. The substituents on the sulfur atom play an important role in determining the conformations. With a single substituent in the 2-position on the phenyl ring, a chair conformation is formed with the two substituents in the seat of the chair in equatorial positions and the remaining sulfur ligand in the axial position. This difference in the structure of the Al_nE_n ring is inherently associated with the difference in geometry of the heteroatom bridging the metal centers; that is, the geometry around the oxygen atom in alkoxide bridged dimers tends to be planar whereas in sulfur bridged species it is pyramidal. These results suggest that the Al_2S_2 ring has the potential for conformational equilibria which is seen on examination of **4** for $[Me_2Al(\mu\text{-}SC_6F_5)]_2$ and will be true for all pyramidal sulfur, selenium, and tellurium derivatives. Exploration of the phenomenon will be discussed in the following section.

To date, the selenium and tellurium analogs of organoaluminums are the least studied. X-ray structural data on selenolates and tellurolates is very scarce. The few structural studies that have been reported are listed in Tables 5.3 – 5.5. The structural data showed $[t\text{-}Bu_2AlTe\text{-}t\text{-}Bu]_2$ to be a dimer with a planar Al_2Te_2 ring and a Al–Te bridge bond distance of 2.732 Å.[33] The structure of $[Me_3AlSeMe]^-$ is based on an "ate" type with an Al–Se bond distance of 2.578 Å.[81]

5.4. NMR Studies

The rotational isomerism in M_2O_2 rings in four-coordinate dimeric organoaluminum derivatives arises from the orientation of the unsymmetrical alkoxide group which can assume the *syn* or *anti* conformation. Symmetrical groups do not yield *syn* and *anti* conformations in the alkoxides because of the planarity of the ring and of the oxygen atom. In the sulfur derivatives rotation about the C–S bond does not lead to equivalence of the groups, even when symmetrical groups are present, because of the stereochemistry around sulfur. To convert from the *anti* into the *syn* conformation in the oxygen derivative requires only rotation about the C–O bond, but in the sulfur derivatives this transformation requires inversion at sulfur or a break in the Al–S bonding. For oxygen derivatives, both *syn* and *anti* forms have been observed in the solid state in the oxygen-bridged *l*-menthol derivatives, $[Me_2Al(l\text{-mentholato})]_2$ (*syn*) and $[i\text{-}Bu_2Al(l\text{-mentholato})]_2$ (*anti*). In the sulfur and heavier chalcogen derivatives, only the *anti* form has been observed in the solid state. The existence of two possible conformations raises the question of their behavior in solution. Recent variable temperature NMR studies on a series 2-methyl-6-allylphenoxide derivatives of aluminum clearly established the presence of both *syn* and *anti* conformation in solution.[62] The equilibrium between these conformations and the barrier to this process have been reported. The rotational barrier is in the range 30 to 50 kJ/mol,[31] which is clearly less than the energy required for disruption of the bridge bond, but of the same order of magnitude as reported for rotation of the bridging groups in tricyclopropylaluminum.[98] Previous studies on the sulfur derivatives did not reveal the *syn* and *anti* conformations in solution, but recent work in our laboratory has shown that in $[Me_2Al(\mu\text{-}SMe)]_2$, two species exist at low temperature which have been interpreted in terms of the *syn* and *anti* configuration. The failure to observe this in other studies may be the result of only the *anti* conformation being present or because rapid rearrangement occurs under the conditions used for observation. Further studies will be required to establish which interpretation is correct.

Another interesting NMR feature which was deducted from the free rotation of the phenoxide around the O–C bond as well as from the planarity of the M_2O_2 ring was ^1H–^{19}F coupling in $(R_2AlOC_6F_5)_2$ derivatives.[70] In solution, the ^1H NMR of $(Me_2AlOC_6F_5)_2$ showed a quintet and a complex multiline pattern for the related ethyl and isobutyl derivatives. These results have been attributed in all cases to the interaction of the four ortho-fluorine atoms with the protons on the α-carbon atom bound to the metal. The ^1H–^{19}F coupling constant in each case studied was 2 Hz. Furthermore, it was observed that the ^1H–^{19}F coupling remained at elevated temperatures, and in fact the spectra showed sharper lines. These facts suggest that the Al–O–Al bridge bonds remain intact or at least do not undergo rapid exchange throughout the temperature range, and the equivalence of coupling between the *ortho*-fluorine atoms and the protons indicated that the pentafluorophenyl rings are rapidly rotating about the C–O bond. If the latter were not the case, one might expect different magnitudes for the coupling between the fluorine atoms on the rings parallel and perpendicular to the M–M axis. If the Al–O–Al bridge bond is reacted with 4-methylpyridine, where the equilibrium between a phenoxide bridged dimer and the monomeric adduct favors the adduct, the ^1H–^{19}F coupling also disappears. Furthermore, no such splitting pattern was observed in thiolato $(R_2AlSC_6F_5)_2$ compounds. These NMR results can be explained if the solid state structures of two similar organoaluminum compounds, $[Me_2Al(\mu\text{-}OC_6F_5)]_2$ and $[Me_2Al(\mu\text{-}SC_6F_5)]_2$ are compared. The Al_2O_2 ring is planar whereas the Al_2S_2 ring is puckered with the angle between the two planes described by Al–S1–Al and Al–S2–Al atoms near 150°. This difference in the structure of the Al_2E_2 ring can be attributed to the variation in geometry of the bridging oxygen and sulfur. Typically, in oxygen-bridged species, the oxygen atom is trigonal planar; in the sulfur-bridged systems, the sulfur atom is in a pyramidal environment. This can be illustrated by considering the sum of the angles around the bridging atom. In the oxygen derivative, this angle is 360°, as required in a planar system. For the sulfur derivative, the sum of the angles around the sulfur is in the range 300–310°.

The magnetic environment surrounding the alkyl groups on the metal center in puckered systems should be non-equivalent for both *syn* and *anti* conformation and, therefore, should give two signals for the proton NMR spectrum in $[Me_2Al(\mu\text{-}SC_6F_5)]_2$. This is further complicated by the fact that inversion at sulfur can occur but typically has a substantial (15-30 kcal/mol) barrier to inversion. If the puckered ring remains rigid in solution, then the alkyl groups bound to the metal must remain non-equivalent. Previous studies on sulfur-bridged systems, which have been found to be planar, indicate that the conversion from the puckered configuration into planar should be a low energy process. The rotation/inversion must be rapid on the NMR time scale to lead to equivalence of the alkyl groups.

The tetrameric molecule, $[Me_2Al(\mu\text{-}SC_6H_3\text{–}2,6\text{-}Me_2)]_4$, forms dimers and trimers in solution, as determined by cryoscopic molecular-weight NMR studies. The ^1H NMR spectra of the solution shows two resonance lines associated with the 2,6-methyl groups. These results suggest slow exchange between the two species in solution but rapid internal rearrangement which leads to equivalence of all the sites within the aggregates. The Me_2Al region of $[Me_2Al(\mu\text{-}SC_6H_3\text{–}2,6\text{-}Me_2)]_3$ is shown in Figure 5.1 as a function of temperature in toluene-d_8. The intensities of the two resonance lines of the Me_2Al groups **a** and **b**, change substantially between −40 and +80°C. The relative peak intensities obtained from ^1H NMR studies on several concentrations of $[Me_2Al(\mu\text{-}SC_6H_3\text{-}2,6\text{-}Me_2)]_n$ as a function of temperature were used to gain information about the equilibrium between the aggregates in solution and were used to calculate the following parameters: $K_{298} = 0.48$, $\Delta G° = -1.84$ kJ/mol, $\Delta H° = 25.9$ kJ/mol, and $\Delta S° = 80.6$ J/Kmol. for the process: $\mathbf{b} \rightleftharpoons \frac{3}{2}\,\mathbf{a}$.

The conclusions drawn from these results are that in solution $[Me_2Al(\mu\text{-}SC_6H_3\text{–}2,6\text{-}Me_2)]_n$ exists in two aggregate states of nearly equal energy which are in equilibrium with one another undergoing slow interchange on the NMR time scale. The observation of only two Me_2Al resonance lines, one associated with all the Me_2Al groups in each aggregate, is most readily interpreted in terms of two conformations which convert from one into the other slowly on the NMR time scale. At the same time, rapid intramolecular motion occurs within each conformation, leading to the equivalence of all the Me_2Al methyl groups and all of the 2,6-dimethylphenylthiolate moieties in each of the aggregates.

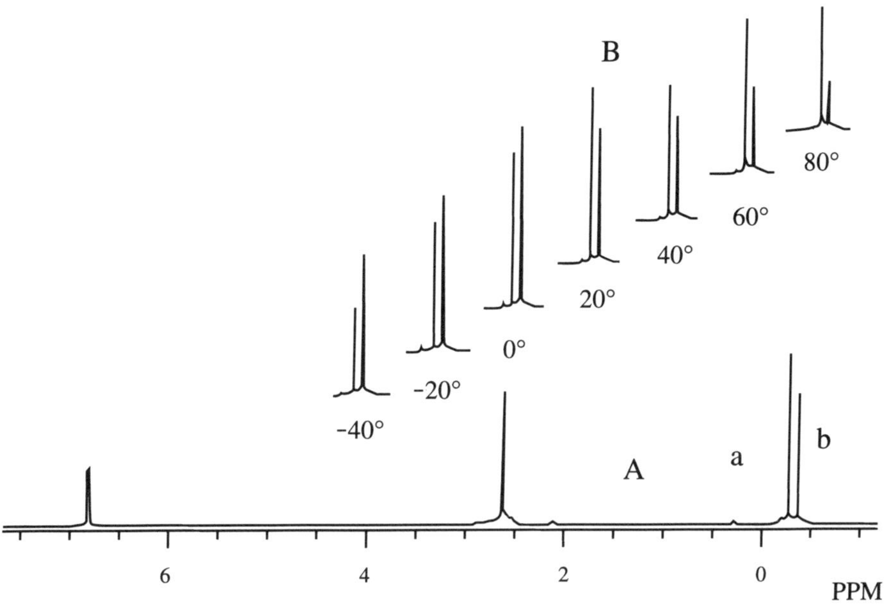

Figure 5.1. ^{1}H NMR of $[Me_2Al(\mu\text{-}SC_6H_3\text{-}2,6\text{-}Me_2)]_n$

NMR studies on the trimeric systems, $[Me_2Al(\mu\text{-}SC_6H_4\text{-}2\text{-}t\text{-}Bu)]_3$ and $[Me_2Al(\mu\text{-}SC_6H_4\text{-}2\text{-}SiMe_3)]_3$, show that the ring structure is maintained in solution but undergoes conformational changes. The 2-*t*-Bu derivative maintains its structure on the NMR time scale at room temperature, but at elevated temperatures undergoes a reversible ring inversion or exchange. The 2-$SiMe_3$ derivative shows the same behavior, but the 2-*i*-Pr derivative undergoes rapid inversion at room temperature. This motion is slowed at low temperatures and, again, gives a spectrum consistent with the chair structure observed in the solid state at low temperature. The 2,4,6-triisopropyl derivative undergoes rapid exchange/inversion under all conditions studied. The rate and energy associated with the conformational changes are dependent on the substituent on sulfur and likely on aluminum, but the latter effects have not been fully explored. The types of change which appear to be occurring are chair $\rightleftharpoons$ chair and/or chair $\rightleftharpoons$ boat.

There is a fine balance between steric effects which inhibit these interchanges and interactions which enhance the exchange rates, with fastest exchange occurring when the sulfur atom is in a planar configuration. This is observed in $[Me_2Al(\mu\text{-}SC_6H_2\text{-}2,4,6\text{-}i\text{-}Pr_3)]_3$ where the inversion cannot be slowed enough to allow the observation of non-equivalent methyl groups. These conclusions are based on the observed ^{13}C and ^{1}H spectra of the compounds as a function of temperature. The room temperature spectra of the 2-*t*-Bu and 2-$SiMe_3$ derivatives are consistent with the trimeric structure observed in the solid state. This indicates slow reorganization on the NMR time scale. The reorganization rate is substantially faster when the 2-*i*-Pr derivative is studied, with observation of the "slow exchange" spectrum only at −70°C. For the 2,4,6-triisopropyl derivative, exchange could not be stopped. These observations are consistent with the structure which shows that in the 2,4,6-(i-Pr$)_3$ derivative, the sulfur atom is in a planar configuration (See **7** and Table 5.3), that is, inversion at sulfur is no longer a factor in the reorganization process. It is anticipated that many additional studies will be required to explore fully the effects of the substituent on ring motion and to determine the effect of the chalcogen on the stability and structure of the aggregates.

5.5. $R_n(ER)_{3-n}$ Al·Base Addition Compounds

The $R_n(ER)_{3-n}$Al·Base addition compounds can be represented by the generalized structure **9**. The Al·Base bond described in terms of a simple donor-acceptor bond with the metal atom serving as the Lewis acid site and E serving as the Lewis base containing an oxygen, nitrogen or other donor atom. The $R_2(ER)$Al·Base addition compounds are simply prepared by reaction of a base with a three-coordinate $Al(ER)_3$ aluminum derivative or by the disruption of the Al−E−Al bridge bond of the $(R_2AlER')_n$ structures by a base. The latter reaction depends on the stability of the Al−E−Al bridge bond which is largely governed by the steric and electronic nature of the ligands attached to the aluminum and the bridging ligand. Thus, the equilibrium for the reaction shown in Equation 5.14 typically lies far to the left with the alkoxide bridged dimer favored for simple alkoxides, whereas it lies to the right if sufficiently sterically demanding ligands are present on the aluminum and the bridge atom.

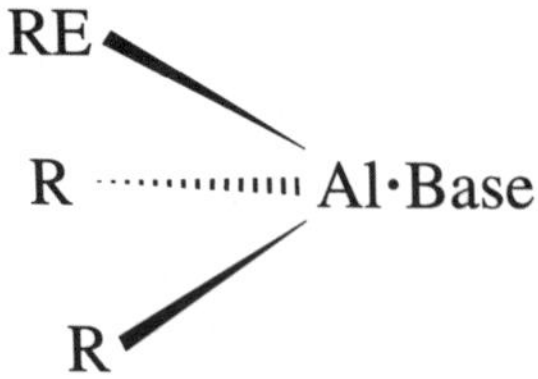

9

With sulfur in the bridging site, reaction occurs more frequently because of the lower metal−sulfur bond energy.

The stability of the addition compounds $R_n(ER)_{3-n}$Al·Base, and their dissociation/exchange have not been studied in detail. The most significant factor is the nature of E, the bridging element, and the strength and steric requirements of the base.

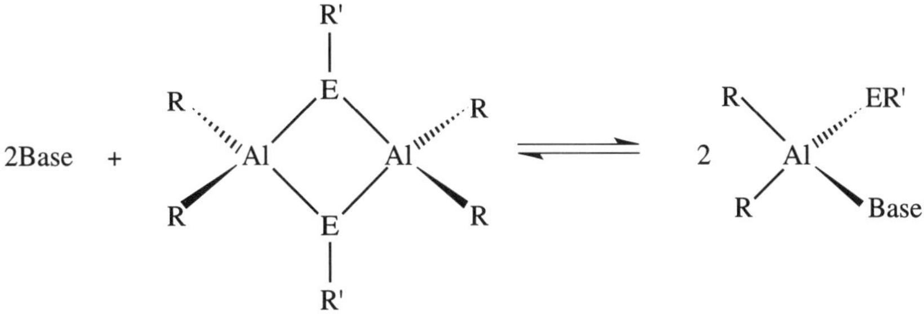

(5.14)

The structures of a variety of simple $R_n(OR)_{3-n}Al{\cdot}Base$ adducts have been determined and are listed in Table 5.6. In general, the aluminum center in $R_n(OR)_{3-n}Al{\cdot}ER'_2$ adducts is in a distorted tetrahedral environment. The data in Table 5.6 show that the average Al–O alkoxide bond distance observed in these derivatives is approximately 1.74 Å. This bond distance is substantially shorter than those found in ether and polyether adducts of simple trialkylaluminums, $R_3Al{\cdot}ER'_2$ and slightly longer than in three-coordinate complexes. There appears to be some modest Al–O alkoxide bond lengthening when alkoxide oxygen also participates in adduct formation with another aluminum center. The Al–O–C angles in these alkoxides are in the range 130–150°, far larger than the 110–130° range expected. The presence of a relatively short Al–O bond distance together with a large Al–O–C angle in $R_n(OR)_{3-n}Al{\cdot}ER'_2$, where neither a vacant p-orbital nor an energetically accessible aluminum d-orbital is available for π-bonding, led to the proposal that π-donation from the oxygen lone pair orbital to the Al–C and Al–E σ antibonding orbitals exists. This fact was later substantiated using *ab initio* molecular orbital calculations and photoelectron spectra on some model and known compounds.[108–110]

There are some noteworthy features of these systems. First, the Al–S–Al bridge bond distances range from 2.35 to 2.41 Å and are much shorter than the Al–S bond lengths observed for the adducts. Second, the Al–S–Al bridge bonds are substantially weaker than the Al–O–Al bridge bonds. Thus, they can be readily cleaved or disrupted by ethers and amines as indicated in Equation 5.15. For example, the aluminum thiolates may react with aldehydes or esters

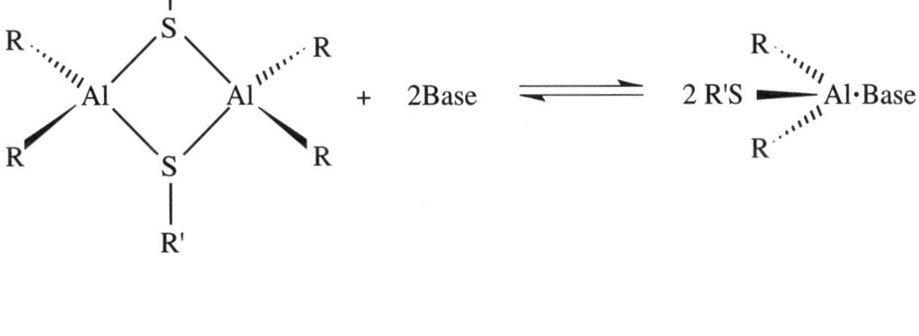

(5.15)

Table 5.6. Selected average bond distances and angles in monomeric four-coordinate organoaluminum alkoxides and related compounds

Compound	Al–O(Å)	Al–C(Å)	C–Al–C(°)	C–Al–O(°)	Al–O–C(°)	Ref.
$Et_2AlBHT \cdot MeC_6H_4CO_2Me$	1.749	1.964	116.4	115.7	145.6	45
	1.887(ester)	–	–	101.8	143.0	
$MeAl(BHT)_2 \cdot C_6H_5CO_2Me$	1.717	1.96	–	114.4	154.8	59
	1.851(ester)	–	–	103.7	174.0	
$Me_2AlBHT \cdot PMe_3$	1.736	1.970	111.7	117.5	164.5	58
$[\{Me_3NH\}(Me(Cl)_2AlBHT\}]$	1.713	1.995	–	117.2	164.0	58
$K[Me_2Al(OPh)_2]$	1.800	1.968	119.4	108.7	125.0	99
[K·dibenzo-18-crown-6][Al_2Me_6OPh]	1.89	2.01	112.1	108.9	116.3	99
$[(AlMe_3)_2\{N,O\text{-}(MeN{=}C(Ph)C\text{-}(Ph){=}O\}]$	1.983	1.954	115.4	101.5		100
$[(Al_2Me_5)\{(t\text{-}Bu)N{=}CHC(Me)_2O\}]$	1.868	1.951	115.1	108.7		100
$[Me_2AlOC(Ph){=}N(Ph)] \cdot ONMe_3$	1.770/1.826	1.962	117.3	108.1		101
$MeAl(BHT)_2 \cdot O{=}CPh_2$	1.727	1.942	–	115.4	152.0	57
	1.903(ketone)	–	–	110.3	144.0	
$Me_2Al(BHT) \cdot O{=}CPh_2$	1.731	1.95	113.1	117.2	157.7	59
	1.907(ketone)	–	–	101.4	153.8	
$[MeAl(BHT)_2 \cdot O{=}C(H)t\text{-}Bu]$	1.728	1.955	–	116.2	135.9	59
	1.920(aldehyde)	–		106.0	136.0	
$Al(DIP)_3 \cdot Py$	1.705	–	–		142.7	102
	1.993(Al–N)					
$MeAl(OMes)_2 \cdot (3,5\text{-}Me_2Py)$	1.718	1.946	–	117.1	143.2	102
	1.961(Al–N)			109.4(C–Al–N)		
$Me_2AlBHT \cdot Py$	1.740	1.960	114.3	116.2	156.1	102
	1.993(Al–N)			102.9(C–Al–N)		
$MeAl(dbmp)_2(bhmap)$	1.735	1.933	–	116.1	158.0	103
	1.836(ketone)		110.0	128.8		
$Me_2Al(BHT) \cdot NH_3$	1.743	–	–	–	149.6	104
	2.015(Al–N)					
$[EtAl(BHT)\{O{=}C(t\text{-}Bu)CH_2C(Me)(t\text{-}Bu)O\}]$	1.726	1.962	–	117.0	136.0	107
	1.911(ketone)	–		107.8	127.5	

$[Me_2Al(Chirald)]$	1.726	1.961	113.2	113.1	132.4	105
	2.031(Al–N)			108.9(C–Al–N)		
$[\{Me_2NH_2\}\{Al(OSiMe_3)_4\}]$	1.752	–	–	–		69
$Me_2Al(BHT)\cdot(3,5\text{-}Me_2Py)$	1.744	1.968	–	116.1	158.0	56
	2.074(Al–N)			102.9(C–Al–N)		
$Et_2Al(BHT)\cdot(H_2N\text{-}t\text{-}Bu)$	1.748	1.967	–	113.7	151.3	56
	2.074(Al–N)			104.7(C–Al–N)		
$Et_2Al(BHT)\cdot(OPy)$	1.754	1.974	–	111.7	144.1	56
	1.868			104.5		
$MeAl(BHT)_2\cdot(OPy)$	1.728	1.956	–	116.4	171.3	56
	1.854			102.3		
$Me_2Al(OC_6F_5)\cdot[N(C_2H_4)_3CH]$	1.787	1.951	–	111.2	128.8	56
	2.002(Al–N)			107.5(C–Al–N)		

*BHT = dbmp = 2,6-di(*t*-butyl)-4-methylphenoxide.
Chirald = (2S,3R)-(+)-4-dimethylamino-1,2-diphenyl-3-methyl-2-butanol.
Hbhmap = 3-tert-butyl-2-hydroxy-5-methylacetophenone.

via Equation 5.15 to yield the simple adduct followed by a rearrangement to yield the dialkylaluminum thioalkoxide shown in Equation 5.16.[29]

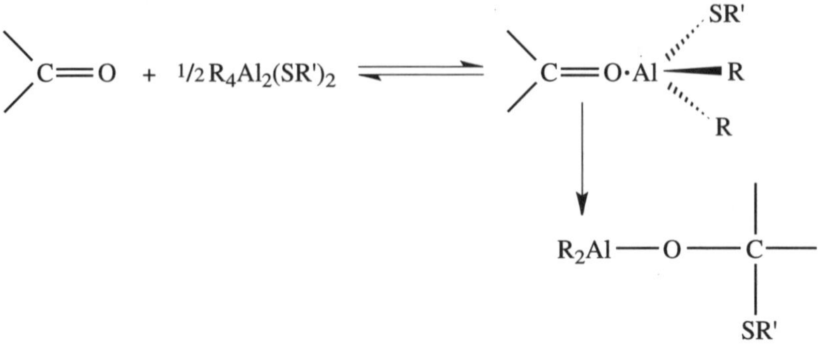

(5.16)

5.6. Bidentate and Polydentate Ligands

The chemistry of bidentate and polydentate ligands with Group III metal alkyls is complex. The introduction of a second (or third) active site in the organic ligand makes it possible for a variety of new structures to be formed. A number of these have been described previously[7] and only a limited discussion of them will be presented here.

Five-coordinate aluminum derivatives have been described in some detail.[7] Recent studies have shown that the structures of species with donor atoms in rigid ligands are often determined by interaction between the donor and the

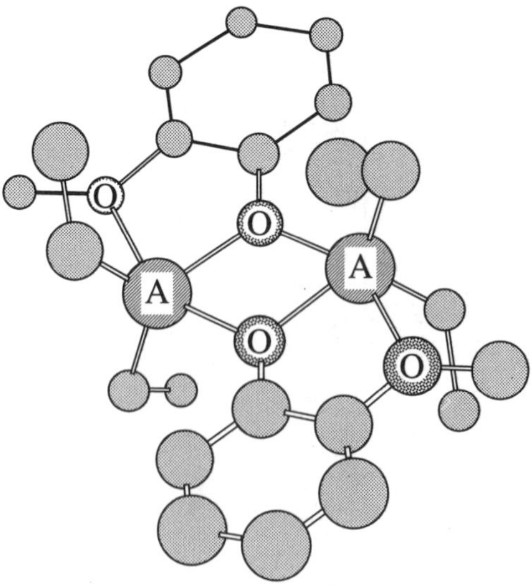

10

metal center, **10**. In this and related structures the molecules have the bridging group positioned to give "closest contact" of the donor atom with the metal center.

Bridging groups of this type appear to occur only in the *anti* configuration with two five-coordinate aluminum centers. This is similar to the studies reviewed earlier[7] which have nitrogen bound to the aluminum as the second coordinating atom. One major difference is observed in the NMR spectra of these systems. No interaction barriers are observed in these systems whereas, in the nitrogen systems, dissociation processes were observed and found to be a function of the bridging ligand. This will likely change as more systems containing coordination to oxygen or sulfur in the fifth coordination site are characterized.

In addition to the simple bridge derivatives discussed above, recent studies have shown that reaction of a bifunctional pyridine-2-thiol with alkylaluminum compounds yields the dimeric molecule, **11**.[77] The central core of this compound contains an $Al_2S_2C_2N_2$ ring with the bridging 2-thiolatopyridine ligands arranged in an unexpected head-to-head fashion. This is in sharp contrast to the structure of the 2-substituted phenyl derivatives and shows the wide range of products possible for these systems. Other bridged systems which contain unusual bridging arrangements are likely to be observed. This area of chemistry, the chemistry of polydentate ligands with Group III metal alkyls, should prove very fruitful in the future.

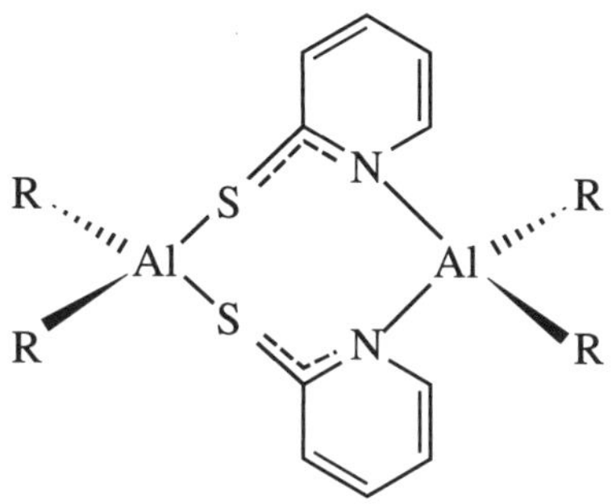

11

References

1. Mole, T.; Jeffrey, E. A. *Organoaluminum Compounds;* Elsevier: Amsterdam, 1972.

2. Eisch, J. J. In *Aluminum;* Wilkinson, G.; Stone, F. G. A.; Abel, E. W., Eds.; Pergamon Press: Oxford, 1982; pp 555–683.

3. Bradley, D. C. *Adv. Inorg. Radio Chem.* **1972**, *15*, 259.

4. Bradley, D. C.; Mehrotra, R. C.; Gear, D. P. *Metal Alkoxides;* Academic Press: New York, 1978.

5. Oliver, J. P. *Adv. Organomet. Chem.* **1970**, *8*, 167.

6. Oliver, J. P. *Adv. Organomet. Chem.* **1977**, *15*, 235.

7. Oliver, J. P.; Kumar, R. *Polyhedron* **1990**, *9*, 409.

8. Matteson, D. S. *Organometallic Reaction Mechanisms*; Academic Press: New York, 1974.

9. Almenningen, A.; Halvorsen, S.; Haaland, A. *Acta Chem. Scand.* **1971**, *25*, 1937.

10. Pazik, J. C.; George, C.; Berry, A. *Inorg. Chim. Acta* **1991**, *187*, 207.

11. Beachley, Jr, O. T.; Victoriano, L. *Organometallics* **1988**, *7*, 63.

12. Beachley, Jr, O. T.; Pazik, J. C. *Organometallics* **1988**, *7*, 1516.

13. Beachley, Jr, O. T.; Spiegel, E. F.; Kopasz, J. P.; Rogers, R. D. *Organometallics* **1989**, *8*, 1915.

14. Pitt, C. G.; Higa, K. T.; McPhail, A. T.; Wells, R. L. *Inorg. Chem.* **1986**, *25*, 2483.

15. Pitt, C. G.; Purdy, A. P.; Higa, K. T.; Wells, R. L. *Organometallics* **1986**, *5*, 1266.

16. Purdy, A. P.; Wells, R. L.; McPhail, A. T.; Pitt, C. G. *Organometallics* **1987**, *6*, 2099.

17. Byrne, E. K.; Parkanyi, L.; Theopold, K. H. *Science* **1988**, *241*, 332.

18. Sauls, F. C.; Interrante, L. V.; Jiang, Z. *Inorg. Chem.* **1990**, *29*, 2989.

19. Cowley, A. H.; Jones, R. A. *Angew. Chem. Int. Ed. Engl.* **1989**, *28*, 1208.

20. Beachley, Jr, O. T.; Noble, M. J.; Churchill, M. R.; Lake, C. H. *Organometallics* **1992**, *11*, 1051.

21. Cowley, A. H.; Harris, P.R.; Jones, R. A.; Nunn, C. M. *Organometallics* **1991**, *10*, 652.

22. Atwood, D. A.; Jones, R. A.; Cowley, A. H.; Atwood, J. L.; Bott, S. G. *J. Organomet. Chem.* **1990**, 394, C6.

23. Theopold, K. H.; Douglas, T. *Inorg. Chem.* **1991**, *30*, 594.

24. Power, M. B.; Barron, A. R. *Angew. Chem. Int. Ed. Engl.* **1991**, *30*, 1353.

25. Almond, M. J.; Drew, M. G. B.; Jenkins, C. E.; Rice, D. A. *J. Chem. Soc., Dalton Trans.* **1992**, 5.

26. Locke, K.; Weidlein, J.; Scholz, F.; Bouanah, N.; Brianese, N.; Gao, Y. *J. Organomet. Chem.* **1991**, *420*, 1.

27. Tuck, D. G. In *Gallium and Indium:* Wilkinson, G.; Stone, F. G. A.; Abel, E. W., Eds.; Pergamon Press: Oxford, 1982; pp 725–754.

28. Kurosawa, H. In *Thallium*; Wilkinson, G.; Stone, F. G. A.; Abel, E. W., Eds.; Pergamon Press: Oxford, 1982; pp 683–723.

29. de Mel, V. S. J.; Kumar, R.; Oliver, J. P. *Organometallics* **1990**, *9*, 1303.

30. Taghiof, M.; Kumar, R.; Oliver, J. P., unpublished observations.

31. Bailey, M.; Hendershot, D. G.; Kumar, R.; Barber, M.; Oliver, J. P. *J. Am. Chem. Soc.* submitted.

32. Power, M. B.; Barron, A. R. *J. Chem. Soc., Chem. Commun.* **1991**, 1513.

33. Cowley, A. H.; Jones, R. A.; Harris, P. R.; Atwood, D. A.; Contreras, L.; Burek, C. J. *Angew. Chem. Int. Ed. Engl.* **1991**, *30*, 1143.

34. Hoffmann, G. G.; Burschka, C. *J. Organomet. Chem.* **1984**, *267*, 229.

35. Histed, M.; Howard, J. A.; Morris, H.; Mile, B. *J. Am. Chem. Soc.* **1988**, *110*, 5290.

36. Nomura, R.; Inazawa, S.; Kanaya, K.; Matsuda, H. *Polyhedron* **1989**, *8*, 763.

37. Nomura, R.; Fujii, S.; Kanaya, K.; Matsuda, H. *Polyhedron* **1990**, *9*, 361.

38. Ruhlandt-Senge, K.; Power, P. P. *Inorg. Chem.* **1991**, *30* 2633.

39. Ruhlandt-Senge, K.; Power, P. P. *Inorg. Chem.* **1991**, *30*, 3683.

40. Banks, M. A.; Beachley, Jr., O. T.; Gysling, H. J.; Luss, H. R. *Organometallics* **1990**, *9*, 1979.

41. Green, J. H.; Kumar, R.; Seudeal, N.; Tuck, D. G. *Inorg. Chem.* **1989**, *28*, 123.

42. Jerius, J. T.; Hahn, J. M.; Rahman, A. F. M. M.; Mols, O.; Ilsley, W. H.; Oliver, J. P. *Organometallics* **1986**, *5*, 1812.

43. Sheldrick, G. M.; Sheldrick, W. S. *J. Chem. Soc. (A)* **1969**, 2279.

44. Uhl, W. *Z. Naturforsch.* **1988**, *43b*, 1113.

45. Shreve, A. P.; Mulhaupt, R.; Fultz, W.; Calabrese, J.; Robbins, W.; Ittel, S. D. *Organometallics* **1988**, *7*, 409.

46. Benn. R.; Janssen, E.; Lehmkuhl, H.; Rufinska, A.; Angermund, K.; Betz, P.; Goddard, R.; Kruger, C. *J. Organomet. Chem.* **1991**, *411*, 37.

47. Petrie, M. A.; Olmstead, M. M.; Power, P. P. *J. Am. Chem. Soc.* **1991**, *113*, 8704.

48. Maruoka, K.; Itoh, T.; Yamamoto, H. *J. Am. Chem. Soc.* **1985**, *107* 4573.

49. Maruoka, K.; Yamamoto, H. *J. Am. Chem. Soc.* **1989**, *111*, 789.

50. Maruoka, K.; Araki, Y.; Yamamoto, H. *J. Am. Chem. Soc.* **1988**, *110*, 2650.

51. Maruoka, K.; Nonoshita, K.; Banno, H.; Yamamoto, H. *J. Am. Chem. Soc.* **1988**, *110*, 7922.

52. Maruoka, K.; Itoh, T.; Shirasaka, T.; Yamamoto, H. *J. Am. Chem. Soc.* **1988**, *110*, 310.

53. Maruoka, K.; Nagahara, S.; Yamamoto, H. *J. Am. Chem. Soc.* **1990**, *112*, 6115.

54. Maruoka, K.; Concepcion, A. B.; Hirayama, N.; Yamamoto, H. *J. Am. Chem. Soc.* **1990**, *112*, 7422.

55. Shambayati, S.; Crowe, W. E.; Schreiber, S. L. *Angew. Chem. Int. Ed. Engl.* **1990**, *29*, 256.

56. Healy, M. D.; Ziller, J. W.; Barron, A. R. *Organometallics* **1991**, *10*, 597.

57. Power, M. B.; Bott, S. G.; Atwood, J. L.; Barron, A. R. *J. Am. Chem. Soc.* **1990**, *112*, 3446.

58. Healy, M. D.; Wierda, D. A.; Barron, A. R. *Organometallics* **1988** *7*, 2543.

59. Power, M. B.; Bott, S. G.; Clark, D. L.; Atwood, J. L.; Barron, A. R. *Organometallics* **1990**, *9*, 3086.

60. Al-Wassil, A. I.; Hitchcock, P. B.; Sarisaban, S.; Smith, J. D.; Wilson, C. L. *J. Chem. Soc., Dalton Trans.* **1985**, 1929.

61. Haaland, A.; Stokkeland, O. *J. Organomet. Chem.* **1975**, *94*, 345.

62. Sierra, M. L.; Kumar, R.; de Mel, V. S. J.; Oliver, J. P. *Organometallics* **1992**, *11*, 206.

63. Kumar, R.; Sierra, M. L.; de Mel, V. S. J.; Oliver, J. P. *Organometallics* **1990**, *9*, 484.

64. Atwood, J. L.; Zaworotko, M. J. *J. Chem. Soc., Chem. Commun.* **1983**, 302.

65. Wengrovius, J. H.; Garbauskas, M. F.; Williams, E. A.; Going, R. C.; Donahue, P. E.; Smith, J. F. *J. Am. Chem. Soc.* **1986**, *108*, 982.

66. Drew, D. A.; Haaland, A.; Weidlein, J. *Z. Anorg. Allg. Chem.* **1973**, *398*, 241.

67. Apblett, A. W.; Barron, A. R. *Organometallics* **1990**, *9*, 2137.

68. Mulhaupt, R.; Calabrese, J.; Ittel, S. D. *Organometallics* **1991**, *10*, 3403.

69. Chisholm, M. H.; Huffman, J. C.; Wesemann, J. L. *Polyhedron* **1991**, *10*, 1367.

70. Hendershot, D. G.; Kumar, R.; Barber, M.; Oliver, J. P. *Organometallics* **1991**, *10*, 1917.

71. Cayton, R. H.; Chisholm, M. H.; Davidson, E. R.; DiStasi, V. F.; Du, P.; Huffman, J. C. *Inorg. Chem.* **1991**, *30*, 1020.

72. Waymouth, R. W.; Potter, K. S.; Schaefer, W. P.; Grubbs, R. H. *Organometallics*, **1990**, *9*, 2843.

73. Erker, G.; Albrecht, M.; Werner, S.; Kruger, C. *Z. Naturforsch,* **1990**, *45b*, 1205.

74. Cetinkaya, B.; Hitchcock, P. B.; Jasim, H. A.; Lappert, M. F.; Williams, H. D. *Polyhedron*, **1990**, *9*, 239.

75. Brauer, D. J.; Stucky, G. D. *J. Am. Chem. Soc.* **1969**, *91*, 5462.

76. Haaland, A.; Stokkeland, O.; Weidlein, J. *J. Organomet. Chem.* **1975**, *94*, 353.

77. Kumar, R.; de Mel, V. S. J.; Oliver, J. P. *Organometallics* **1989**, *8*, 2488.

78. Boardman, A.; Small, R. W. H.; Worrall, I. J. *Inorg. Chim. Acta.* **1986**, *120*, L23.

79. Noth, H.; Konard, P. *Chem. Ber.* **1983**, *116*, 3552.

80. Shakir, R.; Zaworotko, M. J.; Atwood, J. L. *J. Organomet. Chem.* **1979**, *171*, 9.

81. Atwood, J. L.; Seal, S. K. *J. Organomet. Chem.* **1976**, *114*, 107.

82. Sangokoya, S. A.; Pennington, W. T.; Robinson, G. H. *J. Cryst. Spect. Res.* **1990**, *20*, 53.

83. Sangokoya, S. A.; Pennington, W. T.; Robinson, G. H. *J. Cryst. Spect. Res.* **1989**, *19*, 433.

84. Hendershot, D. G.; Barber, M.; Kumar, R.; Oliver, J. P. *Organometallics* **1991**, *10*, 3302.

85. Robinson, G. H.; Sangokoya, S. A. *J. Am. Chem. Soc.* **1988**, *110*, 1494.

86. Robinson, G. H.; Zhang, H.; Atwood, J. L. *Organometallics* **1987**, *6*, 887.

87. Sangokoya, S. A.; Lee, B.; Pennington, W. T.; Robinson, G. H. *J. Coord. Chem.* **1989**, *19*, 331.

88. Burford, N.; Royan, B. W.; Spence, R. E. v. H.; Rogers, R. D. *J. Chem. Soc., Dalton Trans.* **1990**, 2111.

89. Fernholt, L.; Haaland, A.; Hargittai, M.; Seip, R.; Weidlein, J. *Acta. Chem. Scand., Ser. A* **1981**, *A35*, 529.

90. Robinson, G. H.; Self, M. F.; Pennington, W. T.; Sangokoya, S. A. *Organometallics* **1988**, *7*, 2424.

91. Self, M. F.; Lee, B.; Sangokoya, S. A.; Pennington, W. T.; Robinson, G. H. *Polyhedron* **1990**, *9*, 313.

92. Boardman, A.; Jeffs, S. E.; Small, R. W. H.; Worrall, I. J. *Inorg. Chim. Acta* **1985**, *99*, L39.

93. Hoffmann, G. C.; Burshka, C. *Angew. Chem. Int. Ed. Engl.* **1985**, *24*, 970.

94. Maelia, L. E.; Koch, S. A. *Inorg. Chem.* **1986**, *25*, 1896.

95. Kampf, J.; Kumar, R.; Oliver, J. P., unpublished observations.

96. Annan, T.; Kumar, R.; Mabrouk, H. E.; Tuck, D. G.; Chadha, R. K. *Polyhedron* **1989**, *8*, 865.

97. Day, R. O.; Holmes, R. R. *Inorg. Chem.* **1982**, *21*, 2379.

98. Sanders, D. A.; Scherr, P. A.; Oliver, J. P. *Inorg. Chem.* **1976**, *15*, 861.

99. Zaworotko, M. J.; Kerr, C. R.; Atwood, J. L. *Organometallics* **1985**, *4*, 238.

100. van Vliet, M. R. P.; van Koten, G.; Rotteveel, M. A.; Schrap, M.; Vrieze, K.; Kojic-Prodic, B.; Spek, A. L.; Duisenberg, A. J. M. *Organometallics* **1986**, *5*, 1389.

101. Kai, Y.; Yasuoka, N.; Kakudo, M. *Bull. Chem. Soc. Jpn.* **1972**, *45*, 3388.

102. Healy, M. D.; Ziller, J. W.; Barron, A. R. *J. Am. Chem. Soc.* **1990**, *112*, 2949.

103. Power, M. B.; Bott, S. G.; Bishop, E. J.; Tierce, K. D.; Atwood, J. L.; Barron, A. R. *J. Chem. Soc., Dalton Trans.* **1991**, 241.

104. Healy, M. D.; Leman, J. T.; Barron, A. R. *J. Am. Chem. Soc.* **1991**, *113*, 2776.

105. Kumar, R.; Oliver, J. P.; unpublished observations.

106. van der Steen, F. H.; Van Mier, G. P. M.; Spek, A. L.; Karoon, J.; van Koten, G. *J. Am. Chem. Soc.* **1991**, *113*, 5742.

107. Power, M. B.; Apblett, A. W.; Bott, S. G.; Atwood, J. L.; Barron, A. R. *Organometallics* **1990**, *9*, 2529.

108. Lichtenberger, D. L.; Hogan, R. H.; Healy, M. D.; Barron, A. R. *Organometallics* **1991**, *10*, 609.

109. Lichtenberger, D. L.; Hogan, R. H.; Healy, M. D.; Barron, A. R. *J. Am. Chem. Soc.* **1990**, *112*, 3369.

110. Barron, A. R.; Dobbs, K. D.; Francl, M. M. *J. Am. Chem. Soc.* **1991**, *113*, 39.

CHAPTER

6

Anionic and Cationic Organoaluminum Compounds

Jerry L. Atwood

Department of Chemistry, The University of Alabama, Tuscaloosa, Alabama 35487

6.1. Introduction

Anionic organoaluminum compounds have provided a fertile area of investigation for over thirty years. The initial studies were a part of the survey of organoaluminum chemistry produced by Zeigler and co-workers. Subsequently, interest has been spurred by the discovery of liquid clathrates (initially based on anionic organoaluminum compounds) and the success of olefin polymerizations utilizing aluminoxanes as co-catalysts. In this chapter the early liquid clathrate work, as it relates to aluminum chemistry, will be placed in perspective, and recent discoveries which shed light on aluminoxanes will be related. Some initial studies on cationic organoaluminum complexes will also be discussed.

6.2. Liquid Clathrates Based on Anionic Organoaluminum

6.2.1 Initial Observation of Liquid Clathrate Behavior

In 1972 this laboratory reported a class of compounds which exhibited unusually strong interactions with small aromatic molecules.[1,2] The initial studies indicated that when the anion of $M[Al_2Me_6X]$ (M = alkali metal or tetraalkylammonium ions; X = halide or pseudohalide ions) possesses an angular geometry **1**, the substance will react with molecules such as benzene or toluene to form liquid complexes which contain 2.5 to 5 aromatic molecules per anionic unit. On the other hand, a symmetrical anionic structure **2** was not found to produce this effect.[3] The nature of the interaction was thus related to the shape of the anion in $M[Al_2Me_6X]$; it was proposed that a cage or layer-like structure of oriented anions is set up with counter ions and aromatic molecules trapped inside. An appropriate designation for these substances thus appeared to be *liquid clathrates*.

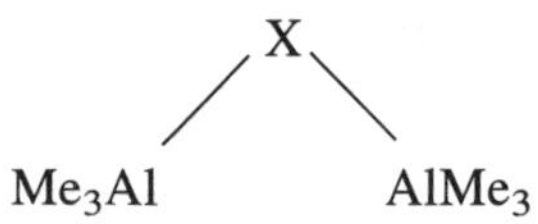

1

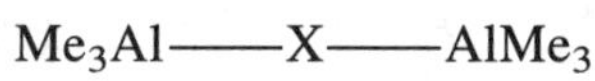

2

Although subsequent investigations have led to the conclusion that lattice energy, rather than an angular anionic geometry, is the overriding consideration, the layer-like liquid structure is still regarded as the best model for the behavior. In the past five years it has been demonstrated that this behavior is not exclusively within the realm of aluminum chemistry.[4–7] However, this solution behavior sparked significant early efforts in this area.

6.2.2 Preparation of the Parent Compounds $M[Al_2R_6X]$

Before further discussion of liquid clathrate behavior, a survey of the structure and chemical behavior of $M[Al_2R_6X]$ will be given.

Zeigler et al.[8] first reported in 1960 that aluminum alkyls react with alkali metal halides to form either 1:1 or 2:1 complexes:

$$AlR_3 + MX \longrightarrow M[AlR_3X]$$

$$2AlR_3 + MX \longrightarrow M[Al_2R_6X]$$

At the time, these substances were of interest for two reasons. In the first place, the very formation of the compounds provided an inorganic study of the relative importance of the lattice energies of MX and $M[AlR_3X]$ or $M[Al_2R_6X]$. Lehmkuhl[9] concluded that complexation occurs most readily for the cases in which the ionic radius of the alkali metal is large, the alkyl chain is short, and the complex ion is small. The second focus of attention came as the result of the X-ray crystallographic investigation of $K[Al_2Et_6F]$ by Allegra and Perego.[10] They found that the anion exhibited a completely linear Al–F–Al linkage, and postulated both sp hybridization of the flourine atom and d orbital participation by the aluminum atom in order to explain this observed bond distance and angle.

Although the logical extension of the preparative method to include alkali metal pseudohalides was hinted in the original paper,[8] no well-defined complexes of this sort were noted until 1971.[11,12–14] Of the compounds prepared thus far, for both the 1:1 and 2:1 substances, those with alkali metal cations are oxygen- and water-sensitive solids. With a tetraalkylammonium ion, the 1:1 compounds are also air-sensitive solids, but the 2:1 moieties are air-sensitive liquids. The reactivity of $M[AlR_3X]$ and $M[Al_2R_6X]$ is, however, in general much less than that of the aluminum alkyls themselves.

6.2.3. Anionic Structure of $M[Al_2R_6X]$

The structures of the 1:1 complexes are not subject to speculation[15–17] as they are certainly quite ionic and at present show no far-reaching synthetic prospects or any indication of liquid clathrate behavior. The configurations of the 2:1 complexes are, however, of utmost importance in understanding the phenomenon, and will be dealt with now and referred to subsequently.

For the 2:1 pseudohalide complexes two clear cases may be differentiated with reference to $[Al_2R_6SCN]^-$:

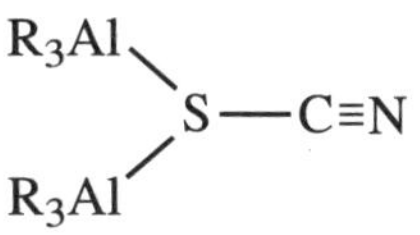

3

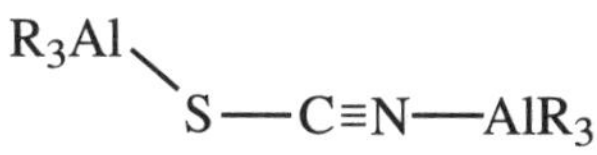

4

For R = Me, structure **3** appeared to be favored in the solid state from the detailed infrared and Raman studies,[14] and by analogy to the structure of $K[Al_2Me_6N_3]$. However, an X-ray crystallographic study later proved that **4** was the correct choice, at least for K^+ as the cation.[18]

A feature of the structure of $K[Al_2Me_6N_3]$ which was originally listed as an interesting sidelight is now viewed as quite significant. The methyl groups could have two possible conformations corresponding to anionic point symmetries of C_{2v} and C_s. Viewed down the nitrogen atom chain they appear as:

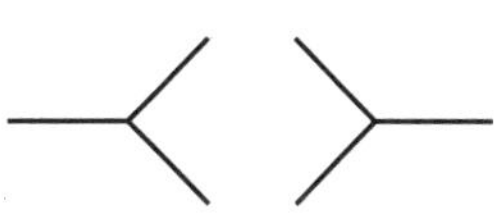

C_{2v}

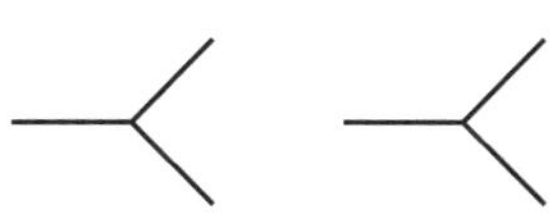

C_s

Weller and Dehnicke[13,14] have determined from spectroscopic studies that the anions in $[NMe_4][Al_2Me_6N_3]$ and $[NMe_4][Al_2Me_6SCN]$ are of C_{2v} symmetry.[13,14] However, in the crystallographic asymmetric unit of $K[Al_2Me_6N_3]$ there are two non-equivalent $[Al_2Me_6N_3]^-$ ions: one of C_{2v} and one of C_s symmetry (Figure 6.1). The bond distances and angles in both forms are quite similar, and the existence of both types may be taken as an indication of the closeness in overall energy of the two conformations in solution. For $Cs[Al_2Me_6N_3]$, only the C_s conformation is found.[19]

As the two independent anions differ geometrically, so also do the two potassium ions differ environmentally (Figure 6.2.). One is most closely associated with the terminal nitrogen atoms of two azide ions, 2.92 (1) and 2.93 (1) Å, with four methyl carbon atoms filling out the coordination sphere. The other potassium ion finds itself in the less favourable electrostatic position of having no portion of the azide ion within 4.90 Å. Its neighborhood consists of six

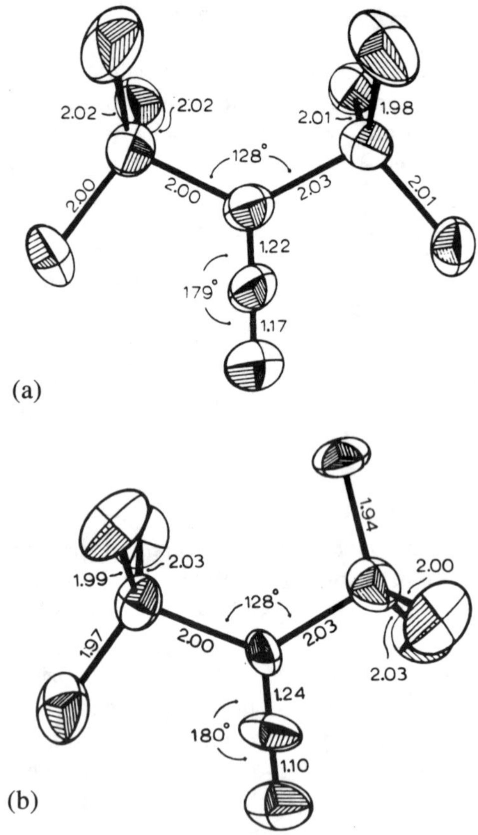

Figure 6.1. (a) Structure of the C_{2v} anion of $K[Al_2(Me_6N_3]$.
(b) Structure of the C_s anion of $K[Al_2(Me_6N_3]$.
Reprinted with permission from ref. 2. Copyright 1974 by Elsevier.

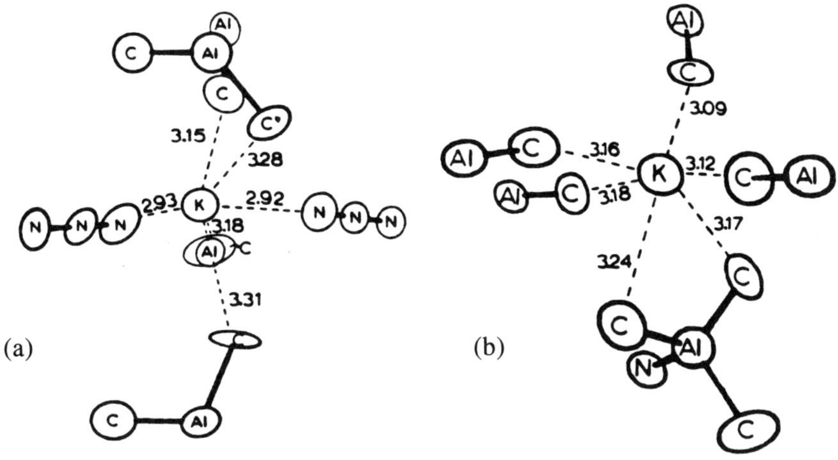

Figure 6.2. Cation environment in the structure of $K[Al_2Me_6N_3]$. Reprinted with permission from ref. 2. Copyright 1974 by Elsevier.

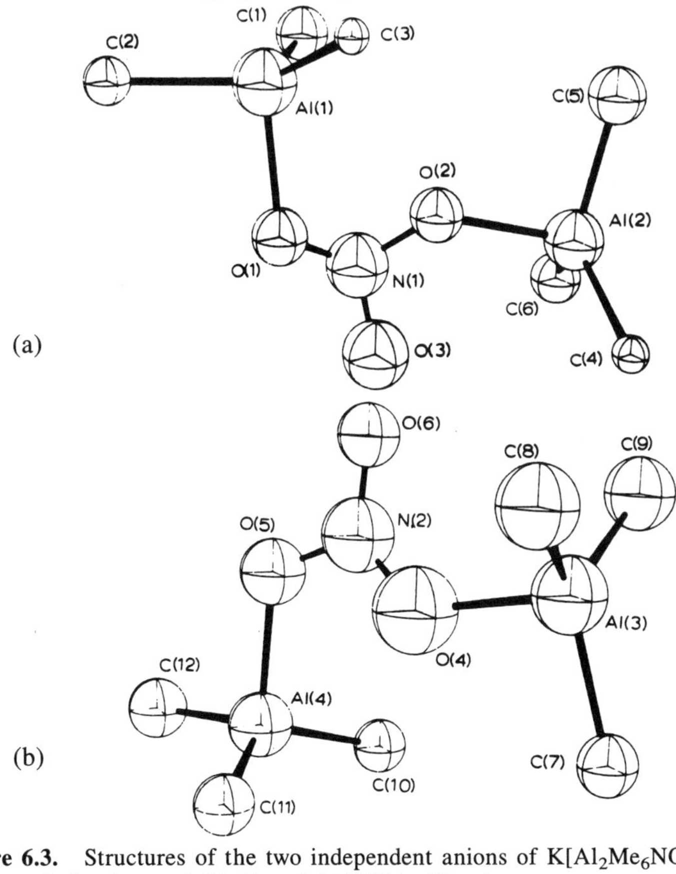

Figure 6.3. Structures of the two independent anions of $K[Al_2Me_6NO_3]$. Reprinted with permission from ref. 20. Copyright 1978 by Elsevier.

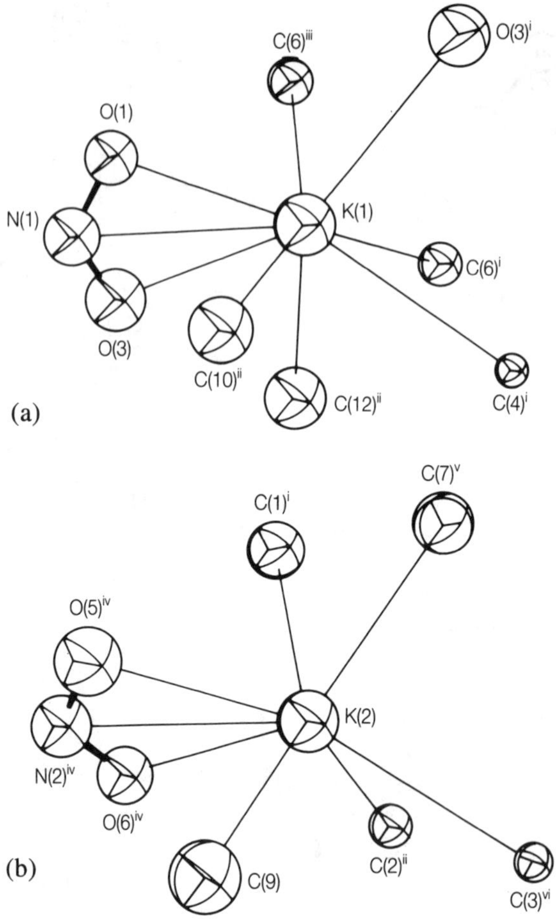

Figure 6.4. Cation environment in the structure of $K[Al_2Me_6NO_3]$. Reprinted with permission from ref. 20. Copyright 1978 by Elsevier.

methyl carbon atoms at distances from 3.08 (1) to 3.24 (1) Å. This anomaly is viewed as extremely important. The existence of two different anions, and, more importantly, two different cation environments in $K[Al_2Me_6N_3]$ is more than an interesting crystallographic artifact. In the second X-ray structure of a parent complex, $K[Al_2Me_6NO_3]$, an equivalent situation was found.[20] The anion containing Al(1) and Al(2) shows a staggered methyl group conformation (Figure 6.3a), with C(2) and C(5) out of the plane of the nitrate ion by 0.31 and 0.16 Å, respectively. The anion with Al(3) and Al(4) shows considerably more distortion of this geometry (Figure 6.3b), and C(7) and C(12) out of the plane of the second nitrate ion by 0.45 and 0.60 Å, respectively. Larger differences are seen for the two different cations (Figures 6.4a and b): K(1) has its closest

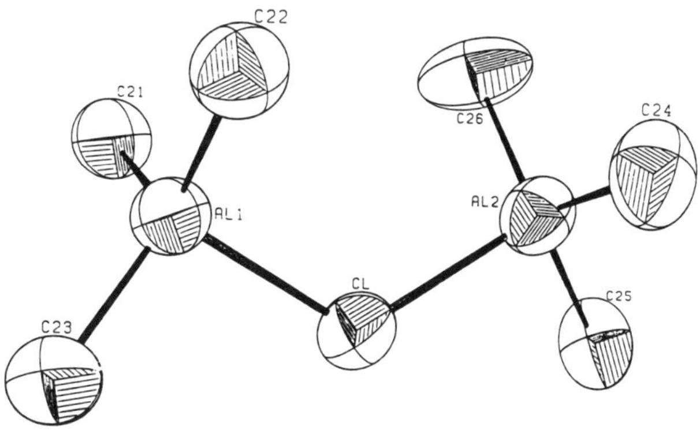

Figure 6.5. Structure of the anion of [K·dibenzo-18-crown-6][Al_2Me_6Cl]·2 C_6H_6. Reprinted with permission from ref. 21.

contacts with three oxygen atoms (at 2.82, 2.92 and 3.11 Å) while K(2) has only two oxygen atoms (2.80 and 2.98 Å) within a sphere of 4 Å radius.

The halide-bridged complexes are also of interest. As was mentioned previously both the angular **1** and the linear **2** structures must be considered. Structure **1** has been demonstrated[21] for the chloride ion (Figure 6.5) for the acetate ion in [NMe_4][$Al_2Me_6CH_3COO$],[22] (Figure 6.6), for the phenoxide anion in K[$Al_2Me_6OC_6H_5$], [23] and is postulated for bromide and iodide bridged anions, whereas **2** has been demonstrated for K[Al_2Et_6F][10] and K[Al_2Me_6F] C_6H_6.[3] The linear structure is also seen for the anion in Na[Al_2Me_6H], (Figure 6.7).[24]

In the previous sections only the 1:1 and 2:1 anionic species have been discussed. There exists one example of a 3:1 anion in K[Al_3Me_9SeMe], (Figure

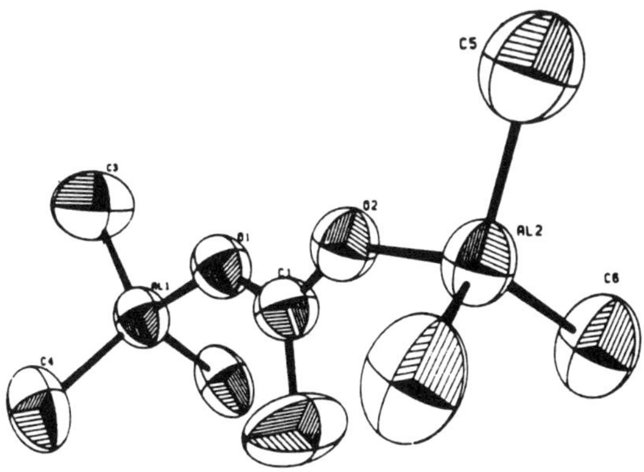

Figure 6.6. Structure of the anion of [NMe_4][$Al_2Me_6(CH_3COO)$]. Reprinted with permission from ref. 22. Copyright 1982 American Chemical Society.

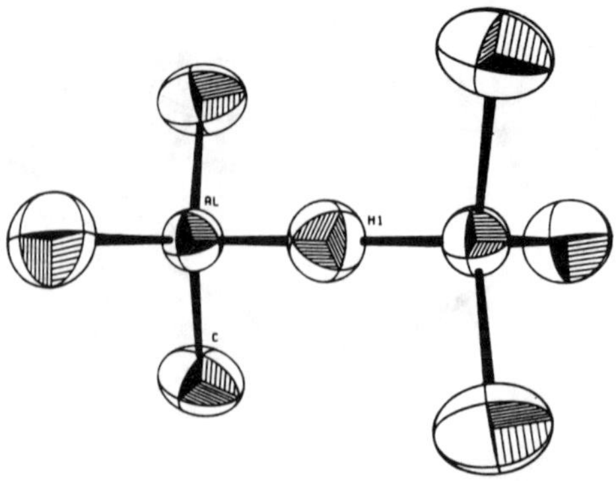

Figure 6.7. Structure of the anion of $Na[Al_2Me_6H]$. Reprinted with permission from ref. 24. Copyright 1981 American Chemical Society.

6.8), the crystal structure of which is discussed in detail later.[25] There is also one example of a 4:1 complex, $K_2[Al_4Me_{12}SO_4]$ (Figure 6.9).[26] In the latter case, only the 4:1 stoichiometry could be isolated. Attempted preparation of the 1:1, 2:1, and 3:1 complexes resulted in the 4:1 complex plus unreacted $K_2SO_4 \cdot K_2[Al_4Me_{12}SO_4]$ forms normal liquid clathrates with aromatic solvents. This presents a clue that the angular geometry is a sufficient, but not mandatory, condition for liquid clathrate formation.

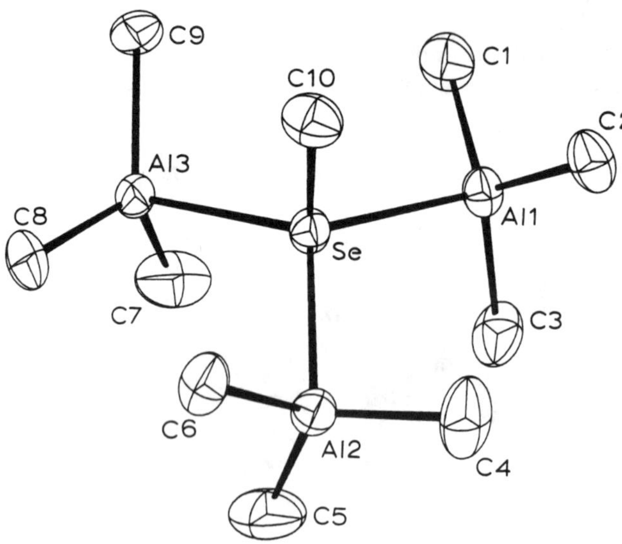

Figure 6.8. Structure of the anion of $K[Al_3Me_9SeMe]\cdot 2\ C_6H_6$. Reprinted with permission from ref. 25. Copyright 1976 Elsevier.

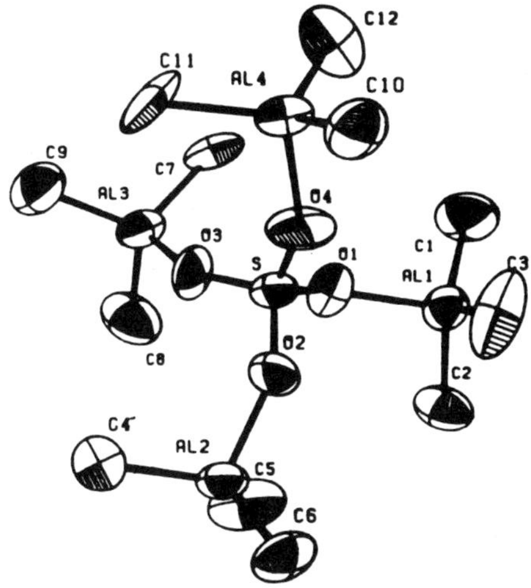

Figure 6.9. Structure of the anion of $K_2[Al_4Me_{12}SO_4]$. Reprinted with permission from ref. 26. Copyright 1984 American Chemical Society.

6.2.4. Definition of Liquid Clathrate Behavior

The term liquid clathrate is used here to designate that group of non-stoichiometric liquid inclusion compounds which form upon the interaction of aromatic molecules with certain $M[Al_2R_6X]$ moieties. A compilation of different combinations of M, R, and X has been given.[27–29] The interaction of

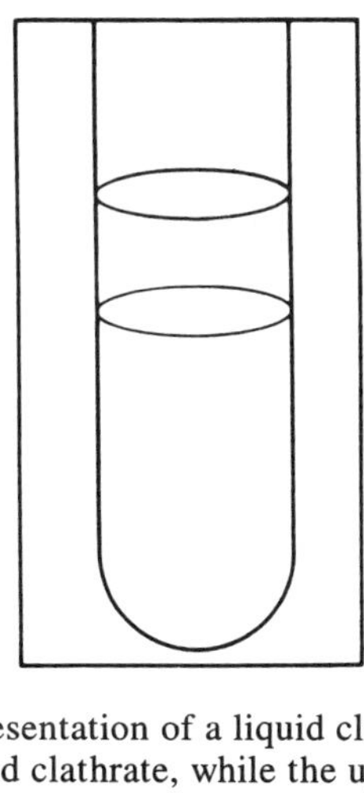

Figure 6.10. Schematic representation of a liquid clathrate. The lower phase contains the salt and aromatic, the liquid clathrate, while the upper phase is the excess aromatic solvent.

$M[Al_2R_6X]$ with an appropiate aromatic substance produces a liquid which is immiscible with excess aromatic. The visually dramatic appearance of the liquid layering is shown in Figure 6.10.

The aromatic molecules in the liquid clathrate are trapped much as they would be in a solid clathrate and can be freed by a change in temperature and reclaimed unchanged. These observations may be summarized in the form of the equilibrium

$$M[Al_2R_6X] + n \text{ aromatic} \rightleftharpoons \underset{\text{(liquid clathrate)}}{M[Al_2R_6X]\cdot n \text{ aromatic}}$$

The exact temperature at which the transformation occurs is strongly dependent on the nature of M, R, X, and the aromatic substance. For example, $K[Al_2Me_6N_3]\cdot 5.8\ C_6H_6$ is metastable at room temperature, and will deposit crystals of $K[Al_2Me_6N_3]$ in a few hours, whereas $Cs[Al_2Me_6N_3]\cdot 7.7\ C_6H_6$ is stable for weeks: $[N(C_5H_{11})_4][Al_2Me_6]\cdot 13.0\ C_6H_6$ is stable at room temperature indefinitely.

6.2.5. Preparation of Liquid Clathrates

6.2.5.1. Parent Compounds with Naked Cations

The two different ways in which these compounds may be synthesized are best illustrated by reference to the $K[Al_2Me_6N_3]$ complex. Using the method previously described,[2] 0.010 mol Me_3Al was added to 0.005 mol KN_3 under an N_2 atmosphere dry box. The mixture was sealed in a bomb tube, removed from the dry box, heated, returned to the dry box, and opened. Another 0.005 mol Me_3Al was then added to the powdered contents. After three cycles of grinding, adding Me_3Al, and heating, the white crystalline product was dried under vacuum. Addition of benzene (~0.01 mol), followed by heating at 60°C for 1 h afforded the liquid clathrate $K[Al_2Me_6N_3]\cdot 5.8\ C_6H_6$.

A much improved method for the production of the compounds involves simply the addition of 0.005 mol KN_3 and ~0.010 mol Me_3Al to 0.10 mol C_6H_6 in the dry box. A liquid clathrate identical in composition to the one prepared by the previous method is obtained in 1 h. As the reaction proceeds, a separation of two liquid layers (liquid clathrate and excess aromatic) becomes obvious; the layers appear upon shaking just as oil and water.

Once a liquid clathrate has reached its maximum composition, it is not possible to cause a further uptake of aromatic molecules. The formation of $K[Al_2Me_6N_3]\cdot 5.8\ C_6H_6$ represents a maximum aromatic:anion ratio.

The liquid clathrates, although water and oxygen sensitive, are much less reactive than the pure parent organometallic compounds. None of the azides have been found to present an explosion hazard, but extreme caution must be exercised in the handling of the formates and acetates.

6.2.5.2 Parent Compounds with Crown Ether-coordinated Cations

Using the methods described above, it was not possible to prepare liquid clathrates based on divalent cations or anions. Reaction temperatures could not practically be raised above ~150°C because of decomposition of either the aluminum alkyl or the resulting complex. In the available thermal range it was not possible to put compounds such as $Ba(NO_3)_2$ or Cs_2CO_3 into solution in aromatic solvents. It was not possible to use solvents of such Lewis basicity as diethyl ether, tetrahydrofuran, or dioxane because these preclude the formation of liquid clathrates. This problem was circumvented by the use of crown ethers. The crown ethers act to effect the transport of the aluminum alkyl to the anion while also providing enhanced solubility of the MX salt.

Not only did this afford a wide range of new liquid clathrates, but also a rich coordination chemistry of aluminum alkyls with crown ethers developed. It could be expected that the steric requirements of the crown ethers might be such as to lead to weak Al–O bonds. However, rather normal bond lengths were found in the first reported complexes; 15-crown-5 and dibenzo-18-crown-6 with Me_3Al.[30] The structure of $[Me_3Al]_2$[dibenzo-18-crown-6] is shown in Figure 6.11, and that of $[Me_3Al]_4$[15-crown-5] in Figure 6.12. Subsequently, the structure of $[Me_3Al]_3$[dibenzo-18-crown-6],[31] $[Me_3Al]_4$[18-crown-6],[31] $[Me_3Al]_2$[dicyclohexano-18-crown-6],[32] and $[Me_3Al]_2$[12-crown-4][33] were reported. The entire area of crown ether coordination complexes with aluminum alkyls has recently been reviewed.[34]

In a typical reaction R_3Al was first complexed by the crown ether as noted earlier. Addition of $Ba(NO_3)_2$, and elevation of the temperature to 100°C, led to the formation of the typical liquid layering.

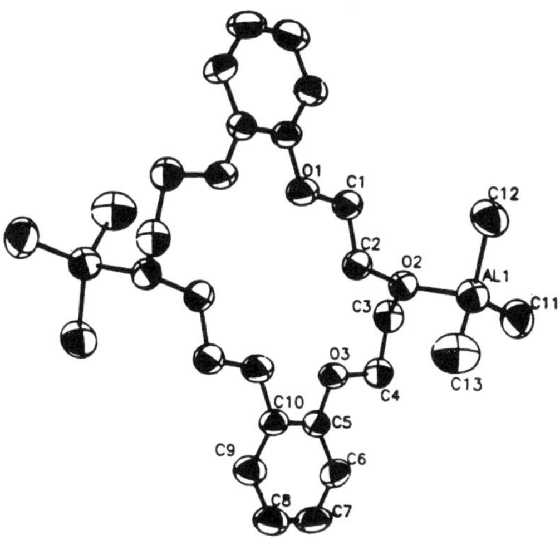

Figure 6.11. Structure of $[AlMe_3]_2$[dibenzo-18-crown-6]. Reprinted with permission from ref. 30. Copyright 1982 American Chemical Society.

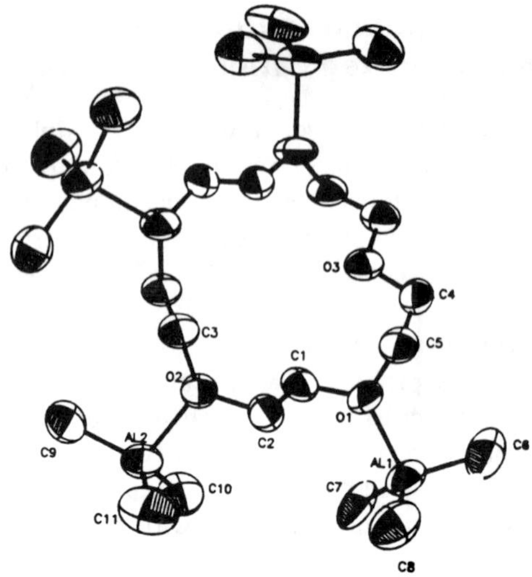

Figure 6.12. Structure of $[Al(CH_3)_3]_4[15\text{-crown-}5]$. Reprinted with permission from ref. 30. Copyright 1982 American Chemical Society.

$$Me_3Al_{(excess)} + 18\text{-crown-}6 \xrightarrow[\text{toluene}]{} [Me_3Al]_4[18\text{-crown-}6]$$

$$[Me_3Al]_4[18\text{-crown-}6] \xrightarrow{Ba(NO_3)_2} Ba[Al_2Me_6NO_3]_2 \cdot n \text{ toluene}$$

This method has led to several new and interesting compounds of the 1:1 and 2:1 type:[K·dibenzo-18-crown-6][$AlMe_3NO_3$],[35] [K·dibenzo-18-crown-6] [$Al_2Me_6N_3$],[19] [Cs_2(18-crown-6)][$Al_3Me_9SO_4$], a new 3:1 complex (Figure 6.13),[36][K·dibenzo-18-crown-6][$Al_2Me_6O_2$] (Figure 6.14),[37] [K·18-crown-6][$Al_2Me_6OCH_3$] (Figure 6.15),[38] and [K·18-crown-6] [$Al_2Me_6OC_6H_5$].[38]

The effect of the crown ether is to generate a liquid clathrate of higher aromatic composition than the naked parent complex would have produced.

6.2.6. Factors Affecting the Composition of the Liquid Clathrate

The number of aromatic molecules in the liquid clathrate is dependent upon the size and type of cation, the nature of the anion, and the type of aromatic molecule. For insight into these factors reference should be made to the data given in reference 29.

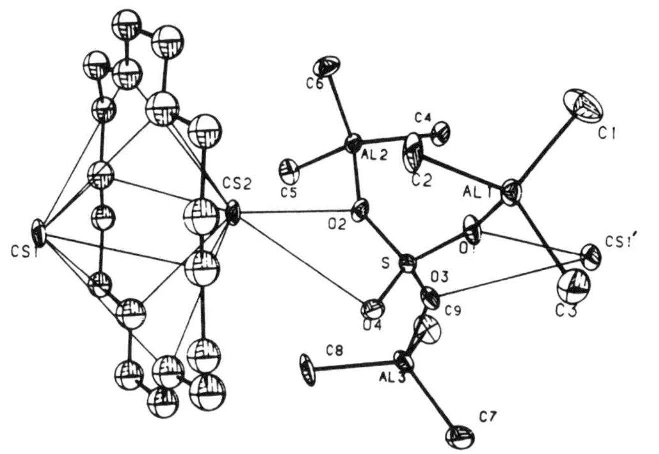

Figure 6.13. Structure of [Cs_2•18-crown-6][$Al_3Me_9SO_4$]. Reprinted with permission from ref. 36. Copyright 1984 American Chemical Society.

Regardless of the aromatic susbstance employed, the effect of the size of the cation is clear-cut: for a given variety of anions, the larger the cation, the greater the number of aromatic molecules in the clathrate. This is true of both the alkali metal and tetraalkylammonium cation series, although at present no meaningful size comparison can be made between the two types of cations.

Apparent anomalies are shown with the higher aluminum alkyls. The smallest numbers of aromatic molecules are taken by NEt_4^+ and NPr_4^+ complexes. No explanation has yet been offered, and the general trend is the same as seen for the trimethylaluminum derivatives. The effect of the anion can be subdivided according to the nature of the alkyl group attached to the aluminum

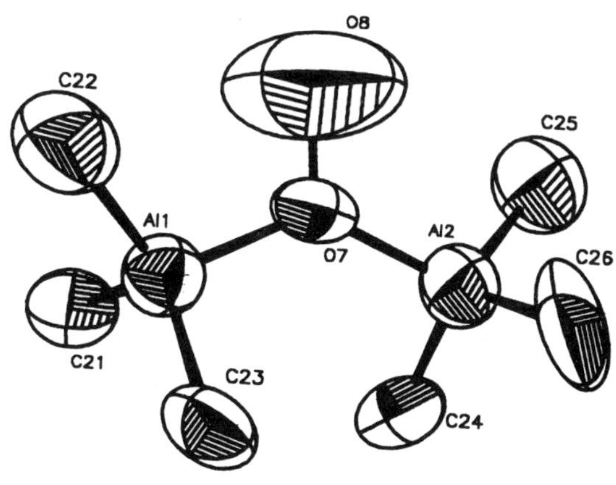

Figure 6.14. Structure of the anion of [K•dibenzo-18-crown-6][$Al_2Me_6O_2$]•1.5 C_6N_6. Reprinted with permission from ref. 37. Copyright 1981 American Chemical Society.

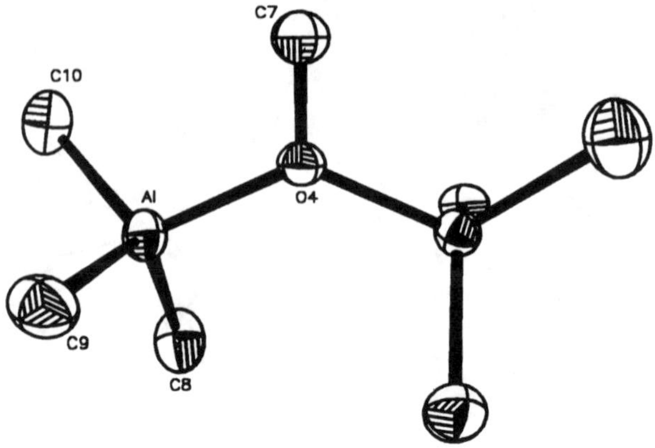

Figure 6.15. Structure of the anion of [K·18-crown-6][$Al_2Me_6CH_3O$].

atom and the type of halide or pseudohalide ion. The larger the anion, the larger the aromatic/anion A/A ratio.

The second effect is illustrated most clearly in the tetraalkylammonium series. The more electronegative the halide ion, the greater the number of aromatic molecules in the clathrate. This could also be a size dependence, but the size of the halide ion probably does not affect the dimensions of the entire anion enough to cause the large observed effects. For example, with the tetra-*n*-butylammonium complexes, the bromide takes on 9.3 toluene molecules per anion, wheras the iodide assumes only 7.0 toluene molecules per anion.

Just as with solid clathrates, the effect of the size of the guest is straight forward: the larger the aromatic molecules, the smaller the number of aromatic molecules in the liquid clathrate. It is possible that subtle electronic effects may also play a role, but the limitations on the accuracy of the determination of the aromatics/anion (A/A) ratio doubtless legislate against the observation of their ramifications. This general trend is seen almost without exception. All the trends shown for the liquid clathrates are just as one would expect for solid state clathrates, and each substance has a maximum A/A ratio. These two observations represent persuasive arguments for the existence of cages or layers of anions and cations with the guest aromatic molecules trapped within.

6.2.7. Solid State Clues to Liquid Clathrate Behavior

In order to understand the formation of liquid clathrates it is necessary to construct an appropiate model for the behavior. Two possibilities fit all the observed properties: a cage-like or layer-like liquid structure. The layer-like structure seems much more likely, since it allows for a great reduction in the electrostatic repulsion of the cations. In two dimensions the model appears as

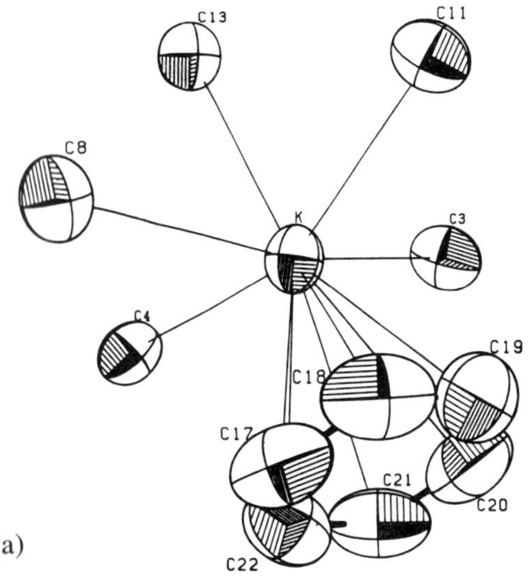

(a)

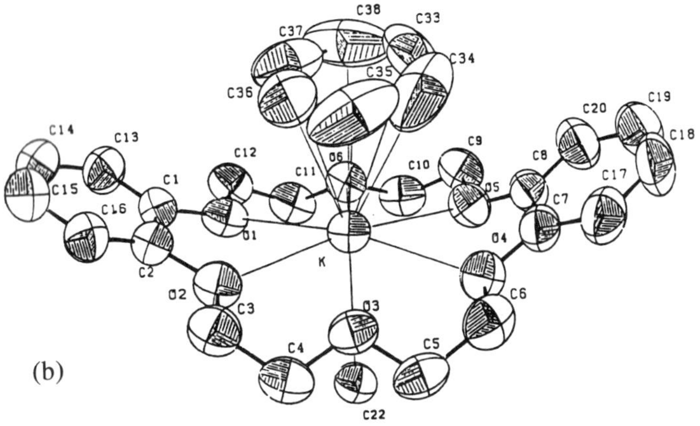

(b)

ure 6.17. (a) Example of a type **5** interaction of K^+ with a benzene ring in the ıcture of $K[Al_7O_6Me_{16}]\cdot C_6H_6$. (b) Type **5** interaction in [K·dibenzo-18-crown-6] $_2Me_6Cl]\cdot 2\ C_6H_6$.

The interaction found in **5** involves a rather symmetrical binding of K^+ to e aromatic ring. Typical examples[21,39] are shown in Figure 6.17.

The interaction of **6** is one in which the aromatic molecule insulates one tassium ion from another. For a given $K^+\cdots K^+$ separation, the insertion of e aromatic clearly improves the electrostatics. Note that this type of interacon is fundamental in the liquid clathrate model. For the two examples[26,37] own in Figure 6.18, the potassium ions are not disposed directly over the ngs.

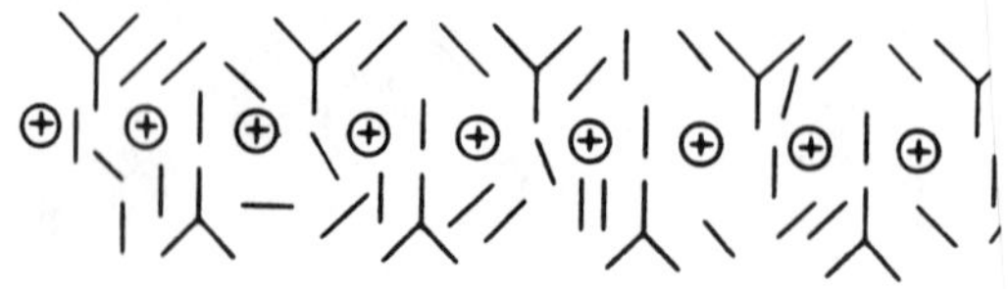

Figure 6.16. Two-dimensional model of liquid clathrate behavic

in Figure 6.16. Although it is difficult to obtain direct experi confirm the layer-like model, several aspects of the model l crystal structures of certain related compounds.

The presence of solvent molecules in the lattice of orga pounds is a common occurrence. An explanation based (packing is often encountered. However, a sufficient data bas of aromatic molecules is in hand to afford some patterns. Ta results of ten crystal structures with potassium cations. There cation aromatic interactions, **5**, **6**, and **7**:

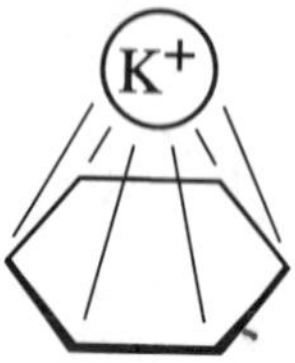

5

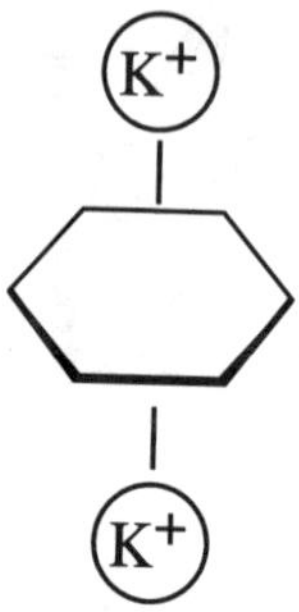

6

Table 6.1. Summary of $K^+\cdots$aromatic interactions.

Compounds	Interaction type	$K^+\cdots C_{(aromatic)}$ Range
[K·DB-18-crown-6][$AlMe_3NO_3$]·$0.5C_6H_6$[a]	7	>3.9
K[$Al_4Me_{12}SO_4$] . $0.5C_6H_6Me_2$-p	6	3.45(2)–3.58(2)
K[Al_2Me_6F]·C_6H_6	7	>3.9
K[$AlMe_3NO_3$]·C_6H_6	5	3.29(1)–3.46(1)
[K·DB-18-crown-6][$Al_2Me_6O_2$]·$1.5C_6H_6$[a]	6 and 7	3.35(3)–3.36(3)[b]
K[$Al_7O_6Me_{16}$]·C_6H_6	5	3.27(1)–3.39(2)
K[Al_3Me_9SeMe]·C_6H_6	6 extended	3.33–3.37
[K·DB-18-crown-6][Al_2Me_6Cl]· $2C_6H_6$[a]	5 and 7	3.29(2)–3.50(3)[b]

[a] DB-18-crown-6 represent dibenxo-18-crown-6.

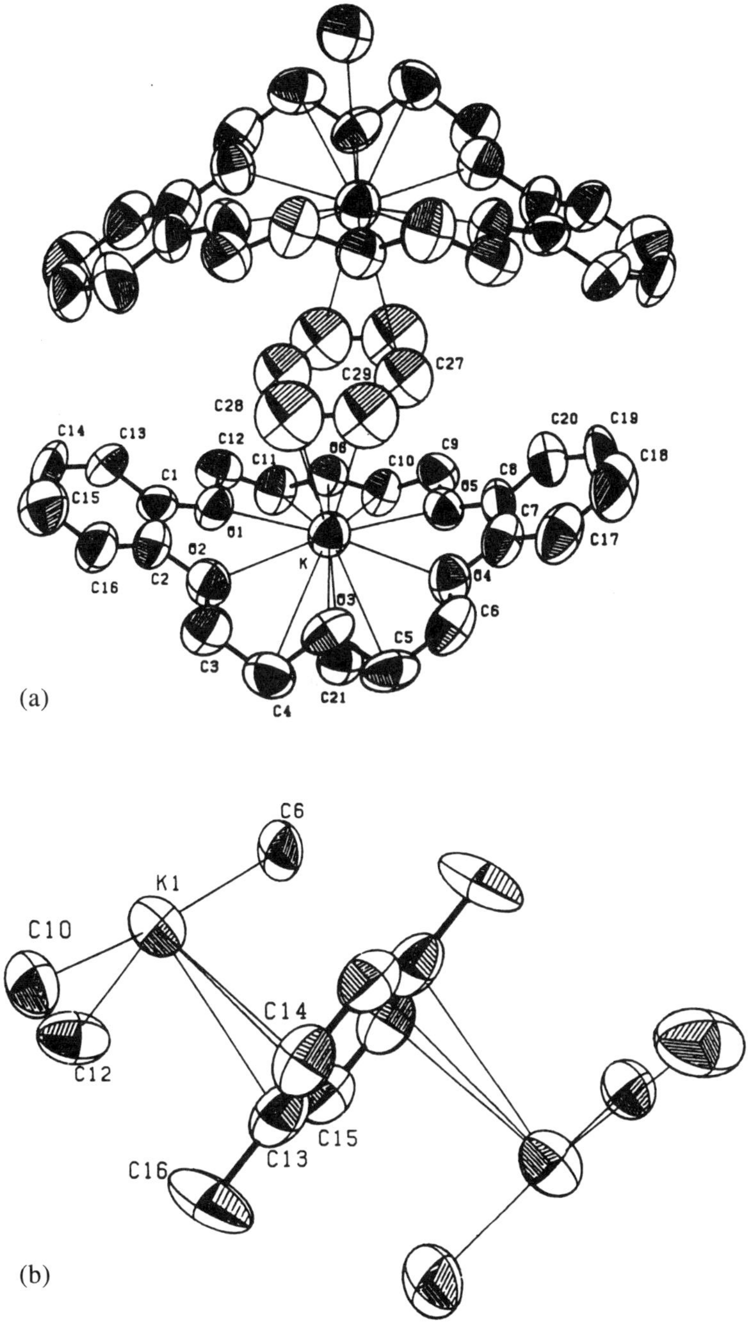

(a)

(b)

Figure 6.18. Type 6 interactions in (a) [K·dibenzo-18-crown-6][$Al_2Me_6O_2$]·1.5 C_6H_6 and (b) [$K_2AlMe_{12}SO_4$]·0.5 $C_6H_4(CH_3)_2$–p. Reprinted with permission from ref. 26. Copyright 1984 American Chemical Society.

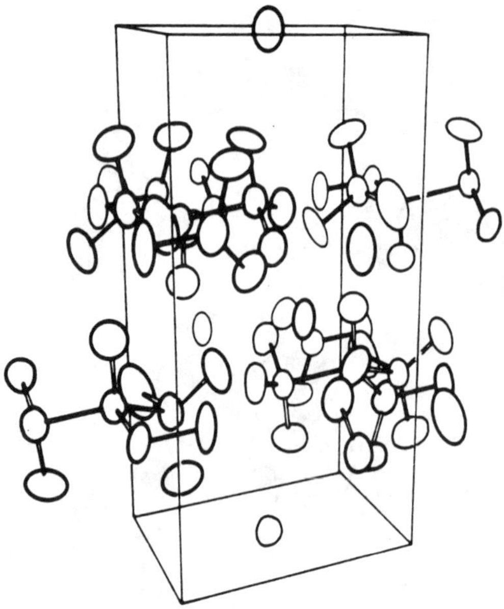

Figure 6.19. Unit cell packing diagram for $K[Al_3Me_9SeCH_3]{\cdot}2\ C_6H_6$. Benzene molecules make little contribution to the X-ray scattering because of high thermal motion or disorder. Reprinted with permission from ref. 27. Copyright 1976 American Chemical Society.

The clearest relation of a crystal structure to the liquid clathrate model is found in the novel selenide, $K[MeSe(AlMe_3)_3]{\cdot}2\ C_6H_6$.[25] From an excellent set of X-ray diffraction data, the conventional agreement index R equals 0.092 for the $K[MeSe(AlMe_3)_3]$ portion of the molecule. (The unit cell packing diagram is given as Figure 6.19.) Even though the atoms of the benzene molecules comprise some 32% of the total electron density, the R factor indicates the structure is solved even in their absence.

In order to explain the lack of importance of the benzene molecules to the total X-ray scattering, two possibilities may be envisioned. First, extremely high thermal motion of the aromatic rings would have the effect of reducing the scattering ability of the carbon atoms. More specifically, such atoms would be expected to contribute strongly to only a few low angle reflections. The same net effect could also be achieved by a highly disordered arrangement of the benzene rings.

By a detailed analysis of electron density difference maps it was possible to locate one of the two independent benzene molecules. The positions of these molecules are shown in Figure 6.20 which shows that the centers of the symmetry-related benzene rings lie close to one of the bc face diagonals. The remaining aromatic molecules must reside with their centers close to the other bc face diagonal. An idealized view of the liquid portion of the unit cell along the a direction is given in Figure 6.21.

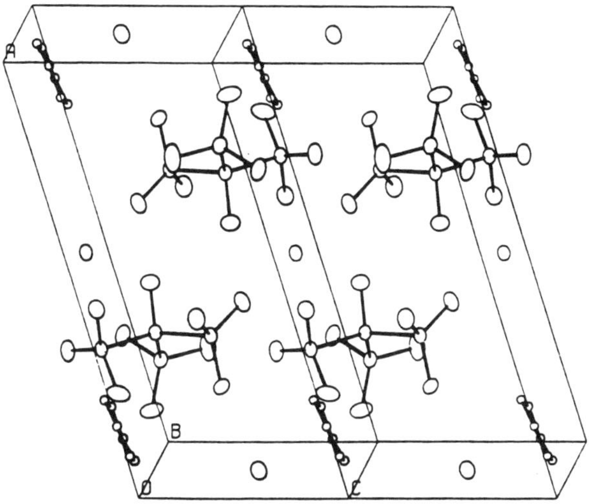

Figure 6.20. Two unit cells with one of the two independent benzene molecules shown in the structure of $K[Al_3Me_9SeCH_3]\cdot 2\ C_6H_6$.

The similarity between the structures given in Figures 6.20 and 6.21 and the proposed liquid clathrate structure is striking. It is possible that in the case of $K[Me_3Se[AlMe_3]_3]$, a solid-state mirror of the liquid ordering has been obtained.

Another interesting feature is that in the asymmetric unit there are also two crystallographically independent potassium ions. Although both reside on centers of symmetry, the environments are quite different. One is packed within the methyl groups of the anions: the K···C distances range from 3.15 to 3.22 Å.

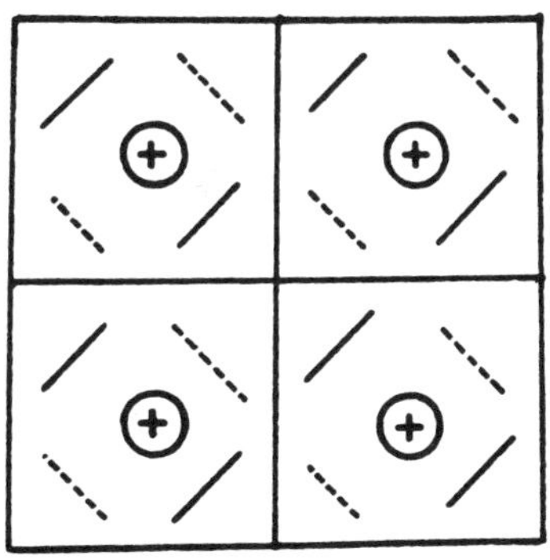

Figure 6.21. Two-dimensional view of the orientation of the benzene molecules with respect to the cations in $K[AlMe_9SeMe]\cdot 2\ C_6H_6$.

(a)

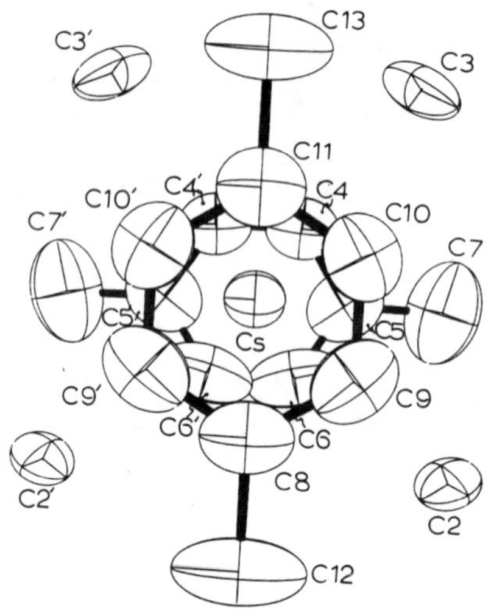

(b)

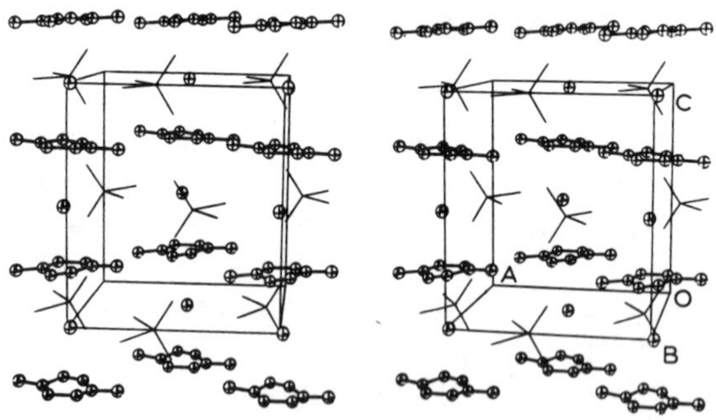

Figure 6.22. (a) Cation environment on $Cs[Al_2Me_6N_3]\cdot 2\ C_6H_5(CH_3)_2$-p. (b) Unit cell packing in $Cs[AlMe_2I_2]\cdot 2\ C_6H_5(CH_3)_2$–p.

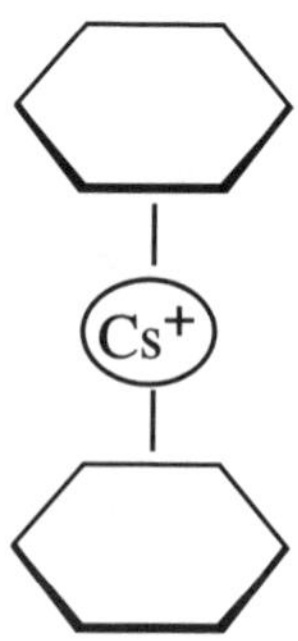

8

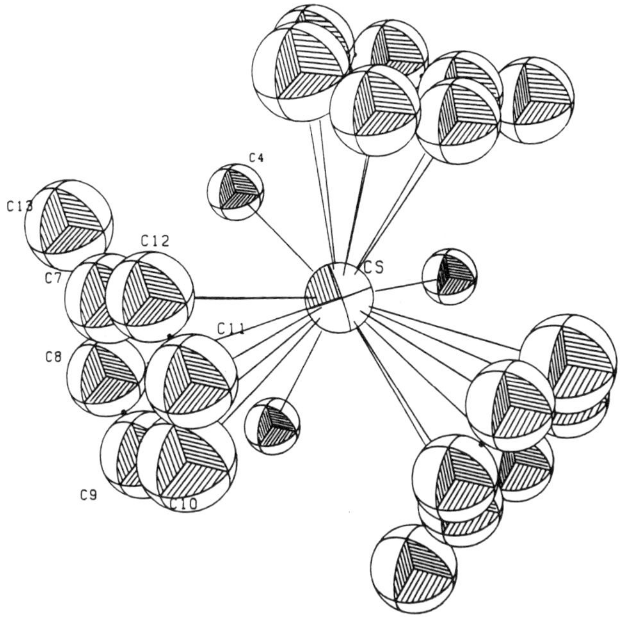

Figure 6.23. Cesium ion environment in $Cs[Al_7O_6Me_{16}]\cdot 3\ C_6H_5CH_3$.

The other potassium ion is associated with two carbon atoms of the anion at 3.29 Å, and the remainder of its contacts must be with the benzene molecules.

There are now several examples in which Cs^+ is sandwiched between aromatic rings **(8)** (Figure 6.22).[19,40] The toluene basket including Cs^+ (Figure 6.23) is also of importance.[39]

6.2.8. Origin of Liquid Clathrate Behavior

In order to expand the scope of the novel solution behavior, it is necessary to formulate an appropriate model which accounts for the experimental observations. We now believe that the parent compound must possess three properties in order to facilitate liquid clathrate formation.

1. The substance must have a relatively low lattice energy.
2. The substance must be capable of exhibiting a very strong cation-anion interaction.
3. Association into tight ion pairs, dimers, or other small units must be prevented.

These factors seem at first glance to be contradictory, but they may be rationalized with reference to an example.

In the solid state the lattice energy of N_3^- is quite high. It is easy for the K^+ ions to pack close to the N_3^- ions. The same situation can also be envisioned for $K[AlMe_3N_3]$, where the Me_3Al group has blocked a portion of the N_3^- ion. However, the crystal structure of $K[Al_2Me_6N_3]$ shows that the presence of two Me_3Al groups causes the K^+ ions to pack in two different environments: one close to the N_3^- group is probably not greatly diminished by the presence of the Me_3Al groups, but the lattice energy is nonetheless relatively low.

In this circumstance the expansion of the structure by addition of the aromatic molecules can have the effect of allowing the K^+ ions packed among methyl groups to migrate to a more favorable electrostatic position, as in Figure 6.16. However, the cation-anion interaction is too strong to allow the ions to move apart as in normal solution. Thus, the structure expands until the layers are filled, but not any further.

Aromatic molecules are especially effective in promoting the effect, primarily because of their ability to insulate the positive charges from each other. It is possible to substitute a moderate percentage of non-aromatic molecules into a liquid clathrate, but a solid parent compound, i.e., $K[Al_2Me_6N_3]$, shows no interaction with molecules such as cyclohexane.

6.2.9. Stability of Liquid Clathrates

In the isolated liquid clathrate there are two principal modes of decomposition. The first is governed by the equilibrium

$$M[Al_2R_6X] + n\ \text{aromatic} \rightleftharpoons M[Al_2R_6X]\cdot n\ \text{aromatic}$$

and forms the basis for the removal of the aromatic from the compound.

The second consideration is affiliated with the equilibrium

$$M[Al_2R_6X]\cdot n\ \text{aromatic} \rightleftharpoons M[Al_2R_6X] + n\ \text{aromatic} + R_3Al$$

With every constituent under discussion, $M[AlR_3X]$ is a solid, and all organoaluminum-based liquid clathrates must be somewhat unstable with respect to this method of decomposition.

All liquid clathrates except those based on the acetate and formate ions are thermally stable to at least 100°C.

6.2.10. Organic Separations

Liquid clathrates have been examined with a variety of separation problems.[27–29,41,42] Two points favor this application over traditional solid state clathrates: high percentages of organic molecules are loaded into the liquid clathrates (typically 60–80% by weight) and techniques of liquid–liquid extraction are applicable. The measure of the effectiveness of a separation is the factor α. For a benzene/hexane separation, α is defined as

$$\alpha = \frac{[\text{benzene/hexane}]_{\text{liquid clathrate phase}}}{[\text{benzene/hexane}]_{\text{hydrocarbon phase}}}$$

For most processes an α of ~10 is considered excellent, whereas a value of 4 is often acceptable. In the benzene/hexane system the average of some 200 runs with $[NMe_4]Al_2Me_6NO_3]$ was ~40, and the optimum for a 24 h run was 240. (An α = 240 roughly means that from a 50:50 mixture 240 parts benzene to one part hexane was obtained in the liquid clathrate layer.)

Other model systems were studied: benzene/toluene, α = 4 for $[NEt_4][Al_2Me_6NO_3]$; benzene/mesitylene, α = 10 for $[NMe_4]$ $[Al_2Me_6]$, α = 4 for $[NEt_4][Al_2Me_6NO_3]$; ethylbenzene/ *o*-xylene, α = 4 for $[NEt_4][Al_2Me_6NO_3]$.

Applications of liquid clathrates in the coal liquefaction arena have also been sought.[43,44]

6.3. Aluminoxanes

It is well known that aluminum alkyls react violently with water, the end product being $Al(OH)_3$ and alkane. However, the controlled reaction of such species has provided an interesting field of chemical endeavor. It was discovered quite early that the products of AlR_3/H_2O mixtures act as catalysts for certain polymerization reactions.[45,46] Other articles had dealt with aspects of the AlR_3/H_2O reaction, but the first detailed report which relied on various spectroscopic methods did not appear until 1968.[47] In this account $[Et_2Al]_2O$ was prepared (from the reaction: 2 Et_3Al + $H_2O \rightarrow [Et_2Al]_2O$ + 2 Et–H) and its structure was postulated to be that of the trimer, **9**.

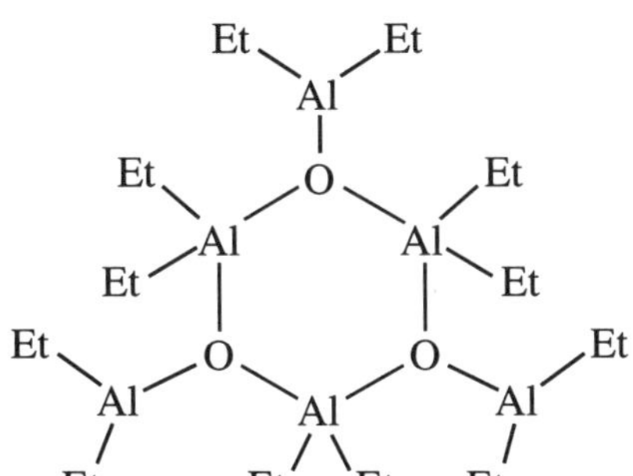

9

In other studies[48,49] a noteworthy feature was the emphasis placed on the method of preparation of the product. Quite complicated NMR spectra were obtained which often varied with the method of preparation. By the mid-1970s these substances were referred to as alumoxanes[50] or aluminoxanes.[51] The lack of definitive structural data on the aluminoxanes was not due to lack of effort. These substances had already found use as co-catalysts in the polymerization of olefins,[52] and had considerable potential value commercially. Indeed, even in the important heterogeneous systems based on $TiCl_4/MgCl_2/AlR_3$, active catalysts were sometimes prepared by the use of hydrated $MgCl_2$.[53] This undoubtedly produced aluminoxanes, but it is not known whether they were active co-catalysts.

The largest area of endeavor has been that of the soluble Ziegler–Natta catalysts. Exposure of active catalytic systems to small amounts of water led to increased activity.[54,55] The structure of an active aluminoxane was postulated to be **10**,[56] and in the absence of crystallographic data this structure is still found in the literature.

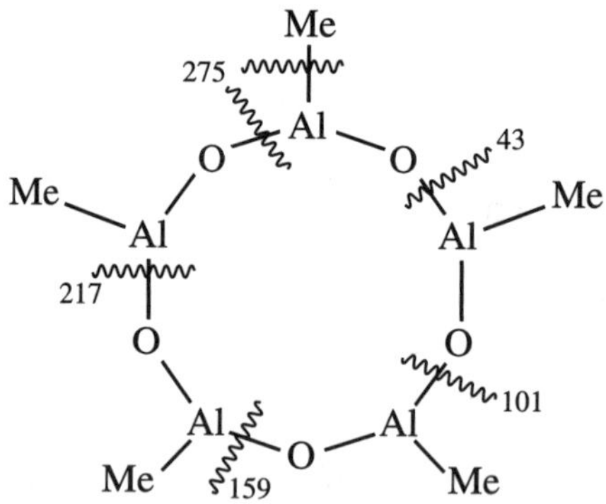

10

Using the general definition of Pasynkiewicz for an aluminoxane,[57] "...species containing an oxygen bridge binding two aluminum atoms," there are still very few crystal structures for such compounds. Since the basic problem with the R_2Al-O-AlR_2 formulation is the coordinative unsaturation at the aluminum atoms, efforts to stabilize such species have centered on the use of bulky R groups and on the generation of anions. Two structures of the former type have been published,[58,59] but is the anionic species that deserve more detailed attention.

Cationic organotitanium species were postulated to play a role in the olefin polymerization behavior of certain soluble catalysts as early as 1967.[60] However, it is only with the work of Jordan,[61] Turner,[62] Ewen,[63] Marks,[64] and others that this idea has received persuasive documentation. Similarly, if the transition metal-containing species is cationic, an anionic aluminoxane seems reasonable.

The best model for an anionic aluminoxane is found in the structure of $K[Al_7O_6Me_{16}]$ (Figure 6.24). This compound was initially prepared by the reaction of KO_2 with Me_3Al in toluene at 80°C. Subsequently, it was found that this structure could also be obtained from the reaction of Cs_2O with Me_3Al in

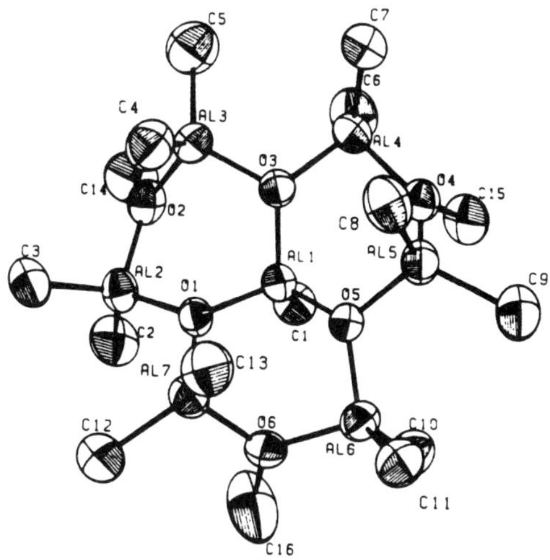

Figure 6.24. Structure of the anion of $K[Al_7O_6Me_{16}]\cdot C_6H_6$. Reprinted with permission from ref. 30. Copyright 1982 American Chemical Society.

toluene at room temperature (via the formation of a liquid clathrate of presumed composition $Cs_2[Al_2Me_6O]\cdot n$ toluene). At the time of publication this result was viewed as significant because (1) the anion presents an important step in the decomposition of high-oxygen content organoaluminum compounds; and (2) the decomposition of the superoxide complex probably passes through the same intermediate as that of the oxide. This structure was viewed as a model for aluminoxanes in the Ziegler–Natta polymerization arena primarily because it was a large structure and it was anionic. However, this anion has recently taken on new significance. The complex [K·18-crown-6]$[Al_7O_6Me_{16}]$ has been isolated from the reaction of K_2SO_4 moistened with two equivalents of H_2O in toluene with Me_3Al and 18-crown-6 according to the following stoichiometry:

$$K_2SO_4 + 2\,H_2O + 18\text{-crown-6} + 4\,Me_3Al \longrightarrow$$

(The target structure is discussed with related compounds in the section on supported aluminoxanes.) It is not likely that SO_4^{2-} decomposed under these conditions. The alternative is that the $[Al_7O_6Me_{16}]^-$ anion is formed by the reaction of Me_3Al with H_2O in toluene.

The anion (Figure 6.24) consists of an open Al_6O_6 framework capped by a seventh aluminum atom which is bonded to three alternate oxygen atoms. The six aluminum atoms are bonded to two terminal methyl groups each, and the unique aluminum atom is attached to only one methyl group. Each oxygen atom is three-coordinate: the ones on the exterior (not bonded to the unique aluminum atom) bridge two aluminum atoms and are also bonded to one methyl group each.

6.3.1 Supported Aluminoxanes

The commercial preparation of aluminoxanes involves a recipe and mixtures are invariably obtained, as noted above. For example, it is appropriate to quote from a Turner patent[65] issued in 1988:

"The alumoxanes are well known in the art and comprise oligomeric linear and/or cyclic alkyl alumoxanes represented by the formula:

$$R\text{——}(\underset{\underset{R}{|}}{Al}\text{–}O)_n\text{——}AlR_2 \quad \text{for oligomeric, linear alumoxanes and}$$

$$(\text{–}\underset{\underset{R}{|}}{Al}\text{–}O)_m \quad \text{oligomeric, cyclic alumoxane.}$$

wherein *n* is 1–40, preferably 10–20, *m* is 3–40, preferably 3–20 and R is a C1–C8 alkyl group and preferably methyl. Generally, in the preparation of alumoxanes from, for example, aluminum trimethyl and water, a mixture of linear and cyclic compounds is obtained."

"The alumoxanes can be prepared in a variety of ways. Preferably, they are prepared by contacting water with a solution of aluminum trialkyl, such as, for example, aluminum trimethyl, in a suitable organic solvent such as benzene or an aliphatic hydrocarbon. For example, the aluminum alkyl is treated with water in the form of a moist solvent. In a preferred method, the aluminum alkyl, such as aluminum trimethyl, can be desirably contacted with a hydrated salt such as hydrated ferrous sulfate. The method comprises treating a dilute solution of aluminum trimethyl in, for example, toluene with ferrous sulfate heptahydrate."

This passage serves to highlight both the uncertain product of the reaction and the preferred method of delivering H_2O to the aluminum alkyl. We had previously published work relating the formation of liquid clathrates based upon calixarenes,[66] **11**, as shown in Scheme 1:

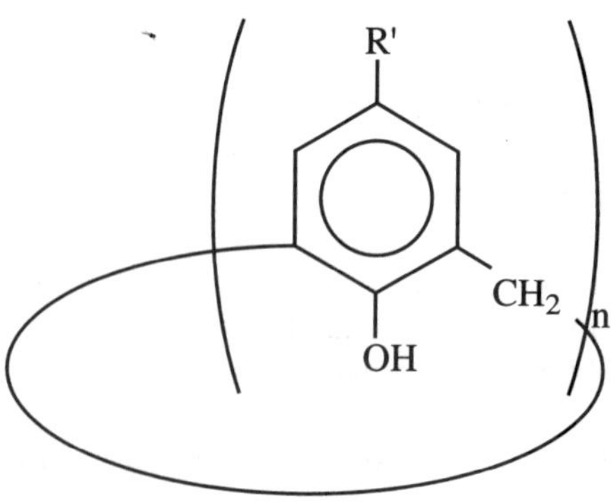

11

Scheme 1

$$CE + nAlMe_3 \xrightarrow[\text{toluene}]{} CE{\cdot}nAlMe_3 \xrightarrow[\text{toluene}]{MX}$$

$$[CE{\cdot}M]^+[(AlMe_3)_2X]^-{\cdot}n\ \text{toluene}$$
(liquid clathrate)

$$CA + nAlMe_3 \xrightarrow[\text{toluene}]{} CA{\cdot}nAlMe_3 \xrightarrow[\text{toluene}]{MX} \text{no reaction}$$

$$(n/2)M[(AlMe_3)_2X] + CA \longrightarrow CA{\cdot}nAlMe_3 + (n/2)MX$$
(liquid clathrate)

CE = crown ether; CA = methoxycalix[4]arene; M = alkali metal ion.

The MX species chosen for this reaction was C_6H_5COONa, but as the monohydrate. A liquid clathrate ensued with the cation as shown in Figure 6.25 and the anion in Figure 6.26. The oxygen atom labelled as O4 in Figure 6.36 had its origin in the H_2O of hydration of the sodium benzoate. Note that the $[C_6H_5COO[AlMe_2]_2O{\cdot}AlMe_3]^-$ contains methyl aluminoxane supported on the benzoate anion. The additional Me_3Al molecule is necessary in this system to produce a liquid clathrate. If the ratio of Me_3Al to benzoate is less than 3 to 1, insoluble material results.

The role of the hydrated ferrous sulfate in the Turner patent[65] (and the $CuSO_4{\cdot}5H_2O$[67] and other hydrated salts from other studies) is viewed as the delivery mechanism for the H_2O to the AlR_3. This has led to work along two

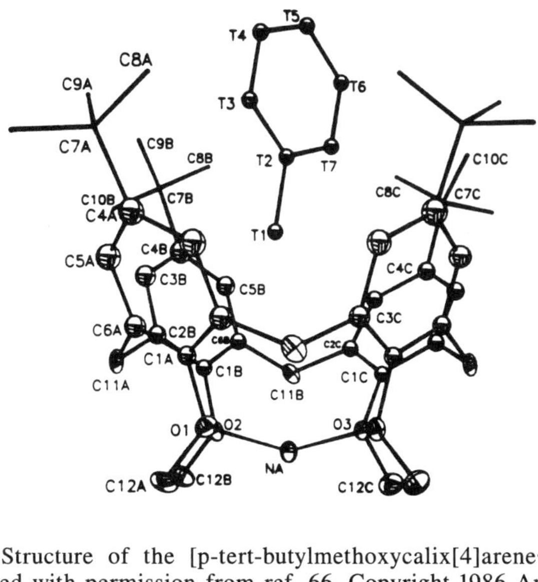

Figure 6.25. Structure of the [p-tert-butylmethoxycalix[4]arene·sodium·toluene]$^+$ cation. Reprinted with permission from ref. 66. Copyright 1986 American Chemical Society.

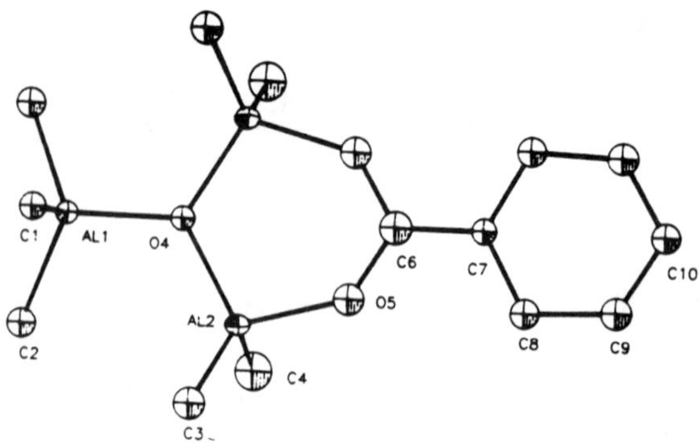

Figure 6.26. Structure of the $[C_6H_5COO\{AlMe_2\}_2OAlMe_3]^-$ anion. Reprinted with permission from ref. 66. Copyright 1986 American Chemical Society.

lines: (1) the support of aluminoxanes on anions such as benzoate or sulfate and (2) the use of non-hydrated salts as carriers of H_2O.

A wide variety of anions can be used to support the $[AlMe_2]_2O$ unit: nitrate, sulfate, benzoate, acetate, etc. Indeed, the key to the support lies in providing the group, **12**, upon which the methyl aluminoxane can be attached.

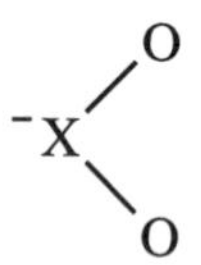

12

It is not necessary to use a salt which is hydrated on a molecular level. For example, C_6H_5COOK can be used as for the sodium salt mentioned earlier if it is moistened with one equivalent of H_2O (Atwood and Zhang, unpublished results). The function of the salt appears to be one of providing a surface from which the adsorbed H_2O can be slowly (on a molecular event time scale) released. NaCl and other simple salts function well for this purpose, although only those which present a template as in **12** lead to well-defined products.

It is certain that aluminoxanes will provide a rich field of investigation for many years. However, the ill-defined nature of the product of the commercial preparation has already led away from aluminoxane anions to such non-coordinating anions as $[B(C_6F_5)_4]^-$ for the stabilization of the organometallic cations in Ziegler–Natta olefin polymerization chemistry.[62,68]

6.4. Cationic Organoaluminum Complexes

Although the liquid clathrates reported earlier were always based on anionic organoaluminum compounds, by 1984 cationic species were shown to exhibit the effect.[69] The first such compounds were prepared as a result of the extension of research into crown ether complexes of Me_3Al. If $EtAlCl_2$ is reacted with 18-crown-6 in toluene, a liquid clathrate immediately results. Based on the known Me_3Al chemistry, a donor–acceptor complex was expected, not a liquid clathrate. X-ray diffraction studies soon revealed that the crown ether has effected an asymmetric split of the bridge in the dimer of $EtAlCl_2$ **13**, leading to $[Cl_2Al\cdot18\text{-crown-6}][Cl_2AlEt_2]$[69] (Figure 6.27). In a similar manner, 12-crown-4 also produces $[Cl_2Al\cdot12\text{-crown-4}]^+$[69] (Figure 6.28). As can be seen, the aluminum atoms in both structures are octahedrally coordinated with *cis*-chloro attachment. This may be referred to as a perching geometry. The *trans*-chloro version, a threaded geometry, was realized for $[Cl_2Al\cdot15\text{-crown-5}]^+$ and $[Cl_2Al\cdot\text{benzeo-5-crown-5}]^+$[70] (Figure 6.29).

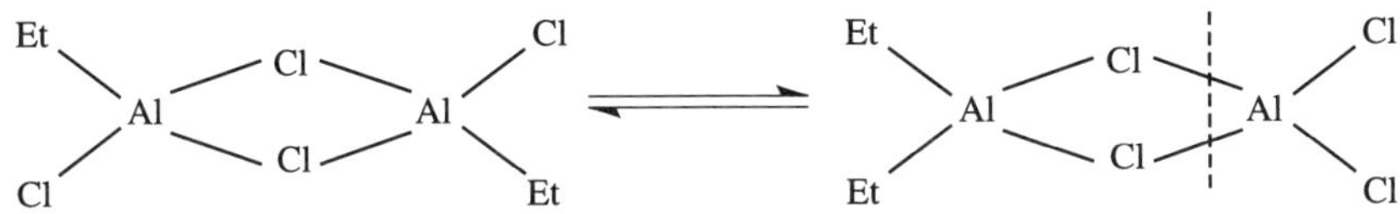

13

It was later discovered that crown ethers were not necessary to produce the asymmetric split of the dimer and thereby a cation-anion pair. Studies of bidentate oxygen donors, such as dimethoxyethane (DME), showed that the related $[Cl_2Al\cdot(DME)_2]^+$ was formed (Elgamal et al., unpublished results). It was therefore of interest to extend the work to bases with only one donor

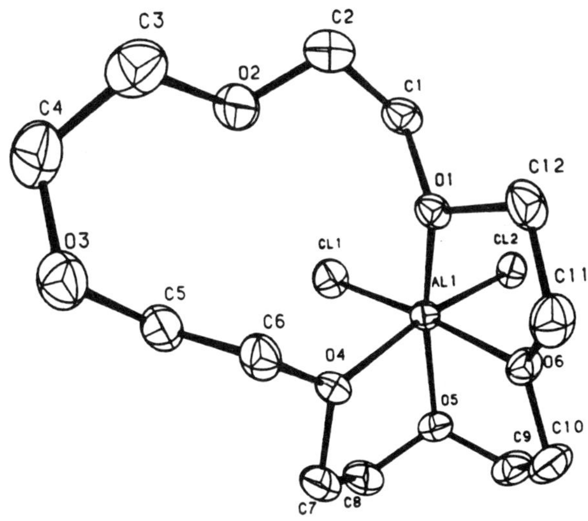

Figure 6.27. Structure of the $[18\text{-crown-6}\cdot AlCl_2]^+$ cation.

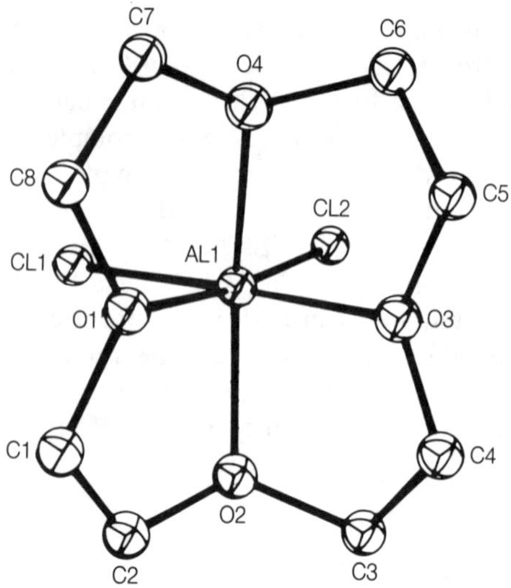

Figure 6.28. Structure of the [12-crown-4·$AlCl_2$] cation.

oxygen atom. The $AlCl_3$/THF system was suitable for study since a number of investigations had already been carried out. The presence of an ionic moiety in the workup of a saturated solution of $AlCl_3$ in THF was deduced from one vibrational spectroscopic study[71] but others found only the molecular $AlCl_3$·2THF in a related system.[72] Our group isolated [$Cl_2Al(THF)_4$][$AlCl_4$]

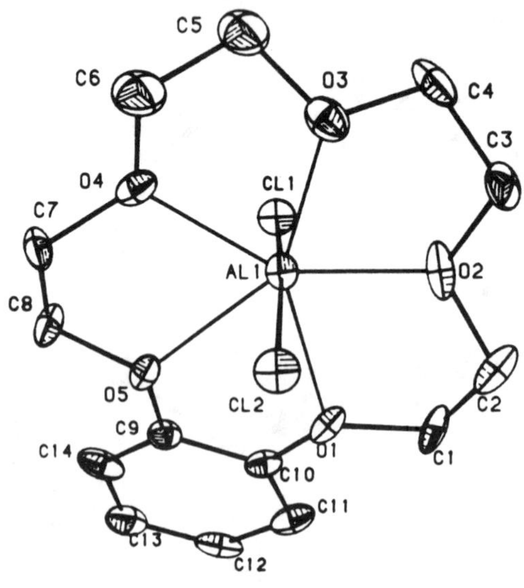

Figure 6.29. Structure of the [benzo-15-crown-5·$AlCl_2$]$^+$ cation. Reprinted with permission from ref. 70. Copyright 1985 American Chemical Society.

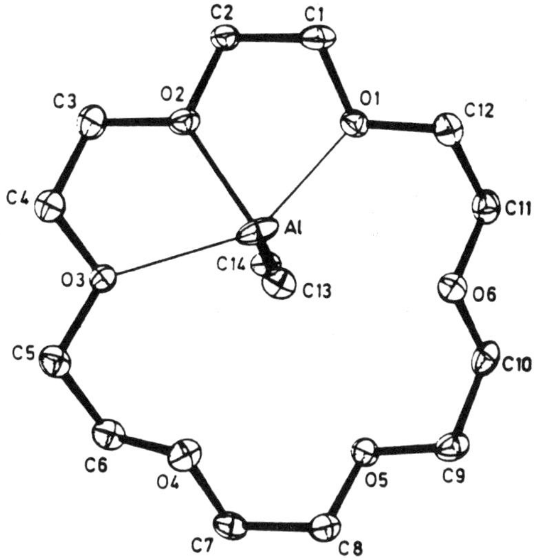

Figure 6.30. Structure of the [18-crown-6·$AlMe_2$]$^+$ cation.

as the only product of the reaction of stoichiometric quantities of the reactants in toluene[73] and similar results were also obtained for pyridine (Elgamal et al., unpublished results). Sangokoya et al. have also found $AlCl_2{}^+$ stabilized by aza-macrocycles.[74]

The first structural verification of organometallic species as cations appeared in 1987,[75,76] although $[R_2Al]^+$ had been implicated in solution studies previously.[77–80] The $[Me_2Al]^+$ cation was discovered during attempts to

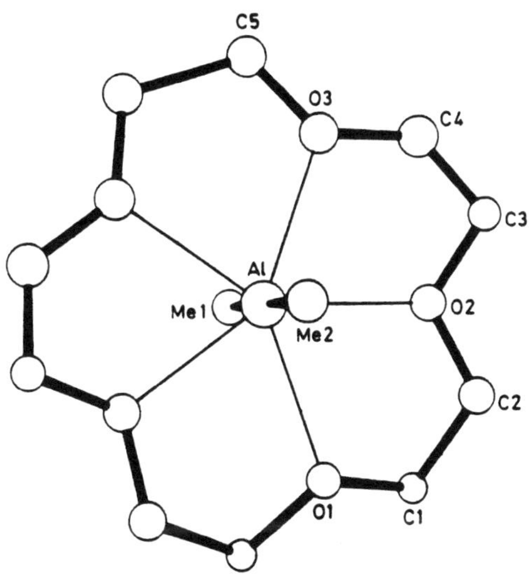

Figure 6.31. Structure of the [15-crown-5·$AlMe_2$]$^+$ cation.

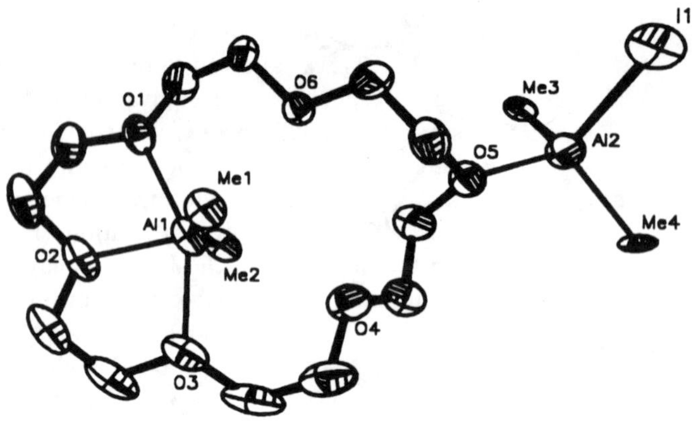

Figure 6.32. Structure of the $[Me_3Al\cdot 18\text{-crown-}6\cdot AlMe_2]^+$ cation.

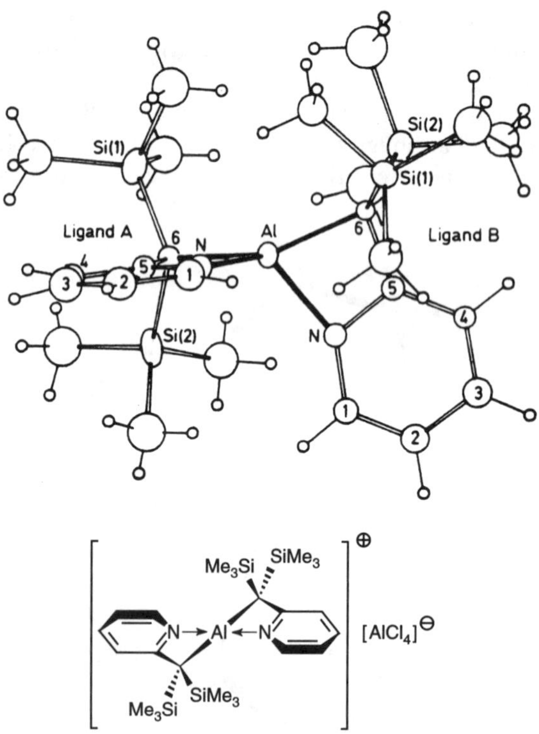

Figure 6.33. Structure of the $[Al\{2\text{-}C(Si(CH_3)_3)_2C_5H_4N_2\}_2]^+$ cation.

trap species related to Ziegler–Natta catalysis with crown ethers. In a typical experiment, one equivalent of Cp_2TiCl_2 was allowed to react with two equivalent of Me_3Al in toluene in the presence of one equivalent of 18-crown-6. Cp_2TiMe_2 and [Me_2Al•18-crown-6][Me_2AlCl_2] were isolated. The Me_2Al•15-crown-5]$^+$ cation was also isolated from the above reaction with Cp_2TiCl_2 and from the reaction of $CoCl_2$ with Me_3Al and 15-crown-5 in toluene in the molar ratio 1:4:1. The structure of the [Me_2Al•18-crown-6]$^+$ cation (Figure 6.30) shows Me_2Al^+ threaded through the crown, with the latter displaying the full symmetry configuration. The aluminum atom is strongly coordinated to one oxygen atom at a distance of 1.929(5) Å, but also interacts with others at 2.181(5) Å and 2.435(5) Å. The $C_{(methyl)}$–Al–$C_{(methyl)}$ bond angle is 140.6(3)°. Figure 6.31 shows Me_2Al threaded through 15-crown-5. Here the Al–O distances range from 2.13(1) Å to 2.26(1) Å and the C-Al-C angle is 178°.

Figure 6.30 shows that there are three donor sites on 18-crown-6 which are not occupied. It has been discovered that reaction of Me_3Al with AlI_3 and 18-crown-6 in toluene afforded the [Me_2Al•18-crown-6•Me_2AlI]$^+$ cation (Atwood et al. unpublished results), (Figure 6.32). Recently, a similar product, [Me_2Al•monoaza-18-crown-6•$AlMe_3$] was obtained as a neutral molecule.[81]

At the same time as the report of Me_2Al^+ stabilized by crown ether coordination, the Raston group noted the formation of an organoaluminum cation stabilized by internal coordination to nitrogen donor groups[76] (Figure 6.33).

Although the chemistry of cationic organoaluminum complexes is not well developed, the reports in the literature indicate that this is a rich area for future development.

Acknowledgments

Support of this research by the National Science Foundation is gratefully acknowledged.

References

1. Atwood, J. L.; Newberry, W. R. *J. Organomet. Chem.* **1972**, *42*, C77
2. Atwood, J. L.; Newberry, W. R. *J. Organomet. Chem.* **1974**, 65, 145
3. Atwood, J. L.; Newberry, W. R. *J. Organomet. Chem.* **1974**, *66*, 15
4. Atwood, J. L.; Bott, S. G.; Coleman, A. W.; Robinson, K. D.; Whetstone, S. B.; Means, C. M. *J. Am. Chem Soc.* **1987**, *109*. 8100.
5. Atwood, J. L.; Bott, S. G.; Means, C. M.; Coleman, A. W.; Zhang, H.; May, M.T. *Inorg. Chem.* **1990**, *29*, 467.
6. Coleman, A.W.; Means, C. M.; Bott, S. G.; Atwood, J. L. *J. Cryst. Spec. Res.* **1990**, *20*, 199
7. Atwood, J. L.; Bott, S. G.; Robinson, K. D.; Bishop, E. J.; May, M.T. *J. Cryst. Spec. Res.* **1991**, *21*, 459.

8. Ziegler, K.; Koster, R.; Lehmkuhl, H.; Reinert, K. *Ann. Chem.* **1960**, *629*, 33
9. Lehmkuhl, H. *Angew. Chem. Int. Ed. Engl.* **1964**, *3*, 107
10. Allegra, G.; Perego, G. *Acta Crystallogr.* **1963**, *16*, 185.
11. Atwood, J. L.; Milton, P. A.; Seale, S. K. *J. Organomet. Chem.* **1971**, *28*, C29.
12. Weller, F.; Wilson, I. L.; Dehnicke, K. *J. Organomet. Chem.* **1071**, *30*, C1.
13. Weller, F.; Dehnicke, K. *J. Organomet. Chem.* **1972**, *35*, 237.
14. Weller, F.; Dehnicke, K. *J. Organomet. Chem.* **1972**, *36*, 23.
15. Atwood, J. L.; Cummings, J. M. *J. Cryst. Mol. Strust.* **1977**, *7*, 257.
16. Atwood, J. L.; Hunter, W. E.; Crissinger, K. D. *J. Organomet. Chem.* **1977**, *127*, 403.
17. Shakir, R.; Zaworotko, M. L.; Atwood, J. L. *J. Cryst. Mol. Struct.* **1979**, *9*, 135.
18. Shakir, R.; Zaworotko, M. J.; Atwood, J. L. *J. Organomet. Chem.* **1979**, *171*, 9.
19 Atwood, J. L.; Hunter, W. E.; Rogers, R. D.; Weeks, J. A. *J. Incl. Phenom.* **1985**, *3*, 113.
20. Atwood, J. L.; Crissinger, K. D.; Rogers, R. D. *J. Organomet. Chem.* **1978**, *155*, 1.
21. Atwood, J. L.; Hrncir, D. C.; Rogers, R.D. *J. Incl. Phenom.* **1983**, *1*, 199.
22. Zaworotko, M. J.; Rogers, R. D.; Atwood, J. L. *Organometallics* **1982**, *1*, 1179.
23. Zaworotko, M. J.; Kerr, C. R.; Atwood, J. L. *Organometallics* **1985**, *4*, 238.
24. Atwood, J. L.; Hrncir, D. C.; Rogers, R. D.; Howard, J. A. K. *J. Am. Chem. Soc.* **1981**, *103*, 6787.
25. Atwood, J. L.; Seale, S. K.; *J. Organomet. Chem.* **1976**, *114*, 107.
26. Rogers, R. D.; Atwood, J. L. *Organometallics* **1984**, *3*, 271.
27. Atwood, J. L.; Atwood, J. D. In *Inorganic Compounds with Unusual Properties;* King, R. B., Ed.; Advances in Chemistry, Vol. 150; American Chemical Society; Washington, **1976**, Chapter 11.
28. Atwood, J. L. *In Recent Developments in Separation Science;* Li, N. N., Ed.; CRC Press; Cleveland, **1977**; Vol 3, pp 195–209.
29. Atwood, J. L. In *Inclusion Compounds*; Atwood, J. L.; Davies, J. E. D.; MacNicol, D. D., Eds,; Academic Press; London, **1984**; Vol. 1, pp. 375–405.
30. Atwood, J. L, Hrncir, D. C.; Shakir, R.; Dalton, M. S.; Priester, R. D.; Rogers, R. D. *Organometallics* **1982**, *1*, 1021.
31. Atwood, J. L.; Priester, R. D.; Rogers, R. D.; Canada, L. C. *J. Incl. Phenom.* **1983**, *1*, 61.
32. Robinson, G. H.; Hunter, W. E.; Bott, S. G..; Atwood, J. L. *J. Organomet. Chem.* **1987**, *326*, 9
33. Robinson, G. H.; Bott, S. G.; Elgamal, H.; Hunter, W. E.; Atwood, J. L. *J. Incl. Phenom.* **1985**, *3*, 65.
34. Robinson, G. H. *Coord. Chem. Rev.* **1992**, *112*, 227.
35. Rogers, R. D.; Atwood, J. L. *J. Cryst. Spec. Res.* **1984**, *14*, 1.
36. Means, C. M.; Means, C. N.; Bott, S. G.; Atwood, J. L. *J. Am. Chem. Soc.* **1984**, *106*, 7627.
37. Hrncir, D. C.; Rogers, R. D.; Atwood, J. L. *J. Am. Chem. Soc.* **1981**, *103*, 4277.
38. Atwood, J. L.; Orr, G. W.; Alvanipour, A.; Zhang, H.; Bott, S. G. *J. Organomet. Chem.* submitted.

39. Atwood, J. L.; Hrncir, D. C.; Priester, R. D.; Rogers, R. D. *Organometallics* **1983**, *2*, 985.

40. Rogers, R.D.; Atwood, J. L. *J. Cryst. Mol. Struct.* **1979**, *9*, 45.

41. Atwood. J. L. In *Chemical Seperations*; King, C. J.; Navratil, J. D., Eds. Litarvan; Denver, CO, **1986**; pp, 335-254.

42. Atwood, J. L. In *Separation Technology*; Li, N. N.; Strathmann, H., Eds. Engineering Foundation; New York, **1988**, 45-56.

43. Atwood, J. L; U. S. Patent 4 321127, **1982**, to the University of Alabama.

44. Bond, D. R.; Jackson, G. E.; Joao, H. C.; Hofmeyr, M. N.; Modro, T. A.; Nassimbeni, L. R. *J. Chem. Soc., Chem. Commun.* **1989**, 1910.

45. Vandenberg, E. J. *J. Polym. Sci.* **1960**, *47*, 486.

46. Saegusa, T.; Imai, H.; Furukawa, *J. Makromol. Chem.* **1962**, *53*, 203.

47. Storr, A.; Jones, K.; Laubengayer, A. W. *J. Am. Chem. Soc.* **1968**, *90*, 3173.

48. Boleslawski, M.; Pasynkiewicz, S. *J. Organomet. Chem.* **1972**, *43*, 81.

49. Ueyama, N.; Araki, T.; Tani, H. *Inorg. Chem.* **1973**, *12*, 2218.

50. Boleslawski, M.; Pasynkiewicz, S.; Kunicki, A.; Serwatowski, J. *J. Organomet. Chem.* **1976**, *116*, 285.

51. Piotrowski, A.; Kunicki, A.; Pasynkiewicz, S. *J. Organomet Chem.* **1980**, *186*, 185.

52. Sinn, H.; Kaminsky, W.; Vollmer, H.-J.; Woldt, R. *Angew Chem. Int. Ed. Engl.* **1980,** *19*, 390.

53. Galli, P.; Susa, E.; Di Drusco, G. U. S. Patent 3 803105, **1974.**

54. Reichert, K. H.; Meyer, K. R. *Makromol. Chem.* **1973**, *169*, 163.

55. Long, W. P.; Breslow, D. S. *Ann. Chem.* **1975**, 463.

56. Sinn, H.-J,; Kaminsky, W. *Adv. Organomet. Chem.* **1980**, *18*, 99.

57. Pasynkiewicz, S. *Polyhedron* **1990**, *9*, 429.

58. Kushi, Y.; Fernando, Q. *J. Chem. Soc., Chem. Commun.* **1969**, 555

59. Lobkovski, E. B.; Sizov, A. I.; Bulychev, B. M.; Sokolova, I. V.; Soloveichik, G. L.; *J. Organomet. Chem.* **1987**, *319*, 69

60. Dyachkovski, F. S.; Shilova, A. K.; Shilov, A. E. *J. Polym. Sci.* **1967**, *16*, 2333.

61. Jordan, R. F.; LaPointe, F. F.; Bradley, P. K.; Baenziger, N. *Organometallics* **1989**, *8*, 2892.

62. Turner, H. W. European Patent Application 0277003, **1988**; 0277004.

63. Ewen, J. A.: Jones, R. L.; Razavi, A. *J. Am. Chem. Soc.* **1988**, *110*, 6255.

64. Sishta, C.; Hathorn, R. M.; Marks, T. J. *J. Am. Chem. Soc.* **1992**, *114*, 1112.

65. Turner, H. W. U. S. Patent 4,791,180, **1988**.

66. Bott, S. G.; Coleman, A. W.; Atwood, J. L. *J. Am. Chem. Soc.* **1986**, *108*, 1709.

67. Sinn, H.; Kaminsky, W.; Vollmer, H. Eur. **1980**, 35242; *Chem. Abstr.* **1980**, *95*, 187927j.

68. Yang, X.; Stern, C. L.; Marks, T. J. *J. Am. Chem. Soc.* **1991**, *113*, 3623.

69. Atwood, J. L.; Elgamal, H.; Robinson, G. H.; Bott, S. G.; Weeks, J. A.; Hunter, W. E. *J. Incl. Phenom.* **1984**, *2*, 367.

70. Bott, S. G.; Elgamal, H.; Atwood, J. L. *J. Am. Chem. Soc.* **1985**, *107*, 1796.

71. Derouault, J.; Forel, M. T. *Inorg. Chem.* **1977**, *16*, 3207

72. Cowley, A. H.; Cushner, M. C.; Davies, R. E.; Riley, P. *Inorg. Chem.* **1981**, *20*, 1179.

73. Means, N. C.; Means, C. M.; Bott, S. G.; Atwood, J. L. *Inorg. Chem.* **1987**, *26*, 1466.

74. Sangokoya, S. A.; Moise, F.; Pennington, W. T.; Self, M. F.; Robinson, G. H. *Organometallics* **1989**, *8*, 2584.

75. Bott, S. G.; Alvanipour, A.; Morley, S. D.; Atwood, D. A.; Means, C. M.; Coleman, A. W.; Atwood, J. L. *Angew. Chem. Int. Ed. Engl.* **1987**, *26*, 485.

76. Engelhardt, L. M.; Kynast, U.; Raston, C. L.; White, A. H. *Angew. Chem. Int. Ed. Engl.* **1987**, *26*, 681.

77. Bonitz, E. *Chem. Ber.* **1955**, *88*, 742.

78. Lehmkuhl, H.; Kobs, H. D. *Tetrahedron Lett.* **1965**, *29*, 2505.

79. Lehmkuhl, H. *Ann. Chem.* **1968**, *719*, 40.

80. Benn, R.; Rufinsky, A. *Angew. Chem. Int. Ed. Engl.* **1986**, *25*, 861.

81. Pajerski, A. D.; Cleary, T. P.; Parvez, M.; Gokel, G. W.; Richey, H. G. *Organometallics* **1992**, *11*, 1400 .

Index